# Neutrosophic Computing and Vague Data Analysis

This edited volume provides a comprehensive exploration of neutrosophic methodologies, emphasizing their computational frameworks, theoretical advancements, and practical applications. Neutrosophic logic, originally developed to address problems involving vague, inconsistent, incomplete, or unclear information, has evolved significantly over the years. Its extensions have found relevance across diverse disciplines, offering innovative solutions to challenges in science, economics, environmental studies, healthcare, and more.

The book highlights the limitations of traditional mathematical and statistical models in handling the complexity and uncertainty inherent in modern systems. While existing frameworks have contributed to uncertainty modeling, they often fail to treat indeterminacy as a distinct and independent component of information. Neutrosophic theory addresses this gap, providing a robust foundation for analyzing and optimizing systems characterized by indeterminacy.

**Key features of the book include:**

- **Computational Toolboxes**: Practical tools for estimating unknown values, simulating results, and making decisions using neutrosophic models. These computational approaches enable practitioners to analyze systems that are too complex for traditional analytical methods.
- **Applications across Disciplines**: Detailed examples and case studies demonstrating the use of neutrosophic models in science, engineering, medical diagnostics, environmental modeling, economics, and decision-making under uncertainty.
- **Accessibility and Adaptability**: By focusing on computational approaches, the book aims to make neutrosophic theory more accessible and applicable to a wider audience, including researchers, practitioners, and students.

This volume serves as a valuable resource for anyone seeking to understand and apply neutrosophic logic to real-world scenarios. It bridges the gap between theoretical advancements and practical applications, empowering readers to tackle complex problems with innovative solutions grounded in neutrosophic methodologies.

# Advances in Applied Mathematics

Series Editor: Daniel Zwillinger

**Engineering Statistics**
A Matrix-Vector Approach with MATLAB®
*Lester W. Schmerr Jr.*

**General Quantum Numerical Analysis**
*Svetlin Georgiev and Khaled Zennir*

**An Introduction to Partial Differential Equations with MATLAB®**
*Matthew P. Coleman and Vladislav Bukshtynov*

**Handbook of Exact Solutions to Mathematical Equations**
*Andrei D. Polyanin*

**Introducing Game Theory and its Applications, Second Edition**
*Elliott Mendelson and Dan Zwillinger*

**Modeling Operations Research and Business Analytics**
*William P. Fox and Robert E. Burks*

**Decision Analysis through Modeling and Game Theory**
*William P. Fox*

**Advances in High-Order Predictive Modelling**
*Dan Gabriel Cacuci*

**Introduction to Financial Mathematics, Second Edition**
*Kevin J. Hastings*

**General Quantum Variational Calculus**
*Svetlin G. Georgiev and Khaled Zennir*

**Advanced Engineering Mathematics with MATLAB®, Sixth Edition**
*Dean G. Duffy*

**Exact Methods for Nonlinear PDEs**
*Andrei D. Polyanin*

**Markov Processes, Second Edition**
*James R. Kirkwood*

**Advances in High-Order Sensitivity Analysis**
*Dan Gabriel Cacuci*

**Mining Complex Networks, Second Edition**
*Bogumił Kamiński, Paweł Prałat, and François Théberge*

**The Math Thought Mindset**
Logic, Reasoning, and Proofs
*James L. Burk*

**Neutrosophic Computing and Vague Data Analysis**
*Edited by Florentin Smarandache and Zahid Khan*

https://www.routledge.com/Advances-in-Applied-Mathematics/book-series/CRCADVAPPMTH

# Neutrosophic Computing and Vague Data Analysis

Edited by

Florentin Smarandache and Zahid Khan

CRC Press
Taylor & Francis Group
Boca Raton London New York

CRC Press is an imprint of the
Taylor & Francis Group, an **informa** business
A CHAPMAN & HALL BOOK

First edition published 2027
by CRC Press
2385 NW Executive Center Drive, Suite 320, Boca Raton FL 33431

and by CRC Press
4 Park Square, Milton Park, Abingdon, Oxon, OX14 4RN

*CRC Press is an imprint of Taylor & Francis Group, LLC*



ISBN: 978-1-041-16202-5 (hbk)
ISBN: 978-1-041-16206-3 (pbk)
ISBN: 978-1-003-68333-9 (ebk)

DOI: 10.1201/9781003683339

Typeset in Latin Modern font
by KnowledgeWorks Global Ltd.

# *Contents*

# Preface

Over the last several decades, the rapid growth of complex systems in science, economics, environmental studies, and healthcare studies has exposed the inherent shortcomings of traditional mathematical and statistical models. The existing frameworks such as probability theory and fuzzy logic variants have contributed significantly to uncertainty modeling. These models remain insufficient to fully capture indeterminacy as an independent component of information. A real-world dataset is rarely a perfect reflection of reality, instead, it is often vague, indeterminate, incomplete, and sometimes contradictory. Neutrosophic logic was originally developed to handle problems where information is vague, inconsistent, incomplete or unclear. Due to the great interest of academicians and analysts, neutrosophic theory has been extended in both theoretical and applied domains across various disciplines. However, understanding its theoretical concepts from a computational and practical perspective remains challenging for practitioners. As computational resources have grown dramatically in recent years, so with the help of computer knowledge, this theory can now be used in many fields to better understand and solve complex problems. These advances no longer simply reduce manual effort; they shift the entire paradigm. Instead of spending time on repetitive tasks, researchers and analysts can explore new ways of understanding problems that better reflect real-life situations.

The primary objective of this edited volume is to present computational frameworks, recent theoretical advancements and real-world applications of neutrosophic methodologies. With the rise of computational capabilities and algorithmic designs, neutrosophic structures like neutrosophic numbers, graphs, calculus, matrices, topology, algebraic systems, and neutrosophic statistical approaches can now be modeled, explored, and optimized efficiently. Computational toolboxes can be used to build and test neutrosophic models that work well with incomplete or vague data. This book presents such tools not only as alternatives but as necessities for modern analysis. These tools facilitate practitioners to estimate unknown values, simulate results, and make decisions based on neutrosophic models that are complex to be analyzed analytically. This book introduces these methods by connecting the theory of neutrosophic logic with hands-on computational approaches. Readers will learn how to apply these models in various fields, including science, engineering, medical diagnostics, environmental modeling, economics, and decision-making under uncertainty.

The chapters are carefully organized to balance theoretical rigor and practical relevance. Initial contributions formalize the algebraic structures necessary for handling indeterminacy, specifically through neutrosophic matrices and interval-valued frameworks. This theoretical base supports the subsequent exploration of complex computational tasks including optimization algorithms, graph systems, correlation measures, control charts, and experimental designs under indeterminacy. A significant portion of the book is devoted to multi-criteria decision-making under vague environments. Advanced frameworks of neutrosophic logic are presented, with applications spanning renewable energy selection, material selection, healthcare diagnostics, groundwater quality assessment, banking location planning, and pandemic preparedness. These chapters validate the superiority of neutrosophic frameworks in scenarios where traditional fuzzy and classical statistical methods prove insufficient.

This book serves as a valuable resource for anyone interested in exploring and applying neutrosophic logic to real-world scenarios characterized by indeterminacy. By adopting a computational approach using computational tools, it seeks to make neutrosophic theory more accessible, adaptable, and widely usable across various disciplines. Ultimately, the book offers a modern framework for analyzing uncertain data and supports complex decision-making in situations involving large volumes of information.

# *Acknowledgments*

The editors would like to express sincere gratitude to all contributing authors for their valuable chapters and for their commitment to maintaining high scholarly standards throughout the preparation of this volume. Their expertise, cooperation, and timely responses made the successful completion of this book possible.

The editors are deeply thankful to their colleagues Prof. Zsolt Tibor Kosztyán and Dr. Attila Imre Katona for their continuous guidance, insightful discussions, and encouragement, which significantly contributed to the intellectual development and successful completion of this work in the field of neutrosophic computing and vague data analysis.

Finally, the editors acknowledge the support and understanding of their families, whose cooperation and encouragement played an important role in the completion of this book.

# About the Editors

**Florentin Smarandache**, PhD, PostDocs, is an Emeritus Professor of Mathematics at the University of New Mexico, United States. He earned his PhD in Mathematics from the State University of Kishinev, and Postdoctoral in Applied Mathematics from Okayama University of Sciences, Japan, and The Guangdong University of Technology, Guangzhou, China. He is the founder of neutrosophy (generalization of dialectics), neutrosophic set, logic, probability, and statistics since 1995 and has published hundreds of papers and books on neutrosophic physics, superluminal and instantaneous physics, unmatter, quantum paradoxes, absolute theory of relativity, redshift and blueshift due to the medium gradient and refraction index besides the Doppler effect, paradoxism, outerart, neutrosophy as a new branch of philosophy, law of included multiple-middle, multispace and multistructure, HyperSoft set, TreeSoft Set, IndetermSoft Set and IndetermHyperSoft Set, SuperHyperGraph, SuperHyperTopology, SuperHyperAlgebra, SuperHyperFunction, Neutrosophic SuperHyperAlgebra, degree of dependence and independence between neutrosophic components, refined neutrosophic set, neutrosophic over-under-off-set, plithogenic set/logic/probability/statistics, symbolic plithogenic algebraic structures, neutrosophic triplet and duplet structures, quadruple neutrosophic structures, extension of algebraic structures to NeutroAlgebra and AntiAlgebra, NeutroGeometry and AntiGeometry, NeutroTopology and AntiTopology, refined neutrosophic topology, refined neutrosophic crisp topology, Dezert-Smarandache Theory, and so on in many peer-reviewed international journals and publishers. Additionally, he also presented papers and plenary lectures to many international conferences around the world.

**Zahid Khan** is currently working as a scientific co-research fellow in the Department of Quantitative Methods at the University of Pannonia, Hungary. He holds a permanent position as an Associate Professor in the Department of Mathematics and Statistics at Hazara University, Pakistan. He earned his PhD from the University Technology PETRONAS, Malaysia, in 2017. His research interests encompass robust estimation in fuzzy probability distributions, neutrosophic statistics, and statistical methods for industrial process control. Dr. Khan received the Talent Award Scholarship from the Higher Education Commission of Pakistan in 2006, a fellowship from Quaid-i-Azam University in 2007, and another fellowship from the University Technology PETRONAS in 2013.

# Contributors

**Razief Perucha Fauzie Afidh**
Department of Informatics
Universitas Syiah Kuala
Banda Aceh, Indonesia

**Liaquat Ahmad**
Department of Statistics and Computer Science
University of Veterinary and Animal Sciences
Lahore, Pakistan

**Abdulrahman AlAita**
Department of Agricultural Economy
Faculty of Agricultural Engineering
Damascus University
Damscus, Syria

**Suriana Alias**
Faculty of Computer and Mathematical Sciences
Universiti Teknologi MARA
Cawangan Kelantan
Machang Campus
Kelantan, Malaysia

**A. H. Ansari**
Kejriwal Institute of Management and Development Studies
Ranchi, India

**Muhammad Aslam**
Department of Statistics
Faculty of Science
King Abdulaziz University
Jeddah, Saudi Arabia

**Amarendra Baral**
Department of Mathematics
Trident Academy of Technology
Bhubaneswar, India

**Said Broumi**
Laboratory of Information Processing
Faculty of Science Ben M'Sik
University of Hassan II
Casablanca, Morocco

**Shariq Aziz Butt**
Department of Computer Science
School of Systems and Technology
University of Management and Technology
Lahore, Pakistan

**Mamoni Dhar**
Department of Mathematics
Science College
Kokrajhar, India

**Paul Augustine Ejegwa**
Department of Mathematics
Joseph Sarwuan Tarka University
Makurdi, Nigeria

**Edrian Hadinata**
Department of Information System
Universitas Harapan
Medan, Indonesia

**Fathiah Hasnizul**
Faculty of Computer and Mathematical Sciences
Universiti Teknologi MARA

Cawangan Kelantan
Machang Campus
Kelantan, Malaysia

**Denny Helzami**
Bank Syariah Indonesia
Banda Aceh, Indonesia

**Irene Hepzibah**
PG & Research Department of Mathematics
T.B.M.L. College affiliated to Annamalai University
Porayar, India

**Irvanizam Irvanizam**
Department of Informatics
Universitas Syiah Kuala
Banda Aceh, Indonesia

**Md. Izhar Khan**
Department of Mathematics
Islamic University of Madinah
Madinah, Saudi Arabia

**Zahid Khan**
Department of Quantitative Methods
University of Pannonia
Pannon Egyetem, Veszprem, Hungary

**Nisha Khanam**
Department of Mathematics and Statistics
Integral University
Lucknow, India

**Murat Kirisci**
Department of Biostatistics and Medical Informatics
Cerrahpasa Faculty of Medicine
Istanbul University-Cerrahpasa
Istanbul, Turkey

**M. Gayathri Lakshmi**
Department of Mathematics
Saveetha Engineering College
Chennai, India

**Devi M**
PG & Research Department of Mathematics
T.B.M.L. College affiliated to Annamalai University
Porayar, India

**Jean-Claude Malela-Majika**
Department of Statistics
College of Science Engineering and Technology
University of South Africa — Florida Campus
South Africa

**Nivetha Martin**
Department of Mathematics
Arul Anandar College (Autonomous)
Karumathur, India

**Norzieha Mustapha**
Faculty of Computer and Mathematical Sciences
Universiti Teknologi MARA
Cawangan Kelantan
Machang Campus
Kelantan, Malaysia

**Sumera Naz**
Department of Mathematics
Division of Science and Technology
University of Education
Lahore, Pakistan

**Taha Yasin Öztürk**
Department of Mathematics
Kafkas University
Kars, Turkey

**Basil Papadopoulos**
Department of Civil Engineering
Democritus University of Thrace
Greece

**Arokia Pratheesha S V**
Department of Mathematics
Dr. N.G.P. Arts and Science College
Coimbatore, India

**Abdul Quddoos**
School of Liberal Arts and Sciences
Era University
Lucknow, India

**R Radha**
Department of Science and Humanities
RVS Technical Campus-Coimbatore
Coimbatore, India

**Fahrul Radhi**
Faculty of Education and Teacher Training
Universitas Syiah Kuala
Banda Aceh, Indonesia

**R. Rajalakshmi**
Department of Mathematics
Panimalar Engineering College
Chennai, India

**Priscilla Rajkumar**
Department of Physics
A.D.M. College for Women (Autonomous)
Nagapattinam affiliated to Bharathidasan University
Tamil Nadu, India

**G. Srinivasa Rao**
Department of Mathematics and Statistics
The University of Dodoma
Dodoma, Tanzania

**Prasanta Kumar Raut**
Department of Mathematics
Trident Academy of Technology
Bhubaneswar, India

**Muhammad Saleem**
Department of Industrial Engineering
Faculty of Engineering-Rabigh
King Abdulaziz University
Jeddah, Saudi Arabia

**K. Saritha**
Department of Mathematics
Panimalar Engineering College
Poonamallee, India

**Ambreen Shafqat**
School of Sciences
Nanjing University of Science and Technology
Nanjing, P.R. China

**Binod Kumar Sharma**
Department of Computer science and Engineering
Birla Institute of Technology
Mesra (Lalpur Unit), Ranchi, India

**Sandile C. Shongwe**
Department of Statistics
College of Science Engineering and Technology
University of South Africa
South Africa

**Florentin Smarandache**
Department of Mathematics
University of New Mexico
Albuquerque, New Mexico

**Apostolos Syropoulos**
Independent Scholar
Greece

**Roliza Md Yasin**
Faculty of Computer and
Mathematical Sciences
Universiti Teknologi MARA
Cawangan Kelantan
Machang Campus
Kelantan, Malaysia

**Adem Yolcu**
Department of Mathematics
Kafkas University
Kars, Turkey

**Sabahat Zafar**
Department of Mathematics
Division of Science and Technology
University of Education
Lahore, Pakistan

**Yasser Al Zaim**
Department of Statistics
Faculty of Science
Damascus University
Damascus, Syria

# Chapter 1

## *Matrices of Neutrosophic Numbers: An Overview*

**Mamoni Dhar**

### 1.1 Introduction

The theory of fuzzy sets introduced by Zadeh [1] showed meaningful applications in many field of studies. To convey the degree of belongingness of the set under consideration, the membership function of a fuzzy set assigns a number from the unit interval [0, 1] to each element of the universe of discourse. A single number between 0 and 1 is considered an element's membership in a fuzzy set according to the fuzzy set theory. The degree of non-membership of an element in a fuzzy set may not always be equal to one minus the membership degree, though, as there may be some hesitation degree exists as well. Thus Atanassov [2] realized a generalization of fuzzy set, which is known as intuitionistic fuzzy set.

Neutrosophic set is a part of neutrosophy which studies the origin, nature and scope of neutralities. The neutrosophic set is the generalization of the concept of fuzzy set and intuitionistic fuzzy sets. Neutrosophic set theory was proposed by Smarandache [3] where each element has three associated defining functions called membership function(T), indeterminacy function (I) and non-membership function (F) defined on the universe of discourse X. To deal with indeterminate and inconsistent information in the real-world problems, this theory has been extensively applied by many scientists in recent past (as for example, [4]–[19]). Many more applications of this theory can be found, all of which cannot be mentioned in this small work.

The idea of fuzzy matrices was initiated by Thomason [20] and the convergence of powers of fuzzy matrices was also taken into consideration. With these particular research works, Kim ([21]–[23]) made significant contributions to the study of determinant theory of square fuzzy matrices and reflected some significant results on the determinant of a square matrices. Dhar ([24]–[26])

DOI: 10.1201/9781003683339-1

extended the definition of determinant of square fuzzy matrices to determinant of square fuzzy matrices defined on the basis of reference function.

Neutrosophic interval matrices werethe concept of Kandaswamy et al [27]. Square neutrosophic fuzzy matrices have been discussed by Dhar et al [28] and multiplication of such neutrosophic fuzzy matrices along with some properties was also discussed in that work. Further study was made by Das et al [29] as an extension of [28]. More researches on this type of matrices can be found in [30, 31].

This work mainly focuses on different types of neutrosophic fuzzy matrices and their associated properties. Further determinant of neutrosophic fuzzy matrices and their properties are also studied.

This chapter is organized as follows: Section1.1 is introductory in nature in which literature review is done. Section 1.2 deals with various existing definitions of the different terms used in later part of the chapter. Section 1.3 defines various types of neutrosophic number matrices. Section 1.4 deals with various operations on neutrosophic number matrices. Section 1.5 defines determinant of neutrosophic number matrices discussed so far. Finally, Section 1.6 deals with conclusions and discussions.

## 1.2 Preliminaries

This section mainly focuses on the definitions of fuzzy matrix, neutrosophic set, neutrosophic fuzzy matrix, etc., which are necessary for the subsequent sections.

**Definition 1.1:** Fuzzy set (Zadeh [1])

If X is a universe of discourse and x is any particular element of X, then a fuzzy set F, defined on X may be written as a collection of ordered pairs $A = \{< \varpi, \Lambda_A(\varpi) >, \varpi \in X\}$, where $\Lambda_A : X \to [0,1]$, is called the membership function or grade of membership of $\varpi$ in X and each pair $< \varpi, \Lambda_A(\varpi) >$ is called a *singleton*.

**Definition 1.2:** Intuitionistic fuzzy set (Atanassov [2])

Let $X$ be a universe of discourse. Then the intuitionistic fuzzy set $A$ is an object having the form $A = \{< \varpi : \Lambda_A(\varpi), \gamma_A(\varpi) >, \varpi \in X\}$, where the function $\Lambda_A(\varpi), \gamma_A(\varpi) : X \to [0,1]$. Here$\Lambda_A(\varpi)$ and $\gamma_A(\varpi)$ define the degree of membership and the degree of non-membership of the element $\varpi \in X$ to the set $A$ assuming that $0 \leq \Lambda_A(\varpi) + \gamma_A(\varpi) \leq 1$.

**Definition 1.3:** Neutrosophic set (Smarandache [3–5])

Let the universe of discourse be $X$. Then the neutrosophic set $A$ will be defined in the form $A = \{< \varpi : \Lambda_A(\varpi), \Omega_A(\varpi), \gamma_A(\varpi) >, \varpi \in X\}$, where functions $\Lambda_A(\varpi), \Omega_A(\varpi), \gamma_A(\varpi) : X \to ]^-0, 1^+[$, define the degree of membership $\Lambda_A$,the degree of indeterminacy $\gamma_A$ and the degree of non-membership, respectively, of the element $\varpi \in X$ to the set A with the condition $0 \leq \Lambda_A(\varpi) + \Omega_A(\varpi) + \gamma_A(\varpi) \leq 3^+$.

**Definition 1.4:** Fuzzy matrix (Thomason [20])

A fuzzy matrix is a matrix in which all the entries belong to [0, 1]. Let $M$ be an $m \times n$ matrix defined by

$$M = \begin{bmatrix} \varepsilon_{11} & \varepsilon_{12} & \cdots & \varepsilon_{1n} \\ \varepsilon_{21} & \varepsilon_{22} & \cdots & \varepsilon_{2n} \\ \vdots & \vdots & \vdots & \vdots \\ \varepsilon_{m1} & \varepsilon_{m2} & \cdots & \varepsilon_{mn} \end{bmatrix}.$$

Then the matrix $M$ is said to be a fuzzy matrix if and only if $\varepsilon_{ij} \in [0, 1]$ for $1 \leq i \leq m, 1 \leq j \leq n$.

**Definition 1.5:** Neutrosophic number matrix (Dhar et al [29])

The entries of neutrosophic number matrix takes the form $a + Ib$(fuzzy number), where $a$, $b$ are elements on [0, 1] and $I$ is an indeterminant such that $I^n = I$,$n$ being a positive integer.

For example,

$$M = \begin{bmatrix} \phi_{11} + I\varphi_{11} & \phi_{12} + I\varphi_{12} \\ \phi_{21} + I\varphi_{21} & \phi_{22} + I\varphi_{22} \end{bmatrix}$$

is a neutrosophic number matrix.

**Definition 1.6:** Square neutrosophic number matrix (Dhar et al [29])

A neutrosophic number matrix is considered to be square neutrosophic number matrix when the number of rows and the number of columns are equal. Thus $M = [\phi_{ij} + I\varphi_{ij}]_{nxn}$ is a square neutrosophic number matrix if $m = n$.

**Definition 1.7:** Rectangular neutrosophic number matrix (Das et al [30])

A neutrosophic number matrix is said to be a rectangular neutrosophic number matrix when the number of rows and the number of columns are unequal. Thus $M = [\phi_{ij} + I\varphi_{ij}]_{nxn}$ is a rectangular neutrosophic number matrix if $m \neq n$.

---

## 1.3 Different Types of Neutrosophic NumberMatrices

In this section different types of neutrosophic number matrices are discussed.

**Definition 1.8:** Row neutrosophic number matrix

A neutrosophic number matrix in which only one row exists is called a row neutrosophic fuzzy matrix.

**Definition 1.9:** Column neutrosophic number matrix

A neutrosophic number matrix in which only one column exists is called a column neutrosophic number matrix.

**Definition 1.10:** Null neutrosophic number matrix

A neutrosophic number matrix is said to be null neutrosophic number matrix when each element of the neutrosophic number matrix is of the form $0 + I0$.

**Definition 1.11:** Equal neutrosophic number matrix

Two neutrosophic number matrices of the form $M = [\phi_{ij} + I\varphi_{ij}]_{mxn}$ and $N = [\varepsilon_{ij} + I\omega_{ij}]_{mxn}$ are said to be equal when the orders of these neutrosophic number matrices are the same and $\phi_{ij} = \varepsilon_{ij} \& \phi_{ij} = \omega_{ij}$.

**Definition 1.12:** Singleton neutrosophic number matrix

If in a neutrosophic number matrix there is only one element, then it is called singleton neutrosophic matrix. Thus $M = [\phi_{ij} + I\varphi_{ij}]_{mxn}$ is a singleton neutrosophic number matrix if $m = n = 1$.

**Definition 1.13:** Identity neutrosophic number matrix

A square neutrosophic number matrix is considered to be identity neutrosophic number matrix when the diagonal elements are of the form $1 + I1$ and others are of the form $0 + I0$.

For example,

$$I = \begin{bmatrix} 1+I1 & 0+I0 \\ 0+I0 & 1+I1 \end{bmatrix}, \begin{bmatrix} 1+I1 & 0+I0 & 0+I0 \\ 0+I0 & 1+I1 & 0+I0 \\ 0+I0 & 0+I0 & 1+I1 \end{bmatrix}$$

are, respectively,identity neutrosophic number matrices of order 2 and 3.

**Definition 1.14:** Diagonal neutrosophic number matrix

If all the elements of the diagonal of a diagonal neutrosophic number matrix are $1 + I1$, then it is considered as diagonal neutrosophic number matrix. Thus a square neutrosophic number matrix $M = [\phi_{ij} + I\varphi_{ij}]_{nxn}$ is a diagonal neutrosophic numbermatrix if $\phi_{ij} + I\varphi_{ij} = 0 + I0$ when $i \neq j$. That is only diagonal has non-zero elements. Two important things to be noted here that

a. A diagonal neutrosophic number matrix is always a square neutrosophic number matrix.

b. The diagonal elements are characterized by a general term $\phi_{ij} + I\varphi_{ij}$ for $i = j$. This means a matrix can have only one diagonal.

For example,

$$\begin{bmatrix} 0.3+I0.3 & 0+I0 \\ 0+I0 & 0.3+I0.5 \end{bmatrix}, \begin{bmatrix} 0.3+I0.4 & 0+I0 & 0+I0 \\ 0+I0 & 0.4+I0.3 & 0+I0 \\ 0+I0 & 0+I0 & 0.2+I0.5 \end{bmatrix}$$

are diagonal neutrosophic matrices of order 2 and 3, respectively.

**Definition 1.15:** Scalar neutrosophic number matrix

If all the entries of the diagonal of a diagonal neutrosophic number matrix are equal, then it is called a scalar neutrosophic number matrix. Thus a square neutrosophic number matrix $M = [\phi_{ij} + I\varphi_{ij}]_{nxn}$ is a scalar neutrosophic numbermatrix if

$$\phi_{ij} + I\varphi_{ij} = \begin{Bmatrix} 0+I0, i \neq j \\ \phi_{ii} + I\varphi_{ii}, i = j \end{Bmatrix}.$$

If all the entries of a diagonal are equal to $1 + I1$, then it will be a scalar matrix and it is obviously a diagonal neutrosophic number matrix. It has got a special name which is called identity neutrosophic number matrix.
For example,

$$\begin{bmatrix} 0.2+I0.3 & 0+I0 \\ 0+I0 & 0.2+I0.3 \end{bmatrix}, \begin{bmatrix} 0.2+I0.3 & 0+I0 & 0+I0 \\ 0+I0 & 0.2+I0.3 & 0+I0 \\ 0+I0 & 0+I0 & 0.2+I0.3 \end{bmatrix}$$

are scalar neutrosophic number matrices of order 2 and 3, respectively.

**Remark 1.1:**

- All identity neutrosophic matrices are scalar neutrosophic number matrices.
- All scalar neutrosophic matrices are diagonal neutrosophic number matrices.

**Definition 1.16:** Bidiagonal neutrosophic number matrix

If there are non-zero entries along the main diagonal and either the diagonal above or below the principal diagonal, then the square neutrosophic number matrix is referred to as a bidiagonal neutrosophic number matrix. There is possibility that these matrices may be of upper and lower bidiagonal form which can be seen in the following way.

For example,

$$\begin{bmatrix} 0.2+I0.4 & 0.2+I0.3 & 0+I0 \\ 0+I0 & 0.2+I0.4 & 0.4+I0.3 \\ 0+I0 & 0+I0 & 0.2+I0.4 \end{bmatrix}$$

is an upper bidiagonal neutrosophic number matrix.

**Definition 1.17:** Triangular neutrosophic number matrix

If the entries above or below the principal diagonal are all zero, a square neutrosophic number matrix is referred to as a triangular neutrosophic number matrix. Triangular neutrosophic number matrices come in two varieties, which are

a. Upper triangular neutrosophic number matrix

   Neutrosophic number upper triangular matrix is a neutrosophic number matrix of the form

$$\begin{bmatrix} 0.2+I0.4 & 0.4+I0.6 & 0.2+I0.5 \\ 0+I0 & 0.2+I0.4 & 0.5+I0.4 \\ 0+I0 & 0+I0 & 0.2+I0.4 \end{bmatrix}.$$

b. Lower triangular neutrosophic number matrix

   Neutrosophic number matrix is a lower triangular neutrosophic number matrix whichis of the form

$$\begin{bmatrix} 0.4+I0.3 & 0+I0 & 0+I0 \\ 0.3+I0.2 & 0.4+I0.3 & 0+I0 \\ 0.5+I0.3 & 0.5+I0.3 & 0.4+I0.3 \end{bmatrix}.$$

---

## 1.4 Operations on Neutrosophic Number Matrices

This section deals with some operations on neutrosophic number matrices.

**Definition 1.18:** Sum of two neutrosophic number matrices (Dhar et al [29])

Let us consider two neutrosophic number matrices $M$ and $N$ such that

$$M = \begin{bmatrix} \phi_{11}+I\varphi_{11} & \phi_{12}+I\varphi_{12} \\ \phi_{21}+I\varphi_{21} & \phi_{22}+I\varphi_{22} \end{bmatrix} \& N = \begin{bmatrix} \varepsilon_{11}+I\omega_{11} & \varepsilon_{12}+I\omega_{12} \\ \varepsilon_{21}+I\omega_{21} & \varepsilon_{22}+I\omega_{22} \end{bmatrix}.$$

Then the sum of two neutrosophic number matrices can be explained when the number of rows and the number of columns of both the matrices are same. Let the sum of $M$ and $N$ is denoted as

$$M + N = [C_{ij}],$$

where

$$\begin{aligned} C_{11} &= max(\phi_{11}, \varepsilon_{11}) + Imax(\phi_{11}, \omega_{11}) \\ C_{12} &= max(\phi_{12}, \varepsilon_{12}) + Imax(\varphi_{12}, \omega_{12}) \\ C_{21} &= max(\phi_{21}, \varepsilon_{21}) + Imax(\varphi_{21}, \omega_{21}) \\ C_{22} &= max(\phi_{22}, \varepsilon_{22}) + Imax(\varphi_{22}, \omega_{22}) \end{aligned}$$

**Definition 1.19:** Subtraction of two neutrosophic matrices (Dhar et al [29])

Let

$$M = \begin{bmatrix} \phi_{11} + I\varphi_{11} & \phi_{12} + I\varphi_{12} \\ \phi_{21} + I\varphi_{21} & \phi_{22} + I\varphi_{22} \end{bmatrix} \& N = \begin{bmatrix} \varepsilon_{11} + I\omega_{11} & \varepsilon_{12} + I\omega_{12} \\ \varepsilon_{21} + I\omega_{21} & \varepsilon_{22} + I\omega_{22} \end{bmatrix}$$

be two neutrosophic fuzzy matrices having order 2 × 2.

The subtraction of two neutrosophic number matrices is defined when the number of rows and columns of both the matrices are the same.

$$M - N = [C_{ij}],$$

where

$$C_{11} = min(\phi_{11}, \varepsilon_{11}) + Imin(\phi_{11}, \omega_{11})$$
$$C_{12} = min(\phi_{12}, \varepsilon_{12}) + Imin(\varphi_{12}, \omega_{12})$$
$$C_{21} = min(\phi_{21}, \varepsilon_{21}) + Imin(\varphi_{21}, \omega_{21})$$
$$C_{22} = min(\phi_{22}, \varepsilon_{22}) + Imin(\varphi_{22}, \omega_{22})$$

**Definition 1.20:** Multiplication of two neutrosophic number matrices (Dhar et al [29])

The multiplication operations of two neutrosophic number matrices is defined when the number of columns of the first matrix is equal the number of rows of the second matrix. Let us consider the two neutrosophic number matrices $M = [\phi_{ij} + I\varphi_{ij}]$, $N = [\varepsilon_{ij} + I\omega_{ij}]$.

Then

$$MN = [maxmin(\phi_{ij}, \varepsilon_{ji}) + Imaxmin(\varphi_{ij}, \omega_{ji})]$$

Let

$$MN = [D_{ij}],$$

where

$$D_{11} = [max\{min(\phi_{11}, \varepsilon_{11}), min(\phi_{12}, \varepsilon_{21})\} + Imax\{min(\varphi_{11}, \omega_{11}), min(\varphi_{12}, \omega_{21})\}]$$
$$D_{12} = [max\{min(\phi_{11}, \varepsilon_{12}), min(\phi_{12}, \varepsilon_{22})\} + Imax\{min(\varphi_{11}, \omega_{11}), min(\varphi_{12}, \omega_{22})\}]$$
$$D_{21} = [max\{min(\phi_{21}, \varepsilon_{11}), min(\phi_{22}, \varepsilon_{21})\} + Imax\{min(\varphi_{21}, \omega_{11}), min(\varphi_{22}, \omega_{21})\}]$$
$$D_{22} = [max\{min(\phi_{21}, \varepsilon_{12}), min(\phi_{12}, \varepsilon_{22})\} + Imax\{min(\varphi_{21}, \omega_{12}), min(\varphi_{22}, \omega_{22})\}]$$

**Definition 1.21:** Transpose of neutrosophic number matrix

Transpose of a neutrosophic number matrix $M = [\phi_{ij} + I\varphi_{ij}]_{mn}$ can be written as $M^T$ and this is defined as $M^T = [\phi_{ji} + I\varphi_{ji}]_{nxm}$.

**Definition 1.22:** Trace of neutrosophic number matrix

The trace of a neutrosophic fuzzy matrix $M = [\phi_{ij} + I\varphi_{ij}]$ is denoted by $tr(M)$ and is defined as $tr(M) = max(\phi_{11}, \phi_{22}, \phi_{33}, ........, \phi_{nn}) + Imax(\varphi_{11}, \varphi_{22}, \varphi_{33}, ...., \varphi_{nn})$.

### 1.4.1 Properties of trace of neutrosophic number matrices

If $M = [\phi_{ij} + I\varphi_{ij}]$ and $N = [\varepsilon_{ij} + I\omega_{ij}]$ are any two neutrosophic number matrices, then the properties which hold for $M$ and $N$ are mentioned below:

i. $tr(M + N) = tr(M) + tr(N)$

ii. $tr(M) = tr(M^T)$

iii. $tr(\mu M) = \mu tr(N)$

**Proof:** Let $M$ and $N$ be two $n \times n$ neutrosophic number matrices such that

$$M = \begin{bmatrix} \phi_{11} + I\varphi_{11} & \phi_{12} + I\varphi_{12} & \dots & \phi_{1n} + I\varphi_{1n} \\ \phi_{21} + I\varphi_{21} & \phi_{22} + I\varphi_{22} & \dots & \phi_{2n} + I\varphi_{2n} \\ \vdots & \vdots & \vdots & \vdots \\ \phi_{n1} + I\varphi_{n1} & \phi_{n2} + I\varphi_{n2} & \dots & \phi_{nn} + I\varphi_{nn} \end{bmatrix},$$

$$N = \begin{bmatrix} \varepsilon_{11} + I\omega_{11} & \varepsilon_{12} + I\omega_{12} & \dots & \varepsilon_{1n} + I\omega_{1n} \\ \varepsilon_{21} + I\omega_{21} & \varepsilon_{22} + I\omega_{22} & \dots & \varepsilon_{2n} + I\omega_{2n} \\ \vdots & \vdots & \vdots & \vdots \\ \varepsilon_{n1} + I\omega_{n1} & \varepsilon_{n2} + I\omega_{n2} & \dots & \varepsilon_{nn} + I\omega_{nn} \end{bmatrix}.$$

Then

$$tr(M) = max(\phi_{11}, \phi_{22}, \phi_{33}, ........, \phi_{nn}) + Imax(\varphi_{11}, \varphi_{22}, \varphi_{33}, ...., \varphi_{nn}),$$
$$tr(N) = max(\varepsilon_{11}, \varepsilon_{22}, \varepsilon_{33}, ........, \varepsilon_{nn}) + Imax(\omega_{11}, \omega_{22}, \omega_{33}, ...., \omega_{nn})$$

$$\begin{aligned} tr(M) + tr(N) &= max\{max(\phi_{11}, \phi_{22}, \phi_{33}, ........, \phi_{nn}), \\ &max(\varepsilon_{11}, \varepsilon_{22}, \varepsilon_{33}, ........, \varepsilon_{nn})\} \\ &+ Imax\{max(\varphi_{11}, \varphi_{22}, \varphi_{33}, ........, \varphi_{nn}), \\ &max(\omega_{11}, \omega_{22}, \omega_{33}, ...., \omega_{nn})\} \\ &= max\{max(\phi_{11}, \phi_{22}, \phi_{33}, ........, \phi_{nn}, \varepsilon_{11}, \varepsilon_{22}, \varepsilon_{33}, ........, \varepsilon_{nn})\} \\ &+ Imax\{max(\varphi_{11}, \varphi_{22}, \varphi_{33}, ........, \varphi_{nn}, \omega_{11}, \omega_{22}, \omega_{33}, ...., \omega_{nn})\}. \end{aligned}$$

Again

$$M + N = \begin{bmatrix} max(\phi_{11}, \varepsilon_{11}) + Imax(\varphi_{11}, \omega_{11}) & max(\phi_{12}, \varepsilon_{12}) + Imax(\varphi_{12}, \omega_{12}) & \dots & max(\phi_{1n}, \varepsilon_{1n}) + Imax(\varphi_{1n}, \omega_{1n}) \\ max(\phi_{21}, \varepsilon_{21}) + Imax(\varphi_{21}, \omega_{21}) & max(\phi_{22}, \varepsilon_{22}) + Imax(\varphi_{22}, \omega_{22}) & \dots & max(\phi_{2n}, \varepsilon_{2n}) + Imax(\varphi_{2n}, \omega_{2n}) \\ \vdots & \vdots & \vdots & \vdots \\ max(\phi_{n1}, \varepsilon_{n1}) + Imax(\varphi_{n1}, \omega_{n1}) & max(\phi_{n2}, \varepsilon_{n2}) + Imax(\varphi_{n2}, \omega_{n2}) & \dots & max(\phi_{nn}, \varepsilon_{nn}) + Imax(\varphi_{nn}, \omega_{nn}) \end{bmatrix}$$

$$\begin{aligned}tr(M+N) &= max\{max(\phi_{11},\varepsilon_{11}), max(\phi_{22},\varepsilon_{22}),\\ &\quad max(\phi_{33},\varepsilon_{33})......., max(\phi_{nn},\varepsilon_{nn})\}\\ &\quad + Imax\{max(\varphi_{11},\omega_{11}), max(\varphi_{22},\omega_{22}),\\ &\quad max(\varphi_{33},\omega_{33}), ........, max(\varphi_{nn},\omega_{nn})\}\\ &= max\{max(\phi_{11},\phi_{22},\phi_{33},........,\phi_{nn},\varepsilon_{11},\varepsilon_{22},\varepsilon_{33},........,\varepsilon_{nn})\}\\ &\quad + Imax\{max(\varphi_{11},\varphi_{22},\varphi_{33},........,\varphi_{nn},\omega_{11},\omega_{22},\omega_{33},....,\omega_{nn})\}.\end{aligned}$$

Hence

$$tr(M+N) = tr(M) + tr(N).$$

Here $M^T = \begin{bmatrix}\phi_{11}+I\varphi_{11} & \phi_{21}+I\varphi_{21} & ... & \phi_{n1}+I\varphi_{n1}\\ \phi_{12}+I\varphi_{12} & \phi_{22}+I\varphi_{22} & ... & \phi_{n2}+I\varphi_{n2}\\ \vdots & \vdots & \vdots & \vdots\\ \phi_{1n}+I\varphi_{1n} & \phi_{2n}+I\varphi_{2n} & ... & \phi_{nn}+I\varphi_{nn}\end{bmatrix}.$

Thus $tr(M^T) = max(\phi_{11},\phi_{22},\phi_{33},........,\phi_{nn}) + Imax(\varphi_{11},\varphi_{22},\varphi_{33},....,\varphi_{nn})$

$$tr(M) = tr(M^T).$$

Let $\mu \in [0,1]$ then

$$\mu M = \begin{bmatrix}\mu\phi_{11}+I\mu\varphi_{11} & \mu\phi_{12}+I\mu\varphi_{12} & ... & \mu\phi_{1n}+I\mu\varphi_{1n}\\ \mu\phi_{21}+I\mu\varphi_{21} & \mu\phi_{22}+I\mu\varphi_{22} & ... & \mu\phi_{2n}+I\mu\varphi_{2n}\\ \vdots & \vdots & \vdots & \vdots\\ \mu\phi_{n1}+I\mu\varphi_{n1} & \mu\phi_{n2}+I\mu\varphi_{n2} & ... & \mu\phi_{nn}+I\mu\varphi_{nn}\end{bmatrix}.$$

Thus

$$\begin{aligned}tr(\mu M) &= max(\mu\phi_{11},\mu\phi_{22},\mu\phi_{33},........,\mu\phi_{nn})\\ &\quad + Imax(\mu\varphi_{11},\mu\varphi_{22},\mu\varphi_{33},....,\mu\varphi_{nn})\\ &= \mu max(\phi_{11},\phi_{22},\phi_{33},........,\phi_{nn}) + I\mu\ max\ (\varphi_{11},\varphi_{22},\varphi_{33},....,\varphi_{nn}).\\ &= \mu\{max(\phi_{11},\phi_{22},\phi_{33},........,\phi_{nn})\\ &\quad + Imax(\varphi_{11},\varphi_{22},\varphi_{33},....,\varphi_{nn})\}.\\ &= \mu tr(M).\end{aligned}$$

---

## 1.5 Determinant of Neutrosophic NumberMatrices

Let $M$ be a square neutrosophic number matrices of order $n \times n$ such that

$$M = \begin{bmatrix}\phi_{11}+I\varphi_{11} & \phi_{12}+I\varphi_{12} & ... & \phi_{1n}+I\varphi_{1n}\\ \phi_{21}+I\varphi_{21} & \phi_{22}+I\varphi_{22} & ... & \phi_{2n}+I\varphi_{2n}\\ \vdots & \vdots & \vdots & \vdots\\ \phi_{n1}+I\varphi_{n1} & \phi_{n2}+I\varphi_{n2} & ... & \phi_{nn}+I\varphi_{nn}\end{bmatrix}.$$

Then the determinant of the above matrix $M$ is denoted by $det(M)$ and can be defined as follows:

$$det(M) = \left\{ \phi_{11} \begin{vmatrix} \phi_{22} & \phi_{23} & \cdots & \phi_{2n} \\ \phi_{32} & \phi_{33} & \cdots & \phi_{3n} \\ \cdots & \cdots & \cdots & \cdots \\ \phi_{n2} & \phi_{n3} & \cdots & \phi_{nn} \end{vmatrix} + I\varphi_{11} \begin{vmatrix} \varphi_{22} & \varphi_{23} & \cdots & \varphi_{2n} \\ \varphi_{32} & \varphi_{33} & \cdots & \varphi_{3n} \\ \cdots & \cdots & \cdots & \cdots \\ \varphi_{n2} & \varphi_{n3} & \cdots & \varphi_{nn} \end{vmatrix} \right\}$$
$$+ \left\{ \phi_{12} \begin{vmatrix} \phi_{21} & \phi_{23} & \cdots & \phi_{2n} \\ \phi_{31} & \phi_{32} & \cdots & \phi_{3n} \\ \cdots & \cdots & \cdots & \cdots \\ \phi_{n1} & \phi_{n3} & \cdots & \phi_{nn} \end{vmatrix} + I\varphi_{12} |..| \right\} + \ldots\ldots\ldots\ldots +$$
$$\left\{ \phi_{1n} \begin{vmatrix} \phi_{21} & \phi_{22} & \cdots & \phi_{2(n-1)} \\ \phi_{31} & \phi_{32} & \cdots & \phi_{3(n-1)} \\ \cdots & \cdots & \cdots & \cdots \\ \phi_{n1} & \phi_{n2} & \cdots & \phi_{n(n-1)} \end{vmatrix} + I\varphi_{1n} \begin{vmatrix} \varphi_{21} & \varphi_{22} & \cdots & \varphi_{2(n-1)} \\ \varphi_{31} & \varphi_{32} & \cdots & \varphi_{3(n-1)} \\ \cdots & \cdots & \cdots & \cdots \\ \varphi_{n1} & \varphi_{n2} & \cdots & \varphi_{n(n-1)} \end{vmatrix} \right\} \quad (1.1)$$

### 1.5.1 Properties of determinant of neutrosophic number matrices

Here in this section some important properties of the determinant of neutrosophic number matrices will be discussed.

**Property 1.1:** If all the entries of a row or column of a neutrosophic numbermatrix are zero, then the value of the determinant becomes zero.

For example, let $M = \begin{bmatrix} 0 + I0 & 0.2 + I0.3 & 0.2 + I0.4 \\ 0 + I0 & 0.4 + I0.3 & 0.2 + I0.3 \\ 0 + I0 & 0.2 + I0.3 & 0.4 + I0.5 \end{bmatrix}$ be the neutrosophic number matrix with zero entries in the first column.

Then expanding the determinant along the first column using (1.1), it can be obtained that

$$|M| = \left\{ 0 \begin{vmatrix} 0.4 & 0.2 \\ 0.2 & 0.4 \end{vmatrix} + I0 \begin{vmatrix} 0.3 & 0.3 \\ 0.3 & 0.5 \end{vmatrix} \right\} + \left\{ 0 \begin{vmatrix} 0.2 & 0.2 \\ 0.2 & 0.4 \end{vmatrix} + I0 \begin{vmatrix} 0.3 & 0.4 \\ 0.3 & 0.5 \end{vmatrix} \right\}$$
$$+ \left\{ 0 \begin{vmatrix} 0.2 & 0.2 \\ 0.4 & 0.2 \end{vmatrix} + I0 \begin{vmatrix} 0.3 & 0.4 \\ 0.3 & 0.3 \end{vmatrix} \right\}$$

and this is clearly zero.

**Property 1.2:** A determinant of a neutrosophic number matrix is expressible as the sum of two or more determinants if any of its rows or columns may be written as the sum of two or more terms.

For example, let $L = \begin{bmatrix} 0.4 + I0.5 & 0.2 + I0.3 & 0.2 + I0.4 \\ 0.2 + I0.4 & 0.4 + I0.2 & 0.2 + I0.3 \\ 0.4 + I0.3 & 0.2 + I0.3 & 0.3 + I0.4 \end{bmatrix}$ be a square neutrosophic number matrix.

To prove

$$\begin{vmatrix} (0.4+I0.5)+(0.3+I0.4) & 0.2+I0.3 & 0.2+I0.4 \\ (0.2+I0.4)+(0.2+I0.3) & 0.4+I0.2 & 0.2+I0.3 \\ (0.4+I0.3)+(0.3+I0.2) & 0.2+I0.3 & 0.3+I0.4 \end{vmatrix} =$$

$$\begin{vmatrix} 0.4+I0.5 & 0.2+I0.3 & 0.2+I0.4 \\ 0.2+I0.4 & 0.4+I0.2 & 0.2+I0.3 \\ 0.4+I0.3 & 0.2+I0.3 & 0.3+I0.4 \end{vmatrix} + \begin{vmatrix} 0.3+I0.4 & 0.2+I0.3 & 0.2+I0.4 \\ 0.2+I0.3 & 0.4+I0.2 & 0.2+I0.3 \\ 0.3+I0.2 & 0.2+I0.3 & 0.3+I0.4 \end{vmatrix}.$$

Let the above equality be expressed as $L' = L_1 + L_2$,
where

$$L' = \begin{vmatrix} (0.4+I0.5)+(0.3+I0.4) & 0.2+I0.3 & 0.2+I0.4 \\ (0.2+I0.4)+(0.2+I0.3) & 0.4+I0.2 & 0.2+I0.3 \\ (0.4+I0.3)+(0.3+I0.2) & 0.2+I0.3 & 0.3+I0.4 \end{vmatrix}$$

$$L_1 = \begin{vmatrix} 0.4+I0.5 & 0.2+I0.3 & 0.2+I0.4 \\ 0.2+I0.4 & 0.4+I0.2 & 0.2+I0.3 \\ 0.4+I0.3 & 0.2+I0.3 & 0.3+I0.4 \end{vmatrix} \text{ and}$$

$$L_2 = \begin{vmatrix} 0.3+I0.4 & 0.2+I0.3 & 0.2+I0.4 \\ 0.2+I0.3 & 0.4+I0.2 & 0.2+I0.3 \\ 0.3+I0.2 & 0.2+I0.3 & 0.3+I0.4 \end{vmatrix}$$

Then by addition of elements, $L'$ becomes

$$\begin{bmatrix} 0.4+I0.5 & 0.2+I0.3 & 0.2+I0.4 \\ 0.4+I0.2 & 0.4+I0.2 & 0.2+I0.3 \\ 0.4+I0.3 & 0.2+I0.3 & 0.3+I0.4 \end{bmatrix}$$

and when the determinant of $L', L_1$ and $L_2$ are evaluated, the values become

$$L' = 0.20 + I0.30$$
$$L_1 = 0.20 + I0.30$$
$$L_2 = 0.15 + I0.20.$$

Again it can be seen that

$$L_1 + L_2 = (0.20 + I0.30) + (0.15 + I0.20) = 0.20 + I0.30.$$

This proves the property $L' = L_1 + L_2$.

**Property 1.3:** If $M$ and $N$ are any two neutrosophic number matrices, then $det(M+N) = detM + detN$.

For example, let

$$M = \begin{bmatrix} 0.4+I0.5 & 0.2+I0.3 & 0.2+I0.4 \\ 0.2+I0.3 & 0.4+I0.3 & 0.2+I0.3 \\ 0.2+I0.4 & 0.2+I0.3 & 0.4+I0.5 \end{bmatrix} \text{ then}$$

$$det\ M = \begin{vmatrix} 0.4+I0.5 & 0.2+I0.3 & 0.2+I0.4 \\ 0.2+I0.3 & 0.4+I0.3 & 0.2+I0.3 \\ 0.2+I0.4 & 0.2+I0.3 & 0.4+I0.5 \end{vmatrix}$$

$$= \{0.4\{min(0.4,0.4) + min(0.2,0.2)\} + 0.2\{min(0.2,0.4) + min(0.2,0.2)\}$$
$$+0.2\{min(0.2,0.2)+min(0.2,0.4)\}\}+I\,\{0.5\{min(0.3,0.5)+min(0.3,0.3)\}$$
$$+\,0.3\{min(0.3,0.5)+min(0.4,0.3)\}+0.4\{min(0.3,0.3)+min(0.4,0.3)\}\}$$
$$= \{0.4(0.4+0.2) + I0.5(0.3+0.3)\} + \{0.2(0.2+0.2) + I0.3(0.3+0.3)\}$$
$$+\,\{0.2(0.2+0.2) + I0.4(0.3+0.3)\}$$
$$= (0.4\text{x}0.6 + I0.5\text{x}0.6) + (0.2\text{x}0.4 + I0.3\text{x}0.6) + (0.2\text{x}0.4 + I0.4\text{x}0.6)$$
$$= (0.24 + I0.30) + (0.08 + I0.18) + (0.08 + I0.24)$$
$$= 0.24 + I0.30. \tag{1.2}$$

Let

$$N = \begin{bmatrix} 0.4+I0.3 & 0.2+I0.5 & 0.2+I0.4 \\ 0.5+I0.3 & 0.4+I0.3 & 0.2+I0.3 \\ 0.2+I0.4 & 0.5+I0.3 & 0.4+I0.5 \end{bmatrix}.$$

Then

$$det\ N = \left\{0.4\begin{vmatrix} 0.4 & 0.2 \\ 0.5 & 0.4 \end{vmatrix} + 0.2\begin{vmatrix} 0.5 & 0.2 \\ 0.2 & 0.4 \end{vmatrix} + 0.2\begin{vmatrix} 0.5 & 0.4 \\ 0.2 & 0.5 \end{vmatrix}\right\}$$
$$+\,I\left\{0.3\begin{vmatrix} 0.3 & 0.3 \\ 0.3 & 0.5 \end{vmatrix} + 0.5\begin{vmatrix} 0.3 & 0.3 \\ 0.4 & 0.5 \end{vmatrix} + 0.4\begin{vmatrix} 0.3 & 0.3 \\ 0.4 & 0.3 \end{vmatrix}\right\}$$

$$= [0.4\{min(0.4,0.4) + min(0.5,0.2)\} + 0.2\{min(0.5,0.4) + min(0.2,0.2)\}]$$
$$+\,0.2\{min(0.5,0.5) + min(0.2,0.4)\}$$
$$+\,I[0.3\{min(0.3,0.5) + min(0.3,0.3)\} + 0.5\{min(0.3,0.5) + min(0.4,0.3)\}$$
$$+\,0.4\{min(0.3,0.3) + min(0.4,0.3)\}]$$
$$= 0.4(0.4+0.2) + 0.5(0.3+0.3) + 0.4(0.3+0.3)$$
$$+\,I[0.3\{min(0.3,0.5), min(0.3,0.3)\}$$
$$+\,0.5\{min(0.3,0.5), min(0.4,0.3)\} + 0.4\{min(0.3,0.3), min(0.4,0.3)\}]$$
$$= \{0.4(0.4+0.2) + I0.3(0.3+0.3)\} + \{0.2(0.4+0.2) + I0.5(0.3+0.3)\}$$
$$+\,\{0.3(0.5+0.2) + I0.4(0.3+0.3)\}$$
$$= (0.24 + I0.18) + (0.12 + I0.30) + (0.21 + I0.24)$$
$$= 0.24 + I0.30. \tag{1.3}$$

Again by summation of two neutrosophic number matrices, it can be seen that

$$M + N = \begin{bmatrix} 0.4+I0.5 & 0.2+I0.5 & 0.3+I0.4 \\ 0.5+I0.3 & 0.4+I0.3 & 0.2+I0.3 \\ 0.2+I0.4 & 0.5+I0.3 & 0.4+I0.3 \end{bmatrix}.$$

Then by definition of determinant of neutrosophic number matrices, it can be obtained that

$$det(M+N) = \left\{0.4\begin{vmatrix}0.4 & 0.2\\0.5 & 0.4\end{vmatrix} + 0.2\begin{vmatrix}0.5 & 0.2\\0.2 & 0.4\end{vmatrix} + 0.3\begin{vmatrix}0.5 & 0.4\\0.2 & 0.5\end{vmatrix}\right\}$$
$$+ I\left\{0.5\begin{vmatrix}0.3 & 0.3\\0.3 & 0.3\end{vmatrix} + 0.5\begin{vmatrix}0.3 & 0.3\\0.4 & 0.4\end{vmatrix} + 0.4\begin{vmatrix}0.3 & 0.3\\0.4 & 0.3\end{vmatrix}\right\}$$

$$\begin{aligned}
&= 0.4(0.4+0.2) + 0.2(0.4+0.2) + 0.3(0.5+0.2) + I[0.5(0.3+0.3) \\
&\quad + 0.5(0.3+0.3) + 0.4(0.3+0.3)] \\
&= (0.4x0.6 + I0.5x0.6) + (0.2x0.6 + I0.5x0.6) + (0.3x0.7 + I0.4x0.6) \\
&= (0.24 + I0.30) + (0.12 + I0.30) + (0.21 + I0.24) \\
&= 0.24 + I0.30. \qquad (1.4)
\end{aligned}$$

Hence from (1.2), (1.3) and (1.4),the following result can be obtained

$$det(M+N) = detM \ + detN.$$

**Property 1.4:** If the column of a neutrosophic number matrix is interchanged, then the value of the resulting determinant remains the same.

For example, let $K$ be any square neutrosophic number matrix such that

$$K = \begin{bmatrix}0.4+I0.5 & 0.2+I0.3 & 0.2+I0.4\\0.2+I0.3 & 0.4+I0.3 & 0.2+I0.3\\0.2+I0.4 & 0.2+I0.3 & 0.4+I0.5\end{bmatrix}$$

Then

$$det(K) = 0.24 + I0.30. \qquad (1.5)$$

Now if the first and second columns are interchanged, then the matrix will become

$$int\ K = \begin{bmatrix}0.2+I0.3 & 0.4+I0.5 & 0.2+I0.4\\0.4+I0.3 & 0.2+I0.3 & 0.2+I0.3\\0.2+I0.4 & 0.2+I0.4 & 0.4+I0.5\end{bmatrix}.$$

Then

$$\begin{aligned}
det(intK) &== \{0.2(0.2+0.2) + I0.3(0.2+0.2)\} \\
&\quad + \{0.4(0.3+0.3) + I0.5(0.3+0.3)\} \\
&\quad + \{0.2(0.2+0.2) + I0.4(0.3+0.3)\} \\
&= (0.08 + I0.12) + (0.24 + I0.30) + (0.08 + I0.24) \\
&= 0.24 + I0.30. \qquad (1.6)
\end{aligned}$$

Hence the result follows from (1.5) and (1.6).

**Theorem 1.1:** If $M$is a square neutrosophic number matrix and $\lambda \in [0,1]$ and if any row or column of $M$ is multiplied by $\lambda$, then the value of the determinant is multiplied by $\lambda$ that is $det(\lambda M) = \lambda \, det \, M$.

**Proof:** Let $M$ be a square neutrosophic number matrix of order $3 \times 3$ such that

$$M = \begin{bmatrix} \phi_{11}+I\varphi_{11} & \phi_{12}+I\varphi_{12} & \phi_{13}+I\varphi_{13} \\ \phi_{21}+I\varphi_{21} & \phi_{22}+I\varphi_{22} & \phi_{23}+I\varphi_{23} \\ \phi_{31}+I\varphi_{31} & \phi_{32}+I\varphi_{32} & \phi_{33}+I\varphi_{33} \end{bmatrix}.$$

Then

$$\begin{aligned} det(M) &= \{\lambda\phi_{11}(min(\phi_{22},\phi_{33}) + min(\phi_{23},\phi_{32}) + I\lambda\varphi_{11}(min(\varphi_{22},\varphi_{33}) \\ &+ min(\varphi_{23},\varphi_{32})\} + \{\lambda\phi_{21}(min(\phi_{12},\phi_{33}) + min(\phi_{13},\phi_{32}) \\ &+ I\lambda\varphi_{21}(min(\varphi_{12},\varphi_{33}) + min(\varphi_{13},\varphi_{32})\} + \{\lambda\phi_{31}(min(\phi_{12},\phi_{23}) \\ &+ min(\phi_{22},\phi_{13})\} + I\lambda\varphi_{31}(min(\varphi_{12},\varphi_{23}) + min(\varphi_{22},\varphi_{13})\} \\ &= \lambda\{\phi_{11}(min(\phi_{22},\phi_{33}) + min(\phi_{23},\phi_{32}) + I\varphi_{11}(min(\varphi_{22},\varphi_{33}) \\ &+ min(\varphi_{23},\varphi_{32})\} + \lambda\{\phi_{21}(min(\phi_{12},\phi_{33}) + min(\phi_{13},\phi_{32}) \\ &+ I\varphi_{21}(min(\varphi_{12},\varphi_{33}) + min(\varphi_{13},\varphi_{32})\} + \lambda\{\phi_{31}(min(\phi_{12},\phi_{23}) \\ &+ min(\phi_{22},\phi_{13})\} + I\varphi_{31}(min(\varphi_{12},\varphi_{23}) + min(\varphi_{22},\varphi_{13})\}. \end{aligned}$$

Thus from above it is clear that $det(\lambda M) = \lambda \, det M$.

**Theorem 1.2:** The value of the determinant remains unchanged when the rows and columns are interchanged. That is

$$det M = det M^T.$$

**Proof:**

$$\text{Let } M = \begin{bmatrix} \phi_{11}+I\varphi_{11} & \phi_{12}+I\varphi_{12} & \phi_{13}+I\varphi_{13} \\ \phi_{21}+I\varphi_{21} & \phi_{22}+I\varphi_{22} & \phi_{23}+I\varphi_{23} \\ \phi_{31}+I\varphi_{31} & \phi_{32}+I\varphi_{32} & \phi_{33}+I\varphi_{33} \end{bmatrix}$$

be a neutrosophic number matrix of order 3. Then the determinant of the neutrosophic number matrix$M$ can be calculated as

$$\begin{aligned} &= \{\phi_{11}(min(\phi_{22},\phi_{33}) + min(\phi_{23},\phi_{32}) + I\varphi_{11}(min(\varphi_{22},\varphi_{33}) \\ &+ min(\varphi_{23},\varphi_{32})\} + \{\phi_{21}(min(\phi_{12},\phi_{33}) + min(\phi_{13},\phi_{32}) \\ &+ I\varphi_{21}(min(\varphi_{12},\varphi_{33}) + min(\varphi_{13},\varphi_{32})\} \\ &+ \{\phi_{31}(min(\phi_{12},\phi_{23}) + min(\phi_{22},\phi_{13})\} + I\varphi_{31}(min(\varphi_{12},\varphi_{23}) \\ &+ min(\varphi_{22},\varphi_{13})\}. \end{aligned}$$

By the definition of transpose of neutrosophic number matrices, it is observed that

$$M^T = \begin{bmatrix} \phi_{11}+I\varphi_{11} & \phi_{21}+I\varphi_{21} & \phi_{31}+I\varphi_{31} \\ \phi_{12}+I\varphi_{12} & \phi_{22}+I\varphi_{22} & \phi_{32}+I\varphi_{32} \\ \phi_{13}+I\varphi_{13} & \phi_{23}+I\varphi_{23} & \phi_{33}+I\varphi_{33} \end{bmatrix}$$

$$= \{\phi_{11}(min(\phi_{22}, \phi_{33}) + min(\phi_{23}, \phi_{32})) + \phi_{21}(min(\phi_{12}, \phi_{33}) + min(\phi_{13}, \phi_{32})) + \phi_{31}(min(\phi_{12}, \phi_{23}) + min(\phi_{22}, \phi_{13})\} + I\{\varphi_{11}(min(\varphi_{22}, \varphi_{33}) + min(\varphi_{23}, \varphi_{32})) + \varphi_{21}(min(\varphi_{12}, \varphi_{33}) + min(\varphi_{13}, \varphi_{32})) + \varphi_{31}(min(\varphi_{12}, \varphi_{23}) + min(\varphi_{22}, \varphi_{13})\}$$

$$= \begin{bmatrix} \phi_{11} + I\varphi_{11} & \phi_{12} + I\varphi_{12} & \phi_{13} + I\varphi_{13} \\ \phi_{21} + I\varphi_{21} & \phi_{22} + I\varphi_{22} & \phi_{23} + I\varphi_{23} \\ \phi_{31} + I\varphi_{31} & \phi_{32} + I\varphi_{32} & \phi_{33} + I\varphi_{33} \end{bmatrix}.$$

This is equal to the value of $det(M)$.
Hence the result $det\ M = det\ M^T$.

## 1.6 Conclusions

The central goal of this chapter is to focus on various types of neutrosophic fuzzy matrices and their associated properties. In order to make the intention clear, all the concepts discussed here are supported by numerical examples. Some operations on neutrosophic fuzzy matrices such as trace and transpose are also discussed. Thereafter the determinant of neutrosophic fuzzy matrices is defined and properties associated with such matrices are discussed. In future this work can be extended to find adjoint of neutrosophic fuzzy matrices as well as the inverse of such matrices considered.

## Bibliography

[1] L.A. Zadeh., Fuzzy Sets, Information and Control, Vol. 8, pp. 338–353, 1965.

[2] K.T. Atanassov, Intuitionistic Fuzzy Sets, Fuzzy Sets and Systems, Vol. 20, No. 1, pp. 87–96, 1986.

[3] F. Smarandache, A Unifying Field in Logics. Neutrosophy: Neutrosophic Probability, Set and Logic, Rehoboth: American Research Press, 1999.

[4] F. Smarandache, Neutrosophy: Neutrosophic Probability, Set, and Logic, Ann Arbor, MI: ProQuest Information &Learning, p. 105, 1998.

[5] F. Smarandache, Neutrosophic Set a Generalization of the Intuitionistic Fuzzy Sets, International Journal of Pure and Applied Mathematics, pp. 287–297, 2005.

[6] W.B.W. Kandasamy and F. Smarandache, Fuzzy Relational Maps and Neutrosophic Relational Maps, HEXIS Church Rock, 2004.

[7] S. Broumi and F. Smarandache, Correlation Co-efficient of Interval Neutrosophic Set, Applied Mechanics and Materials, Vol. 436, pp. 511–517, 2013.

[8] S. Broumi, L.H. Son, A. Bakali, M. Talea, F. Smarandache and G. Selvachandran, Computing Operational Matrices in Neutrosophic Environments: A Matlab Toolbox, Neutrosophic Sets and Systems, Vol. 18, pp. 58–66, 2017.

[9] S. Topal, S.Broumi, A. Bakali, M. Talea and F. Smarandache, A Python Toolfor Implementations on Bipolar Neutrosophic Matrices, Neutrosophic Sets and Systems, Vol. 28, pp. 138–161, 2019.

[10] S. Broumi, A. Bakali, M. Talea and F. Smarandache, A Matlab Toolbox for Interval Valued Neutrosophic Matrices for Computer Applications, Uluslararası Yönetim Bilişim Sistemlerive Bilgisayar Bilimleri Dergisi, Vol. 1, No. 1, pp. 1–21, 2017.

[11] M. Abobala, On Some Neutrosophic Algebraic Equations, Journal of New Theory, Vol. 33, 2020.

[12] H. Sankari and M. Abobala, Neutrosophic Linear Diophantine Equations with Two Variables, Neutrosophic Sets and Systems, Vol. 38, 2020.

[13] H. Khaled, A. Younus and A. Mohammad, The Rectangle Neutrosophic Fuzzy Matrices, Faculty of Education Journal, Vol. 15, 2019 (Arabic version).

[14] F. Smarandache and M. Abobala, n-Refined Neutrosophic Rings, International Journal of Neutrosophic Science, Vol. 6, pp. 83–90 , 2020.

[15] F. Smarandache and M. Abobala., n-Refined Neutrosophic Vector Spaces, International Journal of Neutrosophic Science, Vol. 7, pp. 47–54, 2020.

[16] W.B.V. Kandasamy and F. Smarandache, Neutrosophic Rings, Phoenix: Hexis, Infinite Study, 2006.

[17] M. Abdel-Basset, M. Saleh, A. Gamal and F. Smarandache, An Approach of TOPSIS Technique for Developing Supplier Selection with Group Decision Making under Type-2 Neutrosophic Number, Applied Soft Computing, Vol. 7, pp. 438–452, 2019.

[18] M, Abdel-Basset, M. El-hoseny, A. Gamal and F. Smarandache, A Novel Model for Evaluation Hospital Medical Care Systems Based on Plithogenic Sets, Artificial Intelligence in Medicine, Vol. 100, p. 101710, 2019.

[19] M. Abdel-Basset, M. Mohamed, M. Elhoseny, F. Chiclana and A.E.N.H. Zaied, Cosine Similarity Measures of Bipolar Neutrosophic Set for Diagnosis of Bipolar Disorder Diseases, Artificial Intelligence in Medicine, Vol. 101, p. 101735, 2019.

[20] M.G.Thomason, Convergence of Powers of a Fuzzy Matrix, Journal of Mathematical Analysis and Application, Vol. 57, pp. 476–480, 1977.

[21] J.B. Kim and Baartmans, Determinant Theory for Fuzzy Matrices, Fuzzy Sets and Systems, Vol. 29, pp. 349–356, 1989.

[22] J.B. Kim, Determinant Theory for Fuzzy and Boolean Matrices, Congressus Numerantium Utilitas Mathematica Pub, pp. 273–276, 1978.

[23] J.B. Kim, Idempotents and Inverses in Fuzzy Matrices, Bulletin of Malayasian Mathematical Society, Vol. 6, No. 2, pp. 57–61, 1983.

[24] M. Dhar, Representation of Fuzzy Matrices Based on Reference Function, International Journal of Intelligent Systems and Applications, Vol. 5, No. 2, pp. 84–90, 2013.

[25] M. Dhar, A Note on Determinant of Square Fuzzy Matrix, International Journal of Information Engineering and Electronic Business, Vol. 5, No. 1, pp. 26–32, 2013.

[26] M. Dhar, A Note on Determinant and Adjoint of Fuzzy Square Matrix, International Journal of Intelligent systems and Applications, Vol. 5, No. 5, pp. 58–67, 2013.

[27] W.B.V. Kandaswamy and F. Smarandache, Fuzzy Interval Matrices, Neutrosophic Interval Matrices and Their Applications, Phoenix: Hexis, 2006.

[28] M. Dhar, S. Broumi and F. Smarandache, A Note on Square Neutrosophic Fuzzy Matrices, Neutrosophic Sets and Systems, Vol. 3, pp. 57–41, 2014.

[29] R. Das, F. Smarandache and B.C. Tripathy, Neutrosophic Fuzzy Matrices and Some Algebraic Operations, Neutrosophic Sets and Systems, Vol. 32, pp. 401–409, 2020.

[30] M. Dhar, On K-idempotent Neutrosophic Fuzzy Matrices, Plethogenic Logic and Computation, Vol. 2, pp. 95–106, 2024.

[31] D.S. Purushotam, Secondary Partial Ordering of Neutrosophic Fuzzy Matrices, International Journal of Neutrosophic Sciences, Vol. 24, No. 4, pp. 411–419, 2024.

# Chapter 2

## A Computational Study of Neutrosophic Probability, with Focus on Risk Evaluation

**Prasanta Kumar Raut, M. Gayathri Lakshmi, R. Rajalakshmi, K. Saritha, Said Broumi, and Zahid Khan**

### 2.1 Introduction

Now a days, ambiguity, uncertainty, and insufficient information are common factors in decision-making in business management, science, and engineering. By giving precise numerical probabilities for the occurrence of events, classical probabilistic theory has been a fundamental tool for modeling ambiguity for decades. But in many real-world applications, such as engineering safety analysis, medical diagnostics, and financial risk assessment, uncertainty is brought on not only by chance but also by inconsistent, unpredictable, and conflicting data. Because of this, conventional statistical methods typically fall short in capturing the multifaceted nature of uncertainty. Zadeh [1] developed the concept of fuzzy sets (FSs) to address these issues. By permitting elements to have partial membership values in the range $[0, 1]$, FSs extend classical sets. An additional flexible method for characterizing imprecise or ambiguous data is provided by this theoretical framework. By creating intuitionistic fuzzy sets (IFSs), which comprise both membership and non-membership functions, Atanassov [2] extended this idea. Given their uses, FSs and IFSs are constrained in their ability to express independent indeterminacy by the requirement that the sum of the degrees of truth and false not exceed one. Smarandache [3, 4] introduced the concept of neutrosophic sets (NSs) that overcome these situations by identifying information with three different membership components represented as falsity ($F$), indeterminacy ($I$), and truth ($T$). Each component may vary independently on the non-standard interval $]^{-}0, 1^{+}[$. The enlarged representation allows neutosophic sets to treat highly ambiguous, inconsistent, and missing data in an effective

DOI: 10.1201/9781003683339-2

manner. Neutrophilic logic, set theory, and probability have attracted attention in a number of fields like risk analysis, control systems, information fusion, and decision-making.

The concept of *neutrosophic probability* (NP) generalizes the standard probability in the triplet structure $(T, I, F)$, where $T(E)$ represents the probability of occurrence, $I(E)$ is the degree of indeterminacy, and $F(E)$ is the probability of non-occurrence [4]. The above definition extends classical, fuzzy, and intuitionistic probability to provide a more complete formal structure for assessing vague or conflicting data.

Neutrosophic models have shown their efficacy in many fields like multi-attribute decision-making [5], image processing [6], system reliability [7], and shortest path optimization problems [8–13]. Considering these conceptual developments, the mathematical frameworks and computational tools for the implementation of neutrosophic probability in real-world systems remain scarce. There has been an increasing need for robust computational frameworks capable of assessing neutrosophic probabilities, handling both numerical and interval-valued data inputs, and facilitating aggregation and decision-making methods in uncertain and indeterministic contexts. The rest of the chapter is organized as follows: Section 2.2 provides some basic facts and a literature review of some relevant works in classical, fuzzy, and neutrosophic probability theories. The concepts, notation, and preliminary steps needed for the computational framework are explained in Section 2.3. Section 2.4 presents a framework for neutrosophic probability evaluation through algorithmic techniques and implementation details. Neutrosophic risk assessment model and ways of evaluating and interpreting the unknown risk variables are presented in Section 2.5. Three case studies from the fields of engineering, finance, and medical diagnostics are used to demonstrate the usefulness of the framework in Section 2.6. A summary of the main findings and recommendations for future research are provided in Section 2.7.

### 2.1.1 Objectives and contributions

This chapter's key objectives are as follows:

- To offer a computational methodology for estimating neutrosophic probability combining numerical and interval-valued representations.
- To use the suggested framework in risk assessment issues when standard probability metrics are inadequate.
- Develop algorithms and modeling methodologies for calculating and interpreting $(T, I, F)$ triplets.
- To show the practical value of the suggested strategy via real-world case studies in medical, engineering, and finance.

## 2.2 Preliminaries

In this section, some basic definitions related to neutrosophic sets, neutrosophic numbers, and neutrosophic probability are presented, together with notations and conventions that will be used in this chapter.

### 2.2.1 Neutrosophic sets and neutrosophic numbers

Let $X$ be a nonempty universe of discourse. A *neutrosophic set* $A$ on $X$ is defined by three independent membership functions

$$T_A,\, I_A,\, F_A : X \longrightarrow [0,1],$$

where

- $T_A(x)$ denotes the degree to which $x$ belongs to $A$;
- $I_A(x)$ denotes the degree of indeterminacy regarding the membership of $x$;
- $F_A(x)$ denotes the degree to which $x$ does not belong to $A$.

Thus, each element $x \in X$ is represented by the ordered triplet

$$A(x) = \big(T_A(x),\, I_A(x),\, F_A(x)\big),$$

with no restriction on the sum $T_A(x) + I_A(x) + F_A(x)$, allowing the three components to vary freely and independently.

A *single-valued neutrosophic number* is a triplet

$$\alpha = (T_\alpha,\, I_\alpha,\, F_\alpha), \qquad T_\alpha, I_\alpha, F_\alpha \in [0,1].$$

Such a number represents the truth, indeterminacy, and falsity degrees of an imprecise quantity. For computational purposes, arithmetic operations may be defined componentwise, for example:

$$\alpha + \beta = (T_\alpha + T_\beta,\; I_\alpha + I_\beta,\; F_\alpha + F_\beta),$$

subject to renormalization when necessary, and similarly for subtraction, scalar multiplication, and weighted aggregation.

### 2.2.2 Interval-valued neutrosophic sets

In many practical problems, the three membership components cannot be expressed as precise numbers. To capture this additional level of uncertainty, the concept of an *interval-valued neutrosophic set* (IVNS) is used.

Let $X$ be a universe. An IVNS $A$ on $X$ assigns to every element $x \in X$ three closed intervals:

$$A(x) = \left([T_A^-(x), T_A^+(x)],\ [I_A^-(x), I_A^+(x)],\ [F_A^-(x), F_A^+(x)]\right),$$

where

$$0 \leq T_A^-(x) \leq T_A^+(x) \leq 1,\ \ 0 \leq I_A^-(x) \leq I_A^+(x) \leq 1,\ \ 0 \leq F_A^-(x) \leq F_A^+(x) \leq 1.$$

Each interval represents a range of admissible values for truth, indeterminacy, and falsity. For example, $[T_A^-(x), T_A^+(x)]$ provides the lower and upper bounds on the truth-membership of $x$ with respect to $A$.

Operations on IVNSs are defined using interval arithmetic. For two IVNSs $A$ and $B$, the union and intersection may be written as:

$$T_{A\cup B}^-(x) = \max\left(T_A^-(x), T_B^-(x)\right), \qquad T_{A\cup B}^+(x) = \max\left(T_A^+(x), T_B^+(x)\right),$$

$$T_{A\cap B}^-(x) = \min\left(T_A^-(x), T_B^-(x)\right), \qquad T_{A\cap B}^+(x) = \min\left(T_A^+(x), T_B^+(x)\right),$$

with analogous formulas for the $I$ and $F$ components using suitable interval combinations.

The interval-valued framework is capable of representing both epistemic uncertainty and measurement imprecision, making it particularly useful in computational models that involve fluctuating, incomplete, or conflicting data.

### 2.2.3 Neutrosophic probability and its types

Neutrosophic probability generalizes both classical and fuzzy probabilities for treating indeterminacy and uncertainty. A neutrosophic probability is defined as a function that maps every event $E$ to a tuple of three components: $(T, I, F)$, representing, respectively the probability of the event being true, indeterminate, and false. More precisely, neutrosophic probability can be expressed as This projection of the whole plane into another plane through a point (or a line) is called perspective.

$$P(E) = (T(E), I(E), F(E))$$

where $T(E)$, $I(E)$, and $F(E)$ are the truth, indeterminacy, and falsity components of the probability of event $E$. There are several types of neutrosophic probability, including:

- **Standard Neutrosophic Probability**: It defines a basic form of neutrosophic probability, and the sum of $T$, $I$, and $F$ is not necessarily equal to 1.
- **Neutrosophic Overset**: It occurs when one or more components of neutrosophic probability are greater than 1. This may happen when an occurrence has a degree of truth greater than the usual range, for instance, in overconfirmed cases.

- **Neutrosophic Underset**: When one or more components of neutrosophic probability are less than 0, that indicates the truth or falsity degree of the event is less than the expected situation.

- **Neutrosophic Offset**: A more general case, where some components of neutrosophic probability lie outside the interval [0, 1], both above and below, representing a broader range of uncertainty and inconsistency.

### 2.2.4 Notation and conventions

We will use the following notations and conventions throughout this chapter:

- $T(E), I(E), F(E)$ are the truth, indeterminacy, and falsity components of neutrosophic probability for event $E$.

- $A$ and $B$ will denote neutrosophic sets.

The operators $\oplus$ and $\otimes$ will be used to represent neutrosophic union and intersection, respectively.

- We will refer to the degree of uncertainty (or indeterminacy) as $I$, and the associated components—truth and falsity—will be represented as $T$ and $F$, respectively.

- For interval-valued neutrosophic sets, $[T(x), T^*(x)], [I(x), I^*(x)], [F(x), F^*(x)]$ are interval bounds for the truth, indeterminacy, and falsity components, respectively.

These definitions and notations will form the foundation for the computational models and algorithms developed in subsequent sections.

---

## 2.3 Computational Framework for Neutrosophic Probability

This section articulates the computational structure employed in the evaluation of neutrosophic probability. It is intended to present a well-structured procedure for determining the truth, indeterminacy, and falsity components associated with an event. These foundations also serve to prepare the reader for the risk assessment model developed in Section 2.5, where these same computational principles are applied in practical decision-making settings.

### 2.3.1 Computational procedure

The neutrosophic probability is evaluated in an orderly fashion:

1. **Input Data:** Data on the occurrence of event E are collected from observations, expert estimates, or history.
2. **Determine Components:**
   - $T(E)$ is calculated based on evidence supporting the occurrence of $E$.
   - $I(E)$ is based on ambiguity, conflicting data, or incomplete knowledge.
   - $F(E)$ is obtained from evidence that supports non-occurrence.
3. **Output:** The triplet $(T(E), I(E), F(E))$ represents the neutrosophic probability of $E$.

### 2.3.2 Algorithmic representation

For clarity, the computational steps can be summarized in the following pseudocode:

**Algorithm: Neutrosophic Probability Evaluation**

1. Initialize $T(E)$, $I(E)$, and $F(E)$.
2. Compute $T(E)$ using supportive evidence.
3. Estimate $I(E)$ based on indeterminate or conflicting information.
4. Determine $F(E)$ using evidence against the event.
5. Return $P(E) = (T(E), I(E), F(E))$.

### 2.3.3 Remarks on computational efficiency

The form of input data determines the actual computational effort: interval-valued inputs are handled by interval arithmetic in order to keep the uncertainty; numerical inputs allow for direct computations. In practice, such computations can be conveniently simplified by resorting to standard numerical routines. These results will provide the basis for the risk assessment model to be developed in Section 2.5, where these elements are integrated in order to compute composite risk within a neutrosophic framework.

## 2.4 Proposed Risk Assessment Model

This section develops a neutrosophic probability framework, as discussed in Section 2.3, for a risk assessment model. In doing so, it integrates uncertainty, indeterminacy, and the likelihood of non-occurrence within a unified framework that can better depict real-world risk factors than can classical probability. The formulation in this section will also act as a bridge to connect the computational procedure developed earlier with the case studies introduced in Section 2.6, where the model is applied to practical situations with incomplete or fluctuating information.

### 2.4.1 Risk factor modeling using neutrosophic values

In many decision-making scenarios, individual risk factors contain varying degrees of uncertainty that cannot be fully captured by traditional numerical measures. To reflect this, each risk factor is expressed as a neutrosophic probability. This representation allows the three fundamental components—truth, indeterminacy, and falsity—to be treated independently, offering a clearer view of the uncertain aspects of the assessment. For a given risk factor $R_i$, the neutrosophic probability is written as

$$P_i(R) = (T_i, I_i, F_i),$$

where $T_i$ denotes the degree to which the risk is likely to occur, $I_i$ measures the hesitation or conflict present in the information, and $F_i$ represents the degree of non-occurrence. A typical example is a natural hazard indicator evaluated as $(0.7, 0.2, 0.1)$, where the first component reflects supportive evidence, the second captures uncertainty, and the third summarizes opposing information.

### 2.4.2 Aggregation of multiple risk indicators

In most of the decision-making problems, the uncertainty associated with each individual risk factor cannot be properly conveyed through the traditional numerical representation. For this reason, the neutrosophic probability representation has been adopted for every single risk factor. By employing this representation, all three components—truth, indeterminacy, and falsity—may be handled separately and thus reveal the unclear parts of an evaluation more clearly. Each risk factor $R_i$ is characterized by a neutrosophic probability

$$P_i(R) = (T_i, I_i, F_i),$$

where $T_i$ is the likelihood of the risk occurrence, $I_i$ is the conflict or uncertainty in the information, and $F_i$ is the likelihood of the non-occurrence. A typical instance is a natural hazard indicator estimated as $(0.7, 0.2, 0.1)$, where the first part shows strong support, the second part is about uncertainty, and the last part is about opposition.

### 2.4.3 Evaluation of composite risk under neutrosophic conditions

The next step after combining the individual components is to express the entire composite risk in the following manner:

$$R_{\text{total}} = (T_{\text{total}}, I_{\text{total}}, F_{\text{total}}),$$

where each term results from the summed values. How the terms are understood is determined by the relative sizes of the components. The combination of a large value of $T_{\text{total}}$ and a small value of $F_{\text{total}}$ would indicate a great risk of occurrence, while a high value of $I_{\text{total}}$ would mean that the quality of the information is poor or it is too contradictory. In contrast, if the component of falsity is dominant, then that would be interpreted as indicating a low risk. All these interpretations hand over to the decision-makers a wider comprehension of the structure.

### 2.4.4 Mathematical formulation for risk evaluation

In order to condense the data represented in the neutrosophic triplet, a composite score is introduced. This score is computed based on the weighted contributions of the three components of truth, indeterminacy, and falsity:

$$S_{\text{total}} = w_T T_{\text{total}} + w_I I_{\text{total}} + w_F F_{\text{total}},$$

wherein the weights $w_T$, $w_I$, and $w_F$ are indicative of the relative importance of the corresponding components. These may be determined by criteria specific to the domain, for instance, the degree of uncertainty and the possible impacts of incorrect evaluations. The scoring system operates such that a higher score corresponds to a higher total risk level, considering especially the part played by the indeterminacy.

### 2.4.5 Decision rules and risk categorization

Based on the composite score and the neutrosophic components, risk categories are assigned. A high score corresponds to a high-risk classification, reflecting substantial evidence and limited opposition. A moderate score indicates that the risk is neither clearly high nor clearly low, often due to appreciable indeterminacy. A low score corresponds to a situation in which the evidence of non-occurrence is strong. These categories allow the allocation of resources to be aligned with the assessed level of risk and assist in establishing priorities for further analysis or mitigation.

## 2.5 Case Studies and Illustrative Examples

This section is devoted to a number of examples intended to illustrate how neutrosophic probability can be applied to real situations with incomplete, ambiguous, or even contradictory information. Each has the dual purpose of demonstrating, first, how the computational framework in Section 2.3 can be brought to bear in practical applications, and second, how the proposed risk assessment model in Section 2.5 works when several uncertain indicators interact within the same decision environment.

### 2.5.1 Engineering system reliability

In many engineering analyses, the reliability of a system is determined based on factors whose values cannot be precisely determined. For example, consider a mechanical system whose performance depends on three elements: material strength $M$, load variation $L$, and maintenance quality $Q$. Because of changing operating environments and limitations in measurement processes, each component carries some degree of uncertainty.

Suppose experts estimate the neutrosophic probabilities of the components as follows:

$$P(M) = (0.82, 0.10, 0.08),\ P(L) = (0.65, 0.25, 0.10),\ P(Q) = (0.88, 0.07, 0.05).$$

Assume that the relative importance of these components is represented by the weights:

$$w_M = 0.4, \quad w_L = 0.35, \quad w_Q = 0.25.$$

Using the neutrosophic weighted aggregation rule,

$$P_{\text{agg}} = (T_{\text{agg}}, I_{\text{agg}}, F_{\text{agg}}),$$

where

$$T_{\text{agg}} = 0.4(0.82) + 0.35(0.65) + 0.25(0.88),$$
$$I_{\text{agg}} = 0.4(0.10) + 0.35(0.25) + 0.25(0.07),$$
$$F_{\text{agg}} = 0.4(0.08) + 0.35(0.10) + 0.25(0.05).$$

A direct calculation yields:

$$P_{\text{agg}} \approx (0.77,\ 0.15,\ 0.08).$$

This combined neutrosophic probability suggests that the system has a very good chance of operating properly, but still, some uncertainty exists due to changing load conditions and inadequate maintenance records. The triplet format enables the engineers to tell the difference between guaranteed reliability and remaining doubt.

### 2.5.2 Financial investment risk

The unpredictable nature of the market usually brings uncertainty to the evaluation of financial risks. Imagine there is a portfolio that includes three types of assets: stocks ($E$), government bonds ($B$), and real estate ($R$). Their neutrosophic return-risk estimates are:

$$P(E) = (0.58, 0.30, 0.12),\ P(B) = (0.72, 0.18, 0.10),\ P(R) = (0.81, 0.14, 0.05).$$

Suppose the investor allocates funds according to the weights:

$$w_E = 0.5, \quad w_B = 0.3, \quad w_R = 0.2.$$

Applying neutrosophic aggregation:

$$T_{\text{agg}} = 0.5(0.58) + 0.3(0.72) + 0.2(0.81),$$

$$I_{\text{agg}} = 0.5(0.30) + 0.3(0.18) + 0.2(0.14),$$

$$F_{\text{agg}} = 0.5(0.12) + 0.3(0.10) + 0.2(0.05).$$

This results in:

$$P_{\text{agg}} \approx (0.66,\ 0.24,\ 0.10).$$

The moderate indeterminacy component reflects market fluctuations and inconsistent economic indicators. The investor may classify the portfolio as carrying a moderate risk level, where additional information could reduce the uncertainty component.

### 2.5.3 Medical diagnostic assessment

Medical diagnosis frequently involves conflicting symptoms and incomplete test information. Consider a diagnostic evaluation for a patient suspected of having a particular condition. Three clinical indicators are used: laboratory marker $A$, symptom intensity $S$, and family history $H$. The neutrosophic representations are:

$$P(A) = (0.71, 0.20, 0.09),\ P(S) = (0.67, 0.23, 0.10),\ P(H) = (0.62, 0.30, 0.08).$$

If the diagnostic weights are:

$$w_A = 0.45, \quad w_S = 0.35, \quad w_H = 0.20,$$

then the aggregated neutrosophic probability becomes:

$$P_{\text{agg}} \approx (0.68,\ 0.23,\ 0.09).$$

Although the truth component is moderately high, the indeterminacy component indicates that some medical evidence remains inconclusive. Physicians may therefore recommend further testing to reduce diagnostic uncertainty.

### 2.5.4 Comparison with classical methods

Classical probability assigns a single value to represent the likelihood of an event, which can obscure the distinctions between certainty, ambiguity, and contradiction in the data. In contrast, the neutrosophic approach separates these influences through the triplet $(T, I, F)$, exposing the structure of uncertainty more transparently.

For instance, in the engineering example, the classical reliability might be approximated simply as 0.77, whereas the neutrosophic representation

$$(0.77, 0.15, 0.08)$$

reveals that part of the evaluation stems from incomplete or conflicting information.

Financial analysis requires more than the classical risk estimate of 0.66 to understand what portion of this assessment stems from market behavior and what portion results from economic data unpredictability. The neutrosophic probability system helps decision-makers understand this distinction.

The proposed framework enables decision-makers to understand risk magnitude and evidence-based uncertainty sources through its enhanced uncertainty representation method.

---

## 2.6 Conclusion

This chapter establishes a computational system to assess neutrosophic probability while demonstrating its value for risk evaluation under conditions of incomplete or conflicting or uncertain information. The method distinguishes between truth and indeterminacy and falsity to represent complex risk factors more effectively than traditional classical and fuzzy models. The established procedures together with aggregation rules enable researchers to perform structured uncertainty data analysis for better decision-making. The engineering and financial and medical diagnostic examples show that neutrosophic probability effectively handles uncertainty variations which standard analysis methods fail to detect. The research findings confirm the necessity of including indeterminacy in contemporary risk assessment while creating opportunities to develop new computational models and algorithms.

---

## Bibliography

[1] Zadeh, L.A. (1965). Fuzzy sets. *Information and Control*, 8(3), 338–353.

[2] Atanassov, K.T. (1986). Intuitionistic fuzzy sets. *Fuzzy Sets and Systems*, 20(1), 87–96.

[3] Smarandache, F. (1998). *A Unifying Field in Logics: Neutrosophy. Neutrosophic Probability, Set, and Logic.* American Research Press.

[4] Smarandache, F. (2005). *Neutrosophic Probability, Set, and Logic.* American Research Press.

[5] Biswas, P., Pramanik, S., & Giri, B.C. (2016). TOPSIS method for multi-attribute group decision making under single-valued neutrosophic environment. *Neural Computing and Applications*, 27(3), 727–737.

[6] Ye, J. (2014). A novel technique for image similarity measure using single-valued neutrosophic set. *Applied Mathematical Modelling*, 38(3), 1170–1175.

[7] Ye, J. (2014). Multicriteria decision-making method using the correlation coefficient under single-valued neutrosophic environment. *International Journal of General Systems*, 43(6), 538–553.

[8] Raut, P.K., Behera, S.P., Broumi, S., & Baral, A. (2024). Evaluation of shortest path on multi-stage graph problem using dynamic approach under neutrosophic environment. *Neutrosophic Sets and Systems*, 64, 113–131.

[9] Raut, P.K., Behera, S.P., Broumi, S., & Baral, A. (2024). Evaluation of shortest path by using breadth-first algorithm under neutrosophic environment. *HyperSoft Set Methods in Engineering*, 1, 34–45.

[10] Raut, P.K., Satapathy, S.S., Behera, S.P., Broumi, S., & Sahoo, A.K. (2025). Solving the shortest path problem in an interval-valued neutrosophic Pythagorean environment using an enhanced A* search algorithm. *Neutrosophic Sets and Systems*, 76, 360–374.

[11] Raut, P.K., Behera, S.P., Broumi, S., & Mishra, D. (2023). Calculation of fuzzy shortest path problem using multi-valued neutrosophic number under fuzzy environment. *Neutrosophic Sets and Systems*, 57, 356–369.

[12] Broumi, S., Raut, P.K., & Behera, S.P. (2023). Solving shortest path problems using an ant colony algorithm with triangular neutrosophic arc weights. *International Journal of Neutrosophic Science*, 20(4), 128–138.

[13] Raut, P.K., Pramanik, S., Mohapatra, D.K., & Sahoo, S.K. (2025). Solving the shortest path based on the traveling salesman problem with a genetic algorithm in a Fermatean neutrosophic environment. *Neutrosophic Sets and Systems*, 78, 353–366.

# Chapter 3

## *Optimization Algorithms in Nonstandard Neutrosophic Environments: A Metaheuristic Approach*

**Prasanta Kumar Raut, R. Rajalakshmi, and Said Broumi**

### 3.1 Introduction

Optimization problems are fundamental to many scientific and engineering disciplines. The conventional methods for optimization usually assume that all parameters are precise, consistent, and completely known. Such assumptions are rarely the case in reality since most of the systems have the main characteristics of being *uncertain*, *indeterminate*, and *inconsistent* to different extents. Consequently, these factors severely limit the use and accuracy of traditional optimization methods that rely on deterministic solutions. So, the theoretical methods and classical crisp optimization models have struggled in most cases to represent the vagueness and complexity that are typical of uncertain situations.

In order to deal with these drawbacks, several different extensions of classical set theory have been introduced. One of these extensions is the idea of *fuzzy sets* that was proposed by Zadeh [1], which allows for partial memberships of objects in a range $[0, 1]$ and thus offers a more flexible way of modeling imprecise data. Later, Atanassov [2] came up with the concept of *intuitionistic fuzzy sets*, which include both the degree of membership and the degree of non-membership, to convey the uncertainty or vagueness of the information more accurately.

To maximize the representational capability even more, Smarandache [3] introduced the concept of *neutrosophic sets (NSs)*, which, in practice, apply three separate elements—truth, indeterminacy, and falsity membership degrees. The possible values of these parameters can vary independently within

DOI: 10.1201/9781003683339-3

the range $]^{-}0, 1^{+}[$, thus allowing a broader usage of modelings for uncertainty and incomplete information. A great number of researchers then followed and went on to utilize the neutrosophic theory in different contexts including but not limited to, decision-making, pattern recognition, and optimization [4–14, 16].

At the same time, the theoretical developments led to the emergence of *metaheuristic algorithms* [15] which turned out to be powerful computational tools for dealing with large-scale, nonlinear, and high-dimensional optimization problems. Algorithms like *genetic algorithm (GA)* [14], *particle swarm optimization (PSO)* [17], and *ant colony optimization (ACO)* [13] reflect the greatest possible number of such algorithms that have been placed in the category of being very efficient in overcoming the difficulties of searching in such areas along with expanding the acceptance of convergence. Joining the algorithms with neutrosophy marks a progressive and speedy solution for the optimization issues under uncertainty.

This chapter sets out to examine and evaluate optimization algorithms in nonstandard neutrosophic environments with a metaheuristic framework. It aims to show the positive influence of neutrosophic modeling on the adaptability and robustness of the optimization processes when applied to complex and uncertain systems. The authors have organized the following sections to lead the reader smoothly through the rest of the chapter. The concepts of NSs, neutrosophic numbers, and the necessary mathematical preliminaries for the proposed framework are introduced in Section 3.2. In Section 3.3, the neutrosophic optimization problem is formulated, and the uncertainty category is applied to the structure of decision variables, objective functions, and constraints. Section 3.4 is dedicated to the neutrosophic representation of parameters, which covers membership functions and aggregation mechanisms to model truth, uncertainty, and falsehood. Section 3.5 integrates metaheuristic algorithms into the neutrosophic environment while focusing on the adjustment needed in the genetic operators to fit the neutrosophic space. In Section 3.6, the computational experiments are reported and the performance of the proposed approach on benchmark and real-world datasets is analyzed. To conclude, Section 3.7 summarizes the key points and indicates the directions for future research.

## 3.2 Preliminaries

### 3.2.1 Neutrosophic sets

NSs are the most general forms of the classical, fuzzy, and intuitionistic fuzzy sets. Each element of the NS is characterized by three independent membership degrees.

**Definition 3.1:** *Let U be a universe of discourse. A* ***single-valued neutrosophic set (SVNS)*** *A in U is defined as:*

$$A = \{\langle x, T_A(x), I_A(x), F_A(x)\rangle \mid x \in U\}$$

*where:*

- $T_A(x) : U \to [0,1]$ *is the truth value membership,*
- $I_A(x) : U \to [0,1]$ *is the indeterminacy value membership,*
- $F_A(x) : U \to [0,1]$ *is the falsity value membership,*

*for all* $x \in U$, $0 \leq T_A(x) + I_A(x) + F_A(x) \leq 3$.

The application of this framework to the modeling of uncertain, incomplete, and contradictory information has proven to be more effective than the techniques based on conventional fuzzy logic.

### 3.2.2 Neutrosophic numbers

A **single-valued neutrosophic number (SVNN)** can be defined as $\tilde{a} = (T_a, I_a, F_a)$, where each of $T_a$, $I_a$, and $F_a$ is in the interval $[0,1]$. During the optimization process, SVNN arithmetic operations and ranking functions are used to compare and manipulate the neutrosophic data.

### 3.2.3 Metaheuristic algorithms

Such algorithms are considered to be high-level methods that are put into practice in order to get approximate answers for hard optimization problems. They show excellent performance in problem types that have big, non-convex, or discrete solution spaces.

Following are some metaheuristic algorithms that are very popular and widely used:

- **GA:** The process resembles a natural lottery by virtue of the performing operations like picking, fusing, and changing.
- **PSO:** It imitates the simultaneous shifts of flocks of birds or schools of fish to find the way through the search space by persistently altering the particle's speed and position.
- **ACO:** This method is based on the searching behavior of the ants for food, where the pheromone trails left and the following lead to the discovery of the best routes.

The algorithms are modified in such a way that they work within the limits of the neutrosophic framework, thus making their indeterminate and inconsistency data handling power stronger.

## 3.3 Problem Formulation

Optimization problems in real-world scenarios often involve incomplete, indeterminate, and inconsistent information. To address these issues, we formulate the optimization problem within a neutrosophic framework. Consider an optimization problem:

$$\min_{\tilde{x}\in\tilde{X}} \tilde{f}(\tilde{x}) \tag{3.1}$$

s t:

$$\tilde{g}_i(\tilde{x}) \leq \tilde{b}_i, \quad i = 1, 2, \ldots, m \tag{3.2}$$

$$\tilde{h}_j(\tilde{x}) = \tilde{c}_j, \quad j = 1, 2, \ldots, p \tag{3.3}$$

where:

- $\tilde{x} = (x_1, x_2, \ldots, x_n)$ represents decision variables, each defined as a single-valued neutrosophic number: $\tilde{x}_k = (T_k, I_k, F_k)$,
- $\tilde{f}(\tilde{x})$ is the neutrosophic objective function to be minimized,
- $\tilde{g}_i(\tilde{x})$ and $\tilde{h}_j(\tilde{x})$ are neutrosophic constraint functions,
- $\tilde{b}_i$ and $\tilde{c}_j$ are neutrosophic bounds or constants.

The objective is to obtain the neutrosophic solution $\tilde{x}^*$ that is optimal such that the objective function is minimized and the constraints are met in the neutrosophic way. This formulation makes it possible for the decision-making process to take into account the different levels of uncertainty and imprecision that are characteristic of the real-world applications.

## 3.4 Neutrosophic Representation

We present in this section the neutrosophic representation of the optimization problem's parameters through the definition of the truth, indeterminacy, and falsity membership functions. By using these functions, it becomes possible to capture the decision-making environments' uncertainty, inconsistency, and incompleteness.

### 3.4.1 Membership functions

Let a variable or a parameter involved in the optimization problem be denoted by $x$. Its value is represented as an SVNN:

$$\tilde{x} = (T_x, I_x, F_x)$$

where:

- $T_x \in [0,1]$ is the degree of truth that measures the support or certainty associated with $x$,
- $I_x \in [0,1]$ is the degree of indeterminacy that shows the extent of ambiguity or lack of knowledge about $x$,
- $F_x \in [0,1]$ is the degree of falsity which indicates the extent to which $x$ is not supported or is in contradiction.

The following relationship holds:

$$0 \leq T_x + I_x + F_x \leq 3$$

### 3.4.2 Aggregation of memberships

The parameters or alternatives in neutrosophic form that are analyzed and compared necessitate the use of aggregation operators. The values $\tilde{x}_1, \tilde{x}_2, \ldots, \tilde{x}_n$ are the neutrosophic values in correlation with $n$ different criteria. The aggregated SVNN, $\tilde{x}_{agg}$, is determined as:

$$\tilde{x}_{agg} = \left( \frac{1}{n} \sum_{i=1}^{n} T_{x_i}, \frac{1}{n} \sum_{i=1}^{n} I_{x_i}, \frac{1}{n} \sum_{i=1}^{n} F_{x_i} \right)$$

In a different scenario, weights may be applied for aggregation when the criteria are not of equal importance:

$$\tilde{x}_{agg} = \left( \sum_{i=1}^{n} w_i T_{x_i}, \sum_{i=1}^{n} w_i I_{x_i}, \sum_{i=1}^{n} w_i F_{x_i} \right)$$

with $w_i$, the weight allocated to the $i$th criterion, such that $\sum_{i=1}^{n} w_i = 1$, $w_i \in [0,1]$.

### 3.4.3 Decision-making implication

The neutrosophic model not only represents the uncertain optimization landscape but also provides a more comprehensive picture. This model, by recognizing the three independent viewpoints—truth, indeterminacy, and falsity—allows for the presence of conflicting, vague, and incomplete information, thus enhancing the robustness of the optimization strategy.

---

## 3.5 Metaheuristic Algorithm Integration

This section proposes combining a metaheuristic optimization algorithm with the neutrosophic framework in order to address complex decision-making issues characterized by uncertainty, indeterminacy, and inconsistency. We will first choose a proper metaheuristic technique and then try to make it adapt for dealing with neutrosophic data structures.

### 3.5.1 Choice of metaheuristic algorithm

Out of different metaheuristic methods, we pick GA as one that offers the most pros among all such methods like its strong performance reliability, flexibility, and successful use even in the case of complex, high-dimensional, and nonlinear optimization problems.

### 3.5.2 Adaptation to neutrosophic data

Changes to the genotype's evaluation method will be the first step toward accommodating the neutrosophic representation of decision variables. The **neutrosophic fitness function** which is used for evaluating each and every candidate solution takes into account the degrees of truth, indeterminacy, and falsity as shown below:

$$\text{Fitness}(\tilde{x}) = f(T_x, I_x, F_x)$$

where the function $f$ is determined by the problem at hand and may employ methods like score functions, accuracy functions, or aggregation operators to decode the neutrosophic data.

Moreover, the population individuals will be represented by neutrosophic numbers or structures, wherein the $(T, I, F)$ components of each decision variable will be captured.

### 3.5.3 Algorithm steps

The neutrosophic genetic algorithm follows these main steps:

1. **Initialization:** The first step is to create a population of $N$ candidate solutions. Each solution is represented as a vector of neutrosophic numbers
$$\tilde{X}_i = [(T_1, I_1, F_1), (T_2, I_2, F_2), \ldots, (T_m, I_m, F_m)]$$

2. **Fitness Evaluation:** Determine neutrosophic fitness for everyone applying the tuned fitness function.

3. **Selection:** Choose parent candidates relying on their fitness scores utilizing appropriate selection techniques, e.g., roulette wheel, rank-based selection.

4. **Crossover:** Perform mating between selected parents to create children. In the neutrosophic framework, the crossover is performed on the levels $T$, $I$, and $F$ independently.

5. **Mutation:** Make random changes to the $T$, $I$, or $F$ values of some individuals. This is done in order to maintain diversity in the population.

6. **Replacement:** Refresh the population by substituting low-performing individuals with newly bred offspring.

7. **Termination:** Continue the process until a pre-defined criterion such as a specific number of generations or the convergence of the population is achieved.

   This methodology enables the GA to not only navigate the solution space but also to make the most of it while accounting for the multi-dimensional uncertainty that is represented by the neutrosophic modeling.

## 3.6 Computational Experiments

In order to show the effectiveness of the proposed framework based on the neutrosophic metaheuristics, we do computational experiments on optimization problems of a standard nature and real-world datasets in the presence of uncertainty.

### 3.6.1 Experimental setup

The experimentation was done utilizing Python and MATLAB® on a machine that had an Intel Core i7 processor, 16 GB of RAM, and Windows 10 OS. The following settings were applied to the GA:

- **Population size:** 50
- **Number of generations:** 200
- **Crossover rate:** 0.8
- **Mutation rate:** 0.1
- **Selection method:** Tournament selection
- **Crossover method:** Uniform crossover on $(T, I, F)$ components
- **Mutation method:** Random perturbation on individual neutrosophic components

### 3.6.2 Datasets and problem instances

While the algorithm is being evaluated, we take into account the following:

- **Synthetic benchmark functions** (such as Rastrigin, Ackley) which are accompanied by neutrosophic coefficients allowing for an imprecision to be introduced.

- **Real-world decision-making problems**, for instance, resource allocation and project scheduling, where parameters are represented as neutrosophic sets.

The decision variables and fitness objectives are expressed in neutrosophic terms, meaning that each variable is a triplet $(T, I, F)$ and the aggregated fitness is calculated with the help of suitable scoring functions.

### 3.6.3 Evaluation metrics

Performance is assessed using:

- Best fitness value
- Average fitness value over multiple runs
- Standard deviation
- Convergence rate

### 3.6.4 Results and analysis

Table 3.1 presents the performance of our proposed method compared to classical GA, fuzzy GA, and intuitionistic fuzzy GA.

The findings suggest that the neutrosophic GA is the best choice in terms of accuracy and robustness. By introducing the indeterminacy degree, the model is able to cope with conflicting or incomplete data better and thus to give a higher quality of decision.

**TABLE 3.1**: Comparison of GA variants on a neutrosophic optimization problem.

| **Algorithm** | **Best value** | **Average value** | **Std. dev.** | **Convergence gen.** |
|---|---|---|---|---|
| Classical GA | 92.4 | 89.6 | 4.5 | 150 |
| Fuzzy GA | 95.1 | 92.7 | 3.1 | 132 |
| Intuitionistic GA | 96.3 | 93.9 | 2.8 | 128 |
| **Neutrosophic GA (Proposed)** | **98.7** | **97.5** | **1.2** | **115** |

## 3.7 Conclusion and Future Work

This chapter proposes a novel optimization framework that combines the neutrosophic representation and the metaheuristic algorithm to tackle problems in nonstandard uncertain environments. The primary contributions include a neutrosophic modeling framework able to represent truth, indeterminacy, and falsity separately and the adaptation of a metaheuristic algorithm (like GA) to deal with neutrosophic data. The outcomes validate the proposed method's capability in dealing with the optimization tasks under imprecise and inconsistent information.

## Bibliography

[1] Zadeh, L.A. (1965). Fuzzy sets. *Information and Control*, 8(3), 338–353.

[2] Atanassov, K. (2016). Intuitionistic fuzzy sets. *International Journal Bioautomation*, 20, 95–105. Bulgarian Academy of Sciences.

[3] Smarandache, F. (2006). Neutrosophic set—a generalization of the intuitionistic fuzzy set. In *2006 IEEE International Conference on Granular Computing* (pp. 38–42). IEEE.

[4] Panda, N.R., Raut, P.K., Baral, A., Sahoo, S.K., Satapathy, S.S., & Broumi, S. (2025). An overview of neutrosophic graphs. *Neutrosophic Sets and Systems*, 77, 450–462.

[5] Raut, P.K., Behera, S.P., Broumi, S., & Baral, A. (2024). Evaluation of shortest path on multi-stage graph problem using dynamic approach under neutrosophic environment. *Neutrosophic Sets and Systems*, 64, 113–131.

[6] Raut, P.K., Behera, S.P., Broumi, S., & Baral, A. (2024). Evaluation of shortest path by using breadth-first algorithm under neutrosophic environment. *HyperSoft Set Methods in Engineering*, 1, 34–45.

[7] Raut, P.K., Satapathy, S.S., Behera, S.P., Broumi, S., & Sahoo, A.K. (2025). Solving the shortest path problem in an interval-valued neutrosophic Pythagorean environment using an enhanced A* search algorithm. *Neutrosophic Sets and Systems*, 76, 360–374.

[8] Bhuvaneshwari, S., Sweety, C., Singh, A., Broumi, S., Talea, M., & Raut, P.K. (2024). A novel and an efficient CODAS technique to solve real-life MAGDM problems in Fermatean neutrosophic environment. *Neutrosophic Sets and Systems*, 72(1), 3.

[9] Raut, P.K., & Behera, S.P. (2024). Evaluation of shortest path of network by using an intuitionistic pentagonal fuzzy number with interval values. *International Journal of Reasoning-based Intelligent Systems*, 16(2), 154–159.

[10] Raut, P.K., Behera, S.P., Broumi, S., & Mishra, D. (2023). Calculation of shortest path on Fermatean Neutrosophic Networks. *Neutrosophic Sets and Systems*, 57, 328–341.

[11] Ghadei, S., Panada, A.C., Pramanik, S., Panda, N.R., & Raut, P.K. (2025). Evaluating the minimum spanning trees using Prim's algorithm with undirected neutrosophic graphs. *Neutrosophic Sets and Systems*, 85, 361–379.

[12] Raut, P.K., Pramanik, S., & Mohanty, B.S. (2025). An overview of Fermatean neutrosophic graphs. *Neutrosophic Sets and Systems*, 82, 604–618.

[13] Broumi, S., Raut, P.K., & Behera, S.P. (2023). Solving shortest path problems using an ant colony algorithm with triangular neutrosophic arc weights. *International Journal of Neutrosophic Science*, 20(4), 128–138.

[14] Raut, P.K., Pramanik, S., Mohapatra, D.K., & Sahoo, S.K. (2025). Solving the shortest path based on the traveling salesman problem with a genetic algorithm in a Fermatean neutrosophic environment. *Neutrosophic Sets and Systems*, 78, 353–366.

[15] Gong, J. (2023). An application of meta-heuristic and nature-inspired algorithms for designing reliable networks based on the Internet of Things: A systematic literature review. *International Journal of Communication Systems*, 36(5), e5416.

[16] Raut, P.K., Behera, S.P., Broumi, S., Dey, A., Talea, M., & Baral, A. (2024). Evaluation of the shortest path based on the Traveling Salesman problem with a genetic algorithm in a neutrosophic environment. *Neutrosophic Sets and Systems*, 68(1), 13.

[17] Gafar, M.G., Elhoseny, M., & Manogaran, G. (2020). Modeling neutrosophic variables based on particle swarm optimization and information theory measures for forest fires. *The Journal of Supercomputing*, 76(4), 2339–2356.

# Chapter 4

## An Algebraic Approach to Interval-Valued Neutrosophic Oversets and Undersets

**Prasanta Kumar Raut, R. Rajalakshmi, K. Saritha, and Amarendra Baral**

### 4.1 Introduction

Most of the real-world systems are fundamentally vague and inconsistent, which impact many different kinds of fields, including engineering, applied quantitative research, and decision science. Many mathematical theories have been put forth over time to manage and represent information uncertainty. Among these, the theory of approximate reasoning and its intuitionistic extensions have been important because they provide systematic approaches to dealing with information that is unclear, partial, or incomplete. These methods, however, fall short when dealing with more complex types of indeterminacy, contradicting data, or incomplete knowledge that commonly occur in real-world applications.

To handle such issues, Zadeh [1] proposed the concept of *fuzzy sets* (FSs), wherein each element possesses a degree of membership in the range $[0, 1]$. This provided a concrete mathematical framework to represent gradual truth and uncertainty of real-world systems. Sitara et al. [2] further extended this concept to define *fuzzy graph structures* that served effectively for studying relational ambiguity and uncertainty in complex networks.

In view of the fact that FSs express only one side of uncertainty, Atanassov [3] created the theory of *intuitionistic fuzzy sets* (IFSs), in which every element is assigned both a membership and a non-membership degree. By allowing the representation of hesitancy and incomplete information, this model offers a more complete description of uncertainty.

In order to further develop such ideas, Smarandache [4] suggested the concept of *neutrosophic sets* (NSs), where the parameters of falsity, indeterminacy, and truth are considered independent. This model is more general and

DOI: 10.1201/9781003683339-4

flexible when dealing with contradictory, inconsistent, or incomplete data. This idea was improved by Wang et al. [5] by proposing the *single-valued neutrosophic set* (SVNS) and increasing its applicability to real problems. Thereafter, several works have deeply analyzed its theoretical and applied developments [6–13].

In this regard, the authors Broumi et al. [14] extended this framework by introducing the idea of *single-valued neutrosophic graphs* (SVNGs), which utilizes graph theory combined with NSs for modeling uncertain relationships among interconnected entities. Later, Panda et al. [15] gave a thorough rundown of neutrosophic graphs and how they are used in shortest-path, optimization, and decision analysis problems. Additionally, recent studies have strengthened the applicability of neutrosophic-based models by introducing algorithmic techniques and computational improvements [16–21].

The remainder of this chapter is organized as follows. Section 4.2 introduces the foundational concepts and definitions necessary for the development of interval-valued neutrosophic oversets and undersets. Section 4.3 defines and analyzes key algebraic operations. Section 4.4 illustrates these operations through examples and computational demonstrations. Section 4.5 presents the core algebraic properties and propositions that validate the logical soundness of the proposed model. Section 4.6 discusses potential applications of the developed framework in decision-making and uncertain information analysis. Lastly, Section 4.7 concludes the chapter by summarizing the principal findings and suggesting possible directions for future research.

---

## 4.2 Preliminaries

**Definition 4.1:** *Let us consider $U$ as the universe of discourse. An NS $A$ within $U$ is defined by three independent functions i.e the truth $T_A$, an indeterminacy $I_A$, and a falsity $F_A$. These functions satisfy the following condition:*

$$T_A(x), I_A(x), F_A(x) \subseteq [0,1] \quad \text{for all } x \in U.$$

*Unlike other FS generalizations, the values of $T_A(x)$, $I_A(x)$, and $F_A(x)$ are independent, are not subject to any additive constraint; and each one is allowed to vary independently within the interval* $[0,1]$.

**Definition 4.2:** *An interval-valued neutrosophic set (IVNS) $A$ in $U$ is characterized by three functions:*

$$T_A(x) = [T_A^L(x), T_A^U(x)], \quad I_A(x) = [I_A^L(x), I_A^U(x)], \quad F_A(x) = [F_A^L(x), F_A^U(x)],$$

*where $T_A^L(x), T_A^U(x), I_A^L(x), I_A^U(x), F_A^L(x), F_A^U(x) \in [0,1]$ and*

$$0 \leq T_A^L(x) \leq T_A^U(x) \leq 1, \quad 0 \leq I_A^L(x) \leq I_A^U(x) \leq 1, \quad 0 \leq F_A^L(x) \leq F_A^U(x) \leq 1.$$

**Definition 4.3:** *An IVNS A is called an* **overset** *if at least one of the upper bounds of the intervals* $T_A^U(x)$, $I_A^U(x)$, *or* $F_A^U(x)$ *exceeds 1 for some* $x \in U$. *That is,*

$$T_A^U(x) > 1 \quad or \quad I_A^U(x) > 1 \quad or \quad F_A^U(x) > 1.$$

**Definition 4.4:** *An IVNS A is called an* **underset** *if at least one of the lower bounds of the intervals* $T_A^L(x)$, $I_A^L(x)$, *or* $F_A^L(x)$ *is less than 0 for some* $x \in U$. *That is,*

$$T_A^L(x) < 0 \quad or \quad I_A^L(x) < 0 \quad or \quad F_A^L(x) < 0.$$

---

## 4.3 Algebraic Operations on Interval-Valued Neutrosophic Oversets and Undersets

Let $U$ be a universal set and $A$, $B$ be two IVNSs defined on $U$ with membership functions:

$$A(x) = \left([T_A^L(x), T_A^U(x)], [I_A^L(x), I_A^U(x)], [F_A^L(x), F_A^U(x)]\right),$$
$$B(x) = \left([T_B^L(x), T_B^U(x)], [I_B^L(x), I_B^U(x)], [F_B^L(x), F_B^U(x)]\right).$$

### 1. Union

The union $C = A \cup B$ is defined as:

$$T_C(x) = \left[\max(T_A^L(x), T_B^L(x)), \max(T_A^U(x), T_B^U(x))\right],$$
$$I_C(x) = \left[\min(I_A^L(x), I_B^L(x)), \min(I_A^U(x), I_B^U(x))\right],$$
$$F_C(x) = \left[\min(F_A^L(x), F_B^L(x)), \min(F_A^U(x), F_B^U(x))\right].$$

### 2. Intersection

The intersection $D = A \cap B$ is defined as:

$$T_D(x) = \left[\min(T_A^L(x), T_B^L(x)), \min(T_A^U(x), T_B^U(x))\right],$$
$$I_D(x) = \left[\max(I_A^L(x), I_B^L(x)), \max(I_A^U(x), I_B^U(x))\right],$$
$$F_D(x) = \left[\max(F_A^L(x), F_B^L(x)), \max(F_A^U(x), F_B^U(x))\right].$$

### 3. Complement

The complement $\bar{A}$ of an IVNS $A$ is defined as:

$$T_{\bar{A}}(x) = F_A(x), \quad I_{\bar{A}}(x) = I_A(x), \quad F_{\bar{A}}(x) = T_A(x).$$

### 4. Scalar Multiplication

Let $\lambda \in \mathbb{R}$, the scalar multiplication $\lambda A$ is defined component-wise as:

$$T_{\lambda A}(x) = [\lambda T_A^L(x), \lambda T_A^U(x)],$$
$$I_{\lambda A}(x) = [\lambda I_A^L(x), \lambda I_A^U(x)],$$
$$F_{\lambda A}(x) = [\lambda F_A^L(x), \lambda F_A^U(x)].$$

This operation is especially significant for oversets and undersets, as scalar multiplication may shift membership values beyond the conventional interval $[0, 1]$.

---

## 4.4 Illustrative Examples

Let $U = \{x_1\}$ be a universal set with a single element. Define two IVNSs $A$ and $B$ on $U$ as:

$$A(x_1) = ([1.1, 1.3], [-0.1, 0.2], [0.2, 0.5])$$
$$B(x_1) = ([0.8, 1.2], [0.0, 0.3], [0.1, 0.4])$$

Here, $A$ is an **overset** (since $T_A^U(x_1) = 1.3 > 1$ and $I_A^L(x_1) = -0.1 < 0$), and $B$ is also an **overset**.

**Example 4.1: Union of $A$ and $B$**

$$T_{A\cup B}(x_1) = [\max(1.1, 0.8), \max(1.3, 1.2)] = [1.1, 1.3]$$
$$I_{A\cup B}(x_1) = [\min(-0.1, 0.0), \min(0.2, 0.3)] = [-0.1, 0.2]$$
$$F_{A\cup B}(x_1) = [\min(0.2, 0.1), \min(0.5, 0.4)] = [0.1, 0.4]$$

**Example 4.2: Intersection of $A$ and $B$**

$$T_{A\cap B}(x_1) = [\min(1.1, 0.8), \min(1.3, 1.2)] = [0.8, 1.2]$$
$$I_{A\cap B}(x_1) = [\max(-0.1, 0.0), \max(0.2, 0.3)] = [0.0, 0.3]$$
$$F_{A\cap B}(x_1) = [\max(0.2, 0.1), \max(0.5, 0.4)] = [0.2, 0.5]$$

**Example 4.3: Complement of $A$**

$$\bar{A}(x_1) = ([0.2, 0.5], [-0.1, 0.2], [1.1, 1.3])$$

**Example 4.4: Example 4.4: Scalar Multiplication ($\lambda = 2$) of $A$**

$$T_{2A}(x_1) = [2 \cdot 1.1, 2 \cdot 1.3] = [2.2, 2.6]$$
$$I_{2A}(x_1) = [2 \cdot (-0.1), 2 \cdot 0.2] = [-0.2, 0.4]$$
$$F_{2A}(x_1) = [2 \cdot 0.2, 2 \cdot 0.5] = [0.4, 1.0]$$

**Interpretation**

These examples clearly demonstrate how the algebraic operations behave under the framework of IVNOS/IVNUS. Notably, the union and intersection preserve overset/underset behavior, and scalar multiplication can exaggerate or diminish the degrees beyond classical boundaries.

---

## 4.5 Algebraic Properties and Propositions

This section investigates the key algebraic properties satisfied by the operations defined on IVNOS and IVNUS. Let $A$, $B$, and $C$ be any IVN oversets or undersets defined on a universal set $U$.

### 4.5.1 Commutativity, associativity, and distributivity

- **Commutativity:**

  For two IVN sets $A$ and $B$,

  $$A \cup B = B \cup A, \quad A \cap B = B \cap A.$$

  **Proposition 4.1:** The union and intersection operations on IVN oversets/undersets are commutative.

- **Associativity:**

  For three IVN sets $A$, $B$, and $C$,

  $$(A \cup B) \cup C = A \cup (B \cup C), \quad (A \cap B) \cap C = A \cap (B \cap C).$$

  **Proposition 4.2:** The operations $\cup$ and $\cap$ are associative for IVN sets.

- **Distributivity:**

  For any three IVN sets $A$, $B$, and $C$,

  $$A \cup (B \cap C) = (A \cup B) \cap (A \cup C),$$

  $$A \cap (B \cup C) = (A \cap B) \cup (A \cap C).$$

  **Proposition 4.3:** Union distributes over intersection, and intersection distributes over union.

### 4.5.2 Idempotency and absorption

- **Idempotency:**

  For any IVN set $A$,

$$A \cup A = A, \quad A \cap A = A.$$

**Proposition 4.4:** The IVN set operations $\cup$ and $\cap$ are idempotent.

- **Absorption Laws:**

  For two IVN sets $A$ and $B$,

$$A \cup (A \cap B) = A, \quad A \cap (A \cup B) = A.$$

**Proposition 4.5:** The IVN operations satisfy the absorption laws.

These algebraic properties are important for theoretical analysis and provide a strong foundation for practical implementation in systems dealing with uncertain or extended information.

---

## 4.6 Applications and Discussions

In this section, we explore the practical applicability of IVNOS and IVNUS in real-world problems. These extended models allow better representation of uncertainty, inconsistency, and over-/under-defined information, which are common in complex decision-making systems and information modeling.

### 4.6.1 Decision-making scenarios

Most of the decision-making processes under uncertainty involve vague, inconsistent, and incomplete information. Classical fuzzy or intuitionistic frameworks cannot deal with over-commitment—a truth value greater than 1—or contradiction—a degree of indeterminacy or falsity less than 0—which can be realistically captured using IVN oversets and undersets.

We consider a multi-criteria decision making (MCDM) problem in which the experts may give overly optimistic or overly skeptical evaluations, represented with (truth-membership $> 1$) or (falsity-membership $< 0$), respectively. Using IVNOS/IVNUS, we can:

- Represent each criterion by using interval-valued neutrosophic oversets/undersets.
- Aggregate expert opinions using the algebraic operations defined earlier.

- Rank alternatives based on a defuzzification or score-based function adapted for IVN data.

The method gives a more expressive and flexible methodology for supporting the decision-making process in applications like engineering design, medical diagnosis, risk assessment, and policy analysis.

### 4.6.2 Modeling uncertain information

The real-world systems produce data that is, at times, indeterminate and may possibly exceed the expected bounds due to subjective human judgments, random variables, or malfunctioning equipment. The IVNOS and IVNUS are two concepts that present a coherent method of depicting and evaluating such uncertain or extreme data.

- A sensor network may, for instance, create an overset that indicates a general situation where a sensor has measured unexpected high values thus assigning degrees of truth-membership higher than 1.
- On the other hand, in sentiment analysis the presence of opposing or clashing statements might give rise to negative indeterminacy values that can be effectively illustrated as undersets.
- Lastly, an expert system may show the different degrees of opinions related to various contexts or times by applying intervals that are either above 1 or below 0.

When all these concepts are integrated into the mathematical and computational models, the systems will possess the ability to deal with the more complicated kinds of uncertainty that are typically disregarded by conventional fuzzy, intuitionistic fuzzy, or standard neutrosophic models. Therefore, IVNOS and IVNUS are regarded as particularly valuable in the domains where uncertainty coexists and is usually complex, like artificial intelligence, robotics, pattern recognition, and data fusion.

---

## 4.7 Conclusion

This chapter proposes an algebraic analytical representation of IVNOS and IVNUS. By determining their algebraic structures and basic properties, a generalized multi-valued logical model has been produced, thereby amplifying the traditional theories to include information beyond the usual numerical limits. The suggested framework positively affects the mathematical representation of indeterminacy, ambiguity, and inconsistency that are common in complex information systems. Bringing about formalization of IVNOS and IVNUS, developing their algebraic operations, and demonstrating their relevance in strategic planning and uncertainty evaluation are the most important contributions of the chapter. The analytical formulation developed will be a stronghold for upcoming research in computational intelligence, information analysis, and automated decision-making, having both theoretical and practical significance.

---

## Bibliography

[1] Zadeh, L.A. (1965). Fuzzy sets. *Information and Control*, 8(3), 338–353.

[2] Sitara, M., Akram, M., & Bhatti, M.Y. (2019). Fuzzy graph structures with application. *Mathematics*, 7(1), 63.

[3] Atanassov, K. (2016). Intuitionistic fuzzy sets. *International Journal Bioautomation*, 20, 95–105. Bulgarian Academy of Sciences.

[4] Smarandache, F. (2006). Neutrosophic set—a generalization of the intuitionistic fuzzy set. In *2006 IEEE International Conference on Granular Computing* (pp. 38–42). IEEE.

[5] Wang, H., Smarandache, F., Zhang, Y.Q., & Sunderraman, R. (2010). Single valued neutrosophic sets. *Multispace and Multistructure*, 4, 410–413.

[6] Smarandache, F., & Pramanik, S. (Eds.). (2016). *New trends in neutrosophic theory and applications.* Pons Editions, Brussels.

[7] Smarandache, F., & Pramanik, S. (Eds.). (2018). *New trends in neutrosophic theory and applications, Volume 2.* Pons Editions, Brussels.

[8] Broumi, S., Bakali, A., Talea, M., Smarandache, F., Uluçay, V., Sahin, S., Dey, A., Dhar, M., Tan, R.P., de Oliveira, A., & Pramanik, S. (2018). Neutrosophic sets: An overview. In F. Smarandache & S. Pramanik (Eds.), *New trends in neutrosophic theory and applications, Volume 2* (pp. 403–434). Pons Editions, Brussels.

[9] Pramanik, S., Mallick, R., & Dasgupta, A. (2018). Contributions of selected Indian researchers to multi-attribute decision making in neutrosophic environment. *Neutrosophic Sets and Systems*, 20, 108–131.

[10] Peng, X., & Dai, J. (2020). A bibliometric analysis of neutrosophic set: Two decades review from 1998 to 2017. *Artificial Intelligence Review*, 53(1), 199–255.

[11] Pramanik, S. (2020). Rough neutrosophic set: an overview. In F. Smarandache & S. Broumi (Eds.), *Neutrosophic theories in communication, management and information technology* (pp. 275–311). Nova Science Publishers, New York.

[12] Pramanik, S. (2022). Single-valued neutrosophic set: An overview. In N. Rezaei (Ed.), *Transdisciplinarity. Integrated Science*, Vol. 5, pp. 563–608. Springer, Cham. doi:10.1007/978-3-030-94651-7_26.

[13] Smarandache, F., & Pramanik, S. (Eds.). (2024). *New trends in neutrosophic theories and applications, Volume III.* Biblio Publishing, Grandview Heights, OH.

[14] Broumi, S., Talea, M., Bakali, A., & Smarandache, F. (2016). Single valued neutrosophic graphs. *Journal of New Theory*, 10, 86–101.

[15] Panda, N.R., Raut, P.K., Baral, A., Sahoo, S.K., Satapathy, S.S., & Broumi, S. (2025). An overview of neutrosophic graphs. *Neutrosophic Sets and Systems*, 77, 450–462.

[16] Raut, P.K., Behera, S.P., Broumi, S., & Baral, A. (2024). Evaluation of shortest path on multi-stage graph problem using dynamic approach under neutrosophic environment. *Neutrosophic Sets and Systems*, 64, 113–131.

[17] Raut, P.K., Behera, S.P., Broumi, S., & Baral, A. (2024). Evaluation of shortest path by using breadth-first algorithm under neutrosophic environment. *HyperSoft Set Methods in Engineering*, 1, 34–45.

[18] Raut, P.K., Satapathy, S.S., Behera, S.P., Broumi, S., & Sahoo, A.K. (2025). Solving the shortest path problem in an interval-valued neutrosophic Pythagorean environment using an enhanced A* search algorithm. *Neutrosophic Sets and Systems*, 76, 360–374.

[19] Raut, P.K., Behera, S.P., Broumi, S., & Mishra, D. (2023). Calculation of fuzzy shortest path problem using multi-valued neutrosophic number under fuzzy environment. *Neutrosophic Sets and Systems*, 57, 356–369.

[20] Broumi, S., Raut, P.K., & Behera, S.P. (2023). Solving shortest path problems using an ant colony algorithm with triangular neutrosophic arc weights. *International Journal of Neutrosophic Science*, 20(4), 128–138.

[21] Raut, P.K., Pramanik, S., Mohapatra, D.K., & Sahoo, S.K. (2025). Solving the shortest path based on the traveling salesman problem with a genetic algorithm in a Fermatean neutrosophic environment. *Neutrosophic Sets and Systems*, 78, 353–366.

# Chapter 5

## Neutrosophic Graph Rewriting as a Computational Model

**Apostolos Syropoulos, Basil Papadopoulos, and Binod Kumar Sharma**

### 5.1 Introduction

A graph is a mathematical structure used to visually show the relationship between different items or quantities. It is a fundamental concept in both mathematics and computer science.

**Definition 5.1:** *A graph, often denoted as $G = (V, E)$, consists of two main parts:*

- *Vertices (or Nodes): A set of objects, typically represented as points or circles. The set of all vertices is denoted by $V$.*
- *Edges (or Links): Connections or relationships between pairs of vertices, represented as lines or curves. The set of all edges is denoted by $E$.*

In computer science, graph rewriting, also known as graph transformation, is a method for methodically changing a graph by applying rewrite rules that substitute a specific subgraph pattern with a new, different graph structure. Graph rewriting has found many applications, for example:

- Graph neural networks (GNNs) are neural networks that operate on graphs instead of tabular data. In particular, GNNs are essentially graph transformations [1].
- Origami, the Japanese art of paper folding, has been formalized as graph rewriting [2, 3].
- Compilers can optimize code by manipulating graph-like structures that represent intermediate code, which enables techniques such as loop invariant computation and common subexpression elimination, through the use of graph rewriting [4].

DOI: 10.1201/9781003683339-5

- It has been shown that category theory offers specifications that can be efficiently implemented through imperative algorithms, and this has been applied to the field of graph transformation. It was demonstrated through the use of examples how this software development paradigm facilitates the rapid composition of efficient and accurate code [5].
- It has been demonstrated how graph rewriting can be seen as a computational model [6] and in Section 5.2 we demonstrate this idea.

Pushouts are category theoretic constructs that are used to formalize the changing of a graph by applying rewrite rules—they precisely define how graphs are "glued" together during rewriting. A graph rewriting rule is typically given by a set of morphisms:

$$L \xleftarrow{\;l\;} K \xrightarrow{\;r\;} R$$

where $L$ is the left-hand side (pattern to be replaced), $R$ is the right-hand side (replacement), $K$ is the part preserved during rewriting, $l$ and $r$ graph morphisms. Thus, $K$ is what is common between $L$ and $R$. In category theory, when given a set of arrows $f : A \to B$ and $g : A \to C$ of some category $\mathcal{C}$, a pushout is a triple $(D, \alpha, \beta)$ with $\beta \circ g = \alpha \circ f$ that satisfies the following universal property: for every triple $(E, \alpha', \beta')$ with $\beta' \circ g = \alpha' \circ f$, there exists a unique arrow $\theta : D \to E$ making the following diagram commute:

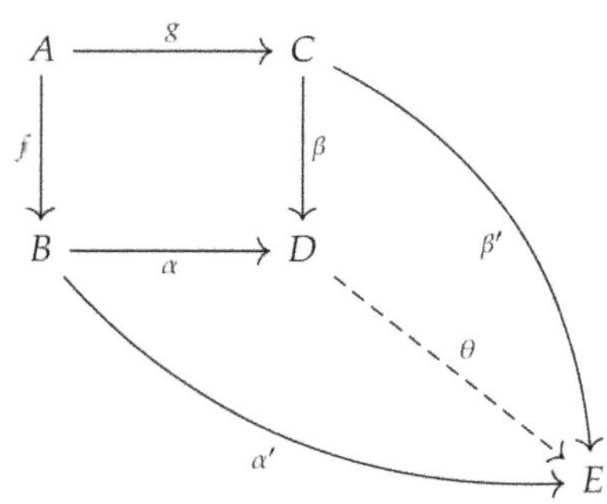


Now, it should be obvious why pushouts are actually graph rewriting rules. More generally, we can use two pushouts to define how to delete and then how to add subgraphs. From this step, we can define *pushout automata* by pushout composition of graphs. For a detailed description of this technique, interested readers should consult the tutorial by Ehrig [7].

A *hypergraph* is a generalized graph where an edge (usually called *hyperedge*) can connect any number of vertices—two or more, or even just one. A hypergraph $H$ is a pair $H = (V, E)$, where $V$ is a set of vertices and $E$ is a set of non-empty subsets of $V$, called hyperedges. Steven Wolfram [8] proposed some models that are some sort of graph rewriting that can be used to understand known fundamental features of physics. In general, the structures on which these models operate consist of collections of relations between identical, labeled, and discrete elements. Wolfram proposed that these structures can be represented by hypergraphs. The elements are the nodes of the hypergraph and the relations are the (directed) hyperedges that connect these elements. For example, the following graph

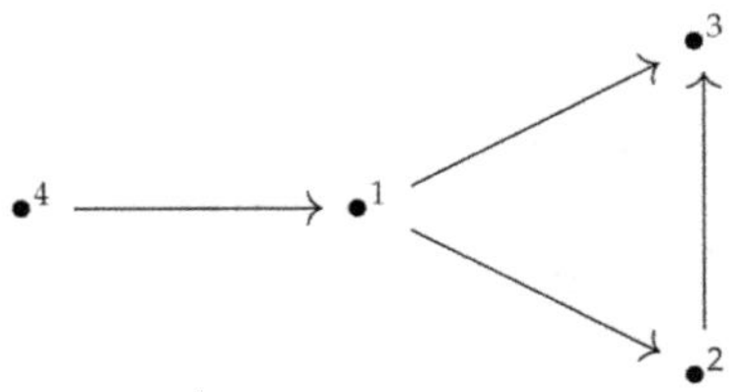


corresponds to the following set of relations:

$$\{\{1, 2\}, \{1, 3\}, \{2, 3\}, \{4, 1\}\}.$$

The order in which these relations are expressed is irrelevant; however, the order in which elements appear within each relation is considered significant (and is evident in the orientations of the hyperedges in the hypergraph). Wolfram's models are rules for rewriting collections of relations. A very simple rule is the following one:

$$\{\{x, y\}\} \to \{\{x, y\}, \{y, z\}\}, \tag{5.1}$$

where $x$, $y$, and $z$ stand for any elements. Note that the following rule

$$\{\{x, y\}\} \to \{\{z, y\}, \{y, x\}\} \tag{5.2}$$

is different from the previous one since order within each relation matters. Suppose we have the set of relations $\{\{1, 2\}\}$ and we apply rule 5.2, then we get the set of relations $\{\{7, 2\}, \{2, 1\}\}$, where 7 stands for a new node whose label is free (i.e., not assigned to some other node). Now, let us apply the same rule 5.2 to the new set of relations. The result will be the set $\{\{8, 2\}, \{2, 7\}, \{9, 1\}, \{1, 2\}\}$. The graph that follows is a depiction of this resulting relation.

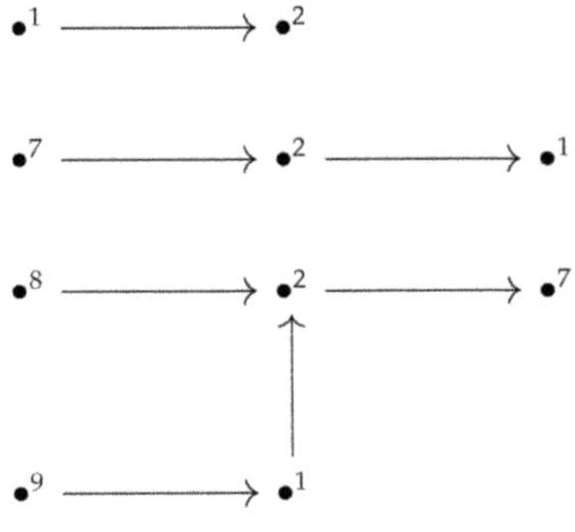


Neutrosophic sets (sometimes called neutrosophic logic or neutrosophy) are a generalization of fuzzy sets and intuitionistic fuzzy sets, introduced by

Florentin Smarandache [9] in the late 1990s. Neutrosophic sets have been used in computer science for the analysis of medical images [10]. In a neutrosophic set, each element $x$ is characterized by a triple of functions:

$$A(x) = \big(T(x), I(x), F(x)\big),$$

where $T(x)$ is the degree of truth (membership), $I(x)$ is the degree of indeterminacy (uncertainty, unknown, inconsistent info), and $F(x)$ is the degree of falsehood (non-membership). Since we are going to study graph rewriting of neutrosophic graphs, we need to provide a definition of these structures:

**Definition 5.2:** *A* neutrosophic graph *[11] is an ordinary graph (i.e., a set of vertices and a set of edges connecting these vertices) accompanied by the following functions:*

1. *a function $T : V \to [0, 1]$ that represents truth;*
2. *a function $I : V \to [0, 1]$ that represents indeterminacy; and*
3. *a function $F : V \to [0, 1]$that represents falsity;*

*such that $0 \le T(v) + I(v) + F(v) \le 3$ for all $v \in V$. Clearly, these functions assign membership values to each vertex $v \in V$, representing the truth, indeterminacy, and falsity degrees, respectively, of the proposition "$v$ is a member of the graph $G$." Similarly, the membership values of an edge $(u, v) \in E$ are defined by the following three functions:*

1. *a membership function $T : E \to [0, 1]$ that represents truth;*
2. *a membership function $I : E \to [0, 1]$ that represents indeterminacy;*
3. *a membership function $F : E \to [0, 1]$ that represents falsity;*

*such that $0 \le T(u, v) + I(u, v) + F(u, v) \le 3$ for all $(u, v) \in E$. These functions represent the truth, indeterminacy, and falsity degrees, respectively, of the proposition "the edge $(u, v)$ is a member of the graph $G$."*

### 5.1.1 Organization of the chapter

First we demonstrate how we can actually compute using graph rewriting, then we present neutrosophic graph rewriting and define neutrosophic double pushout graph rewriting and neutrosophic hypergraph rewriting à la Wolfram.

---

## 5.2 Graph Rewriting as a Model of Computation

In this section we will describe a simple method to compute the successor of (positive) binary number, which is represented by a graph, using graph rewriting. A binary number will be represented by a chain of nodes representing bits and the least significant bit will be at the end. There are two special nodes, one at the beginning and one at the end of the graph. Also, the leftmost node containing a digit is the least significant bit. For example, the number 101 (5 in decimal) might be represented as:

$$\text{start} \longrightarrow 1 \longrightarrow 1 \longrightarrow 1 \longrightarrow \text{end}$$

The rules to compute the successor are:

1. If the current bit is 0, change it to 1 and stop (no carry needed).
2. If the current bit is 1, change it to 0 and propagate carry to the next bit.
3. If carry goes past the last bit, add a new 1 node.

We can express these as rewrite patterns:

**Rule 1:** Increment a 0 bit

$$[0] \rightarrow \text{rest} \Rightarrow [1] \rightarrow \text{rest},$$

when no carry needed.

**Rule 2:** Increment a 1 bit

$$[1] \rightarrow \text{rest} \Rightarrow [0] \rightarrow (\text{carry-to-rest}),$$

propagate carry if necessary.

**Rule 3:** Carry past the end

$$(\text{end with carry}) \Rightarrow [1] \rightarrow \text{end}.$$

Let us apply these rules to number 110 (binary 6) represented as:

$$\text{start} \longrightarrow 0 \longrightarrow 1 \longrightarrow 1 \longrightarrow \text{end}$$

In this case, we simply apply **Rule 1** and so [0] becomes [1] and we stop. Thus, we get

$$\text{start} \longrightarrow 1 \longrightarrow 0 \longrightarrow 1 \longrightarrow \text{end}$$

which is the number 7 in binary.

The previous example showed that we can actually compute with graph rewriting. However, this example just verified that we can compute but it is not a proof that we can compute anything that is computable. Max Dauchet [12] has proved that it is possible to simulate Turing machines with graph rewriting. Thus, graph rewriting is Turing-complete, which means that we can compute with it anything we can compute with a Turing machine.

## 5.3 Neutrosophic Graph Rewriting

Since neutrosophic graphs are a generalization of ordinary graphs, it makes sense to perform graph rewriting to such graphs. However, authors of other works exploring graph rewriting under vagueness (e.g., see [13]), consider only fuzzy graphs whose vertices form a crisp set and not a fuzzy one. First, we will explore graph rewriting of a restricted version of neutrosophic graphs whose vertices do not have truth, falsity, and indeterminacy degrees. Then, we will briefly explain how we can categorically handle neutrosophic graphs whose set of vertices and edges are also neutrosophic sets.

In general, a fuzzy graph can be represented by a fuzzy relation [13]. Thus, we can adopt this idea and we will view neutrosophic graphs as binary neutrosophic relations:

**Definition 5.3:** *A binary neutrosophic relation from a (crisp) set $X$ to a (crisp) $Y$ is a neutrosophic subset in $X \times Y$, that is, a structure $R$ given by*

$$R = \{\langle (x,y), \mu_R(x,y), \sigma_R(x,y), \nu_R(x,y)\rangle \mid (x,y) \in X \times Y\},$$

*where $\mu_R : X \times Y \to [0,1]$, $\sigma_R : X \times Y \to [0,1]$, and $\nu_R : X \times Y \to [0,1]$. For any $(x,y) \in X \times Y$, the value $\mu_R(x,y)$ is called the degree of a membership of $(x,y)$ in $R$, $\sigma_R(x,y)$ is called the degree of indeterminacy of $(x,y)$ in $R$, and $\nu_R(x,y)$ is called the degree of non-membership of $(x,y)$ in $R$.*

Using this definition we define a neutrosophic graph $G = (V, E)$ as a pair $(V, R)$, where $V$ is its set of vertices and $R$ is a binary neutrosophic relation from $V$ to $V$. Mizoguchi and Kawahara [14] defined a *relational calculus* that has been redefined for fuzzy relations by Mori and Kawahara [13] and which we redefine for neutrosophic relations.

**Definition 5.4:** *The (crisp) set of all neutrosophic relations from $A$ to $B$ will be denoted by $\mathbf{Rel}_n(A,B)$. This set contains the relations $\mathbf{0}_{A,B}(a,b) = 0$ and $\mathbf{U}_{A,B}(a,b) = 1$, for all $(a,b) \in A \times B$.*

Assume that $S, R \in \mathbf{Rel}_n(A,B)$, then $S \sqsubseteq R$ if and only if $\mu_S(a,b) \leq \mu_R(a,b)$, $\sigma_S(a,b) \leq \sigma_R(a,b)$, and $\nu_S(a,b) \geq \nu_R(a,b)$ for all $(a,b) \in A \times B$. The set $\mathbf{Rel}_n(A,B)$ equipped with the relation $\sqsubseteq$ is a poset.

**Proof:** *The relation $\sqsubseteq$ is clearly reflexive, transitive, and anti-symmetric. Ergo, $\mathbf{Rel}_n(A,B)$ equipped with the relation $\sqsubseteq$ is a poset.*

Given two relations $S, R \in \mathbf{Rel}_n(A, B)$, we define the operations $S \sqcap R$ and $S \sqcup R$ as follows:

$$\mu_{S\sqcap R}(a,b) = \min\{\mu_S(a,b), \mu_R(a,b)\}, \quad \mu_{S\sqcup R}(a,b) = \max\{\mu_S(a,b), \mu_R(a,b)\}$$

$$\sigma_{S\sqcap R}(a,b) = \min\{\sigma_S(a,b), \sigma_R(a,b)\}, \quad \sigma_{S\sqcup R}(a,b) = \max\{\sigma_S(a,b), \sigma_R(a,b)\}$$

$$\nu_{S\sqcap R}(a,b) = \max\{\nu_S(a,b), \nu_R(a,b)\}, \quad \nu_{S\sqcup R}(a,b) = \min\{\nu_S(a,b), \nu_R(a,b)\}$$

When the operants of "$\sqcap$" and "$\sqcup$" are not finite, we have to replace the operator "min" with the operator "inf" and the operator "max" with the operator "sup," respectively. The quadruple $(\mathbf{Rel}_n(A, B), \sqsubseteq, \sqcap, \sqcup)$ is a lattice with a zero element.

**Proof:** *Any poset is a lattice if and only if all operations $S_1 \sqcap S_2$ and $S_1 \sqcup S_2$ are defined. Obviously for any $S_1, S_2 \in \mathbf{Rel}_n(A, B)$, these operations are defined, therefore $(\mathbf{Rel}_n(A, B), \sqsubseteq, \sqcap, \sqcup)$ is a lattice.*

Now, let us define the relation $S \Rightarrow R$:

$$\mu_{S\Rightarrow R}(a,b) = \begin{cases} 1, & \text{if } \mu_S(a,b) \le \mu_R(a,b) \\ \mu_R(a,b), & \text{if } \mu_S(a,b) > \mu_R(a,b) \end{cases}$$

$$\sigma_{S\Rightarrow R}(a,b) = \begin{cases} 1, & \text{if } \sigma_S(a,b) \le \sigma_R(a,b) \\ \sigma_R(a,b), & \text{if } \sigma_S(a,b) > \sigma_R(a,b) \end{cases}$$

$$\nu_{S\Rightarrow R}(a,b) = \begin{cases} 0, & \text{if } \nu_S(a,b) \ge \nu_R(a,b) \\ \nu_S(a,b), & \text{if } \nu_S(a,b) < \nu_R(a,b) \end{cases}$$

It is clear that $S \Rightarrow R \in \mathbf{Rel}_n(A, B)$ and the operator "$\Rightarrow$" is used to define the relative pseudo-complement in the lattice $\mathbf{Rel}_n(A, B)$. Suppose that $R, S \in \mathbf{Rel}_n(A, B)$. Then, for any $Q \in \mathbf{Rel}_n(A, B)$, it is true that

$$R \sqcap Q \sqsubseteq S \quad \text{if and only if} \quad Q \sqsubseteq R \Rightarrow S.$$

**Proof:** *Suppose that $Q \sqsubseteq R \Rightarrow S$. Then, this means that*

$$\mu_Q(a,b) \le \mu_{R\Rightarrow S}(a,b)$$

$$\sigma_Q(a,b) \le \sigma_{R\Rightarrow S}(a,b)$$

$$\nu_Q(a,b) \ge \nu_{R\Rightarrow S}(a,b)$$

*for any pair* $(a,b)$.

$$\mu_{R\sqcap(R\Rightarrow S)}(a,b)=\begin{cases}R(a,b), & \text{if } \mu_R(a,b)\le\mu_S(a,b)\\ \mu_{R\sqcap S}(a,b), & \text{if } \mu_R(a,b)>\mu_S(a,b)\end{cases}$$

$$\sigma_{R\sqcap(R\Rightarrow S)}(a,b)=\begin{cases}R(a,b), & \text{if } \sigma_R(a,b)\le\sigma_S(a,b)\\ \sigma_{R\sqcap S}(a,b), & \text{if } \sigma_R(a,b)>\sigma_S(a,b)\end{cases}$$

$$\nu_{R\sqcap(R\Rightarrow S)}(a,b)=\begin{cases}R(a,b), & \text{if } \sigma_R(a,b)\ge\sigma_S(a,b)\\ \sigma_{R\sqcap S}(a,b), & \text{if } \sigma_R(a,b)<\sigma_S(a,b)\end{cases}$$

*This means that* $R\sqcap(R\Rightarrow S)=R\sqcap S$. *We have assumed that* $R\sqcap Q\sqsubseteq R\sqcap(R\Rightarrow S)=R\sqcap S\sqsubseteq S$. *Conversely, let us suppose that* $R\sqcap Q\sqsubseteq S$. *If*

$$\mu_R(a,b)\le\mu_Q(a,b),$$
$$\sigma_R(a,b)\le\sigma_Q(a,b),$$

*and*

$$\nu_R(a,b)\ge\nu_Q(a,b),$$

*then*

$$\mu_R(a,b)=\mu_{R\sqcap Q}(a,b)\le\mu_S(a,b),$$
$$\sigma_R(a,b)=\sigma_{R\sqcap Q}(a,b)\le\sigma_S(a,b),$$

*and*

$$\nu_R(a,b)=\nu_{R\sqcap Q}(a,b)\ge\nu_S(a,b).$$

*This means that*

$$\mu_{R\Rightarrow S}(a,b)=1,$$
$$\sigma_{R\Rightarrow S}(a,b)=1,$$

*and*

$$\nu_{R\Rightarrow S}(a,b)=0.$$

*Therefore it holds that*

$$\mu_Q(a,b)\le\mu_{R\Rightarrow S}(a,b),$$
$$\sigma_Q(a,b)\le\sigma_{R\Rightarrow S}(a,b),$$

*and*

$$\nu_Q(a,b)\ge\nu_{R\Rightarrow S}(a,b).$$

*Otherwise, it should be true that*

$$\mu_R(a,b) > \mu_Q(a,b),$$
$$\sigma_R(a,b) > \sigma_Q(a,b),$$

*and*

$$\nu_R(a,b) < \nu_Q(a,b),$$

*but we have assumed that*

$$\mu_Q(a,b) = \mu_{R\sqcap Q}(a,b) \leq \mu_S(a,b) \leq \mu_{R\Rightarrow S}(a,b),$$
$$\sigma_Q(a,b) = \sigma_{R\sqcap Q}(a,b) \leq \sigma_S(a,b) \leq \sigma_{R\Rightarrow S}(a,b),$$

*and*

$$\nu_Q(a,b) = \nu_{R\sqcap Q}(a,b) \geq \nu_S(a,b) \geq \nu_{R\Rightarrow S}(a,b).$$

*Hence,* $Q \sqsubseteq (R \Rightarrow S)$.

Usually, we denote the relatively pseudo-complement $T \Rightarrow \mathbf{0}$ by $\neg T$. The following result can be proved using the previous results.

**Theorem 5.1:** *For any sets* $A$ *and* $B$*, the quadruple* $\left(\mathbf{Rel}_n(A,B), \sqsubseteq, \sqcap, \sqcup\right)$ *is a Heyting algebra.*

**Definition 5.5:** *A neutrosophic relation* $R \in \mathbf{Rel}_n(A,B)$ *is* regular *if and only if* $\mu_R(a,b) = 0$, $\sigma_R(a,b) = 0$, *and* $\nu_R(a,b) = 1$ *or* $\mu_R(a,b) = 1$, $\sigma_R(a,b) = 1$, *and* $\nu_R(a,b) = 0$.

**Definition 5.6:** *Assume that* $R, S \in \mathbf{Rel}_n(A,B)$ *and* $R \sqsubseteq S$. *Then,* $R$ *is* regular with respect to $S$ *if and only if* $\mu_R(a,b) \neq 0$ *implies* $\mu_R(a,b) = \mu_S(a,b)$, $\sigma_R(a,b) \neq 0$ *implies* $\sigma_R(a,b) = \sigma_S(a,b)$, *and* $\nu_R(a,b) \neq 1$ *implies* $\nu_R(a,b) = \nu_S(a,b)$.

**Definition 5.7:** *Assume that* $0 < \epsilon_1 < 1$, $0 < \epsilon_2 < 1$, *and* $0 < \epsilon_3 < 1$. *Then, a neutrosophic relation* $R$ *is* $\varepsilon$*-regular with respect to* $T$ *if for all* $a$ *and* $b$

$$\mu_R(a,b) \neq 0 \Rightarrow \mu_T(a,b) \neq 0 \text{ and } |\mu_R(a,b) - \mu_T(a,b)| \leq \epsilon_1,$$
$$\sigma_R(a,b) \neq 0 \Rightarrow \sigma_T(a,b) \neq 0 \text{ and } |\mu_R(a,b) - \mu_T(a,b)| \leq \epsilon_2, \text{ and}$$
$$\nu_R(a,b) \neq 0 \Rightarrow \nu_T(a,b) \neq 0 \text{ and } |\mu_R(a,b) - \mu_T(a,b)| \leq 1 - \epsilon_3.$$

**Definition 5.8:** *Given a neutrosophic relation* $R : A \to B$*, the domain* $d(R) : A \to A$ *of* $R$ *is a relation defined by*

$$d(R) = R \circ R^{-1} \circ \mathrm{id}_A,$$

*where* $R^{-1}$ *is the* inverse *relation of* $R$*,* $R \circ S$ *is the* composite *relation of* $R$ *and* $S$*, and* $\mathrm{id}_A$ *is the* identity *relation of* $A$ *(the diagonal set of* $A$*).*

**Definition 5.9:** *Given two neutrosophic graphs* $(V_1, R)$ *and* $(V_2, S)$, *a* partial graph morphism *from* $(V_1, R)$ *to* $(V_2, S)$ *is a regular-partial*[1] *function* $f : V_1 \to V_2$ *satisfying* $d(f) \circ R \circ f \sqsubseteq f \circ S$.

It is not difficult to show that all neutrosophic graphs and the partial graph morphisms between them form a category, which will be denoted by **Pfn**(**NGraph**). Using the same arguments presented in [14], we can prove the following:

**Theorem 5.2:** *The category* **Pfn**(**NGraph**) *has pushouts.*

In order to apply a rewriting rule to a graph we need a mechanism that will check whether the rule can be applied to a subgraph. This mechanism is called matching. Let us now formalize rewriting rules and matching:

**Definition 5.10:** *A rewriting rule is a triple* $p = \big((V_1, R), (V_2, S), f : V_1 \to V_2\big)$, *where* $f$ *is a partial function.*

**Definition 5.11:** *An arrow* $g$ *from* $(V, R)$ *to* $(G, \Xi)$ *is a* matching *from* $(V, R)$ *to* $(G, \Xi)$ *if* $g^{-1} \circ R \circ g$ *is regular with respect to* $\Xi$.

In the crisp case, we can construct the pushout from the rewriting rule and matching. For example, the following pushout square exists in category **Pfn** of sets and partial functions between them.

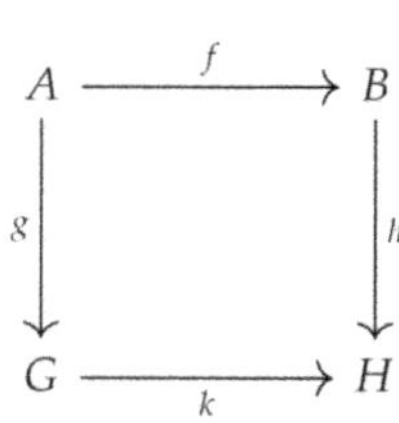


The graph $(G, \Xi)$ is transformed into $(H, \Lambda)$ by applying a rewriting rule $p$ together with a matching $g$ if $\Lambda = k^{-1} \circ T \circ k \sqcup h^{-1} \circ (\Xi - g^{-1} \circ R \circ g) \circ k$. It is customary to denote the previous rewriting square by $(G, \Xi) \Longrightarrow_{p/g} (H, \Lambda)$. The application of a rewriting rule can be viewed as a rewriting square:

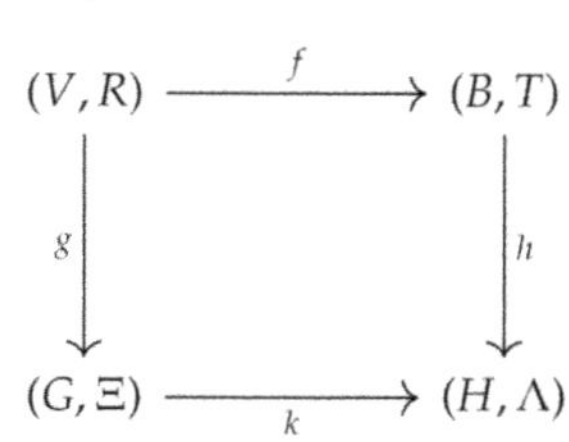


Mori and Kawahara [13] have shown that the rewriting square is a pushout square in **Pfn**(**FGraph**), the category of fuzzy graphs and partial morphisms between them. Thus, we can prove the following theorem for category **Pfn**(**NGraph**).

**Theorem 5.3:** *Suppose that* $p$ *is a rewriting rule* $((V,R),(W,S),f:V\to W)$ *and* $g:(V,R)\to(W,S)$ *is a matching. If* $f:(V,R)\to(W,S)$ *is a partial morphism, then the rewriting square* $(G,\Xi)\Longrightarrow_{p/g}(H,\Lambda)$ *is a pushout in* **Pfn(NGraph)**.

The next thing we need to define is the vague matching condition.

**Definition 5.12:** *Assume that* $\epsilon_1,\epsilon_2,\epsilon_3\in[0,1]$. *Then,* $g:V\to G$ *is an* $\varepsilon$-*matching from* $(V,R)$ *to* $(G,\Xi)$ *if and only if the relation* $g^{-1}\circ R\circ g$ *is* $\varepsilon$-*regular with respect to* $\Xi$.

Two neutrosophic graphs $(V,R)$ and $(W,S)$ are $\varepsilon$-equal if and only if $R$ is $\varepsilon$-regular with respect to $S$ and vice versa.

An alternative approach to the categorical representation of fuzzy graphs has been presented in [15]. Here we adopt the theory developed in [15] to the case of neutrosophic graphs.

**Definition 5.13:** *Assume that* $\mathcal{L}=[0,1]^3$ *and that* $(\mathcal{L},\leq)$ *is a poset. Then, a neutrosophic fuzzy set is a pair* $(A,\alpha)$ *consisting of a set* $A$ *and a membership function* $\alpha:A\to\mathcal{L}$. *A morphism* $f:(A,\alpha)\to(B,\beta)$ *of fuzzy sets is a function* $f:A\to B$ *such that* $\alpha\leq\beta\circ f$, *that is,* $\alpha(a)\leq(\beta\circ f)(a)$ *for all* $a\in A$. *They form the category* **NFuzzySet**$(\mathcal{L})$. *Membership functions are represented in gray. The fact that* $\alpha\leq\beta\circ f$ *is expressed by the following diagram, which is called* $\leq$-*commuting.*

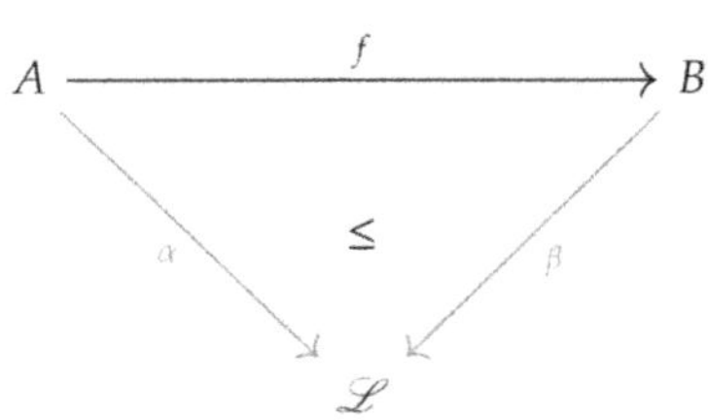


**Definition 5.14:** *A neutrosphic graph* $(A,\alpha)$ *consists of a graph* $A$ *and two membership functions* $\alpha_V:A(V)\to[0,1]^3$, *and* $\alpha_E:A(E)\to[0,1]^3$, *where* $[0,1]^3$ *is a poset. A neutrosophic graph homomorphism* $f:(A,\alpha)\to(B,\beta)$ *is a graph homomorphism* $f:A\to B$ *such that* $\alpha_V\leq\beta_V\circ f_V$ *and* $\alpha_E\leq\beta_E\circ f_E$. *All neutrosphic graphs and the neutrosophic graph homomorphism between them form the category* **NGraph**.

**Definition 5.15:** *A presheaf on a category* $\mathbf{I}$ *is a functor* $F:\mathbf{I}^{\mathrm{op}}\to\mathbf{Set}$. *An arrow of presheaves is a natural transformation between functors. The category of presheaves on* $\mathbf{I}$ *is denoted by* $\mathbf{Set}^{\mathbf{I}^{\mathrm{op}}}$.

Suppose that $\mathsf{incl}:\mathbf{Set}\to\mathbf{Poset}$ the inclusion functor that gives the coarsest possible partial order, that is, the reflexive relation.

**Definition 5.16:** *Given a category* $\mathbf{I}$ *and a functor* $L:\mathbf{I}^{\mathrm{op}}\to\mathbf{Poset}$, *a (poset)* $L$-*fuzzy presheaf is a pair* $(A,\alpha)$ *consisting of a presheaf* $A:\mathbf{I}^{\mathrm{op}}\to\mathbf{Set}$ *and a* *lax natural transformation* $\alpha : \mathsf{incl} \cdot A \Rightarrow L$, *that is, a natural transformation whose naturality squares* $\leq$*-commute, as shown in the diagram below right for each* $\iota : j \to i$. *A morphism of L-fuzzy presheaves* $f : (A, \alpha) \to (B, \beta)$ *is a natural transformation* $f : A \to B$ *such that* $\alpha \leq \beta \circ f$, *that is,* $\alpha_i \leq \beta_i \circ f_i$ *for all* $i \in \mathbf{I}$. *The result is the category* $\mathbf{FuzzyPresheaf}(\mathbf{I}, L)$.

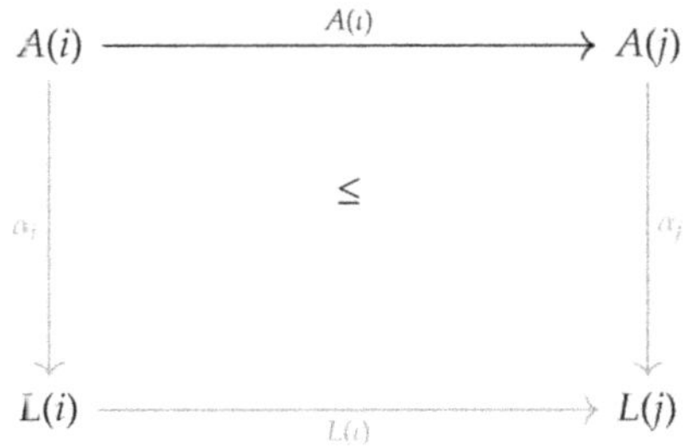


**Theorem 5.4:** *Suppose that* $\mathbf{I}$ *is a small category and a functor* $L : \mathbf{I}^{\mathrm{op}} \to$ **CompHeytAlg**. *Then,* $\mathbf{FuzzyPresheaf}(\mathbf{I}, L)$ *is a quasitopos.*

Since a quasitopos is a category for which every finite limit and finite colimit exists, this means that pushouts exists. Therefore, graph rewriting rules can be represented as pushouts. In order to avoid lax natural transformation and other such constructs, we can adopt a definition of fuzzy categories such as those defined in [16].

---

## 5.4 Neutrosophic Hypergraph Rewriting

Neutrosophic hypergraphs of the kind we are interested in are defined as follows [17]:

**Definition 5.17:** *Assume that* $V = \{v_1, v_2, \ldots, v_n\}$ *be a finite set and* $E = \{E_1, E_2, \ldots, E_m\}$ *be a finite family of non-trivial single-valued neutrosophic subsets of* $V$ *such that*

$$V = \bigcup_i \mathrm{supp}(E_i), i = 1, 2, \ldots, m,$$

*where the edges* $E_i$ *are single-valued neutrosophic subsets of* $V$,

$$E_i = \{(v_j, T_{E_i}(v_j), I_{E_i}(v_j), F_{E_i}(v_j)) \mid j = 1, 2, \ldots, k\},$$

$E_i \neq \emptyset$, *for* $i = 1, 2, \ldots, m$. *Then, the pair* $H = (V, E)$ *is a single-valued neutrosophic hypergraph on* $V$, $E$ *is the family of single-valued neutrosophic hyperedges of* $H$, *and* $V$ *is the crisp vertex set of* $H$.

We can use the neutrosophic sets $E_i$ to form the cartesian product $\mathcal{D} = D_1 \times D_2 \times \cdots \times D_m$. Then, we can define a neutrosophic hypergraph by $(V, R)$, where $R \subseteq \mathcal{D}$. Thus, we have managed to transform a neutrosophic hypergraph to a neutrosophic relation. This means that we can generalize the relational calculus of the previous section for neutrosophic hypergraphs and so we can define hypergraph rewriting as a pushout in some category. In addition, it is possible to reformulate Wolfram's models as rules for rewriting collections of neutrosophic relations. Of course this is useful once we accept that vagueness is a fundamental property of this world.

---

## 5.5 Conclusions

We have explained why graph rewriting is a model of computation and defined categorically neutrosophic graph rewriting. In addition, we have showed how an alternative approach to categorical neutrosophic graph rewriting can be used for "full" neutrosophic graphs. Moreover, we have briefly examined neutrosophic hypergraph rewriting and its possible applications. We believe this is an exciting field of study and research as there are many things to be examined and studied.

---

## Bibliography

[1] Adam Machowczyk and Reiko Heckel. Graph rewriting for graph neural networks. In Maribel Fernández and Christopher M. Poskitt, editors, *Graph Transformation*, pages 292–301. Springer Nature Switzerland, Cham, Switzerland, 2023.

[2] Tetsuo Ida. Graph rewriting in computational origami. In *13th International Symposium on Symbolic and Numeric Algorithms for Scientific Computing*, pages 20–27, Los Alamitos, CA, USA, 2008. IEEE Computer Society.

[3] Tetsuo Ida and Hidekazu Takahashi. Origami fold as algebraic graph rewriting. *Journal of Symbolic Computation*, 45(4):393–413, 2010. The paper can be downloaded from `https://www.sciencedirect.com/science/article/pii/S0747717109001588`.

[4] Manfred Nagl. Application of graph rewriting to optimization and parallelization of programs. In Walter Knödel and Hans Jürgen Schneider, editors, *Parallel Processes and Related Automata / Parallele Prozesse und

*damit zusammenhängende Automaten*, pages 105–124. Springer Vienna, Vienna, 1981.

[5] Kristopher Brown, Evan Patterson, Tyler Hanks, and James Fairbanks. Computational category-theoretic rewriting. In Nicolas Behr and Daniel Strüber, editors, *Graph Transformation*, pages 155–172. Springer International Publishing, Cham, Switzerland, 2022.

[6] D. L. McBurney and M. R. Sleep. Graph rewriting as a computational model. In Akinori Yonezawa and Takayasu Ito, editors, *Concurrency: Theory, Language, and Architecture*, pages 235–256. Springer Berlin Heidelberg, Berlin, Heidelberg, 1991.

[7] Hartmut Ehrig. Tutorial introduction to the algebraic approach of graph grammars. In Hartmut Ehrig, Manfred Nagl, Grzegorz Rozenberg, and Azriel Rosenfeld, editors, *Graph-Grammars and Their Application to Computer Science*, pages 1–14. Springer Berlin Heidelberg, Berlin, Heidelberg, 1987.

[8] Stephen Wolfram. A class of models with the potential to represent fundamental physics. *Complex Systems*, 29(2):107–536, 2020. The paper can be downloaded from `https://www.complex-systems.com/abstracts/v29_i02_a01/`.

[9] Florentin Smarandache. Neutrosophic set—a generalization of the intuitionistic fuzzy set. In *2006 IEEE International Conference on Granular Computing*, pages 38–42, May 2006.

[10] Deepika Koundal and Bhisham Sharma. 15 - challenges and future directions in neutrosophic set-based medical image analysis. In Yanhui Guo and Amira S. Ashour, editors, *Neutrosophic Set in Medical Image Analysis*, pages 313–343. Academic Press, 2019.

[11] Nihar Ranjan Panda, Prasanta Kumar Raut, Amarendra Baral, Srikanta Kumar Sahoo, Sakya Singh Satapathy, and Said Broumi. An overview of neutrosophic graphs. *Neutrosophic Sets and Systems*, 77: 450–462, 2025. The paper can be downloaded from `https://fs.unm.edu/NSS/NeutrosophicGraphs22.pdf`.

[12] Max Dauchet. Simulation of turing machines by a regular rewrite rule. *Theoretical Computer Science*, 103(2):409–420, 1992.

[13] Masao Mori and Yasuo Kawahara. Fuzzy graph rewritings. *Proceedings of the Research Institute for Mathematical Analysis, Kyoto University Research Institute for Mathematical Sciences*, 918:65–71, 1995. The paper can be downloaded from `http://hdl.handle.net/2433/59676`.

[14] Yoshihiro Mizoguchi and Yasuo Kawahara. Relational graph rewritings. *Theoretical Computer Science*, 141(1):311–328, 1995. Paper freely available from ScienceDirect.

[15] Aloïs Rosset, Roy Overbeek, and Jörg Endrullis. Fuzzy presheaves are quasitoposes. In Maribel Fernández and Christopher M. Poskitt, editors, *Graph Transformation*, pages 102–122. Springer Nature Switzerland, Cham, 2023.

[16] Apostolos Syropoulos. Categories of fuzzy structures and fuzzy categories. In José Acacio de Barros and Décio Krause, editors, *A True Polymath: A Tribute to Francisco Antonio Doria*, pages 230–249. College Publications, Rickmansworth, England, 2020.

[17] Muhammad Akram, Sundas Shahzadi, and Arsham Borumand Saeid. Single-valued neutrosophic hypergraphs. *TWMS Journal of Applied and Engineering Mathematics*, 8(1):122–135, 2018. Paper freely available from `https://jaem.isikun.edu.tr/web/`.

---

## Note

1. Recall that any function is a relation and a (crisp) relation is a neutrosophic relation.

# Chapter 6

## A Novel Correlation Coefficient between Single-Valued Neutrosophic Sets and Its Use in Renewable Energy Source Selection

**Paul Augustine Ejegwa and Murat Kirisci**

### 6.1 Introduction

Energy is essential to human life and significantly influences a nation's economic development. However, the overuse of fossil fuels causes serious environmental issues, including rising greenhouse gas emissions that greatly contribute to global warming. Most nations are actively developing sustainable or renewable energy sources (RESs) to address these environmental challenges. Many are committed to reducing carbon emissions and promoting a green economy, recognizing the grave threat posed by climate change, as evidenced by the Kyoto Protocol negotiations in 1997 and the Paris Climate Change Conference held in 2015. Thus, for many countries, transitioning from fossil fuels to sustainable energy is crucial [24]. Today, nations have an ever-increasing need for energy to sustain their development.

On the other hand, while producing the required energy, protecting natural systems is also critical. Since the 1990s, environmentally friendly energy production has gained prominence on the agendas of countries, international institutions, and organizations. With increased environmental awareness, renewable energy (RE) has become significantly more important. RESs continuously renew themselves or can be regenerated naturally within a short period during their natural cycles. Until recently, the integrated utilization of RE was hindered by objective conditions and the high cost of equipment for large-scale applications. However, as solar cell prices continue to drop and tariffs on fossil fuels and energy services rise, the challenge of improving the efficiency of renewable resource utilization has reached a new level. The policy

DOI: 10.1201/9781003683339-6

of a gradual transition to a carbon-neutral economy also contributes to this progress [33].

According to the REN21 Global Status Report 2025, the installed capacity of RE is 15.1 GW, with total investment in 2024 exceeding 728 billion US dollars worldwide. The average annual growth rate is 0.88% [17]. Notably, 38% of the total increase in global energy supply came from renewable sources. The global economy's energy intensity rose by only 1%, continuing the recent slowdown. Compared to 2023, when it increased by 1.2%, the rise in energy-related $CO_2$ emissions fell to 0.8% [23]. Due to the decline in technological costs, IRENA predicts that by 2030, RE will account for about 40% of global energy production. RE is clean and inexhaustible, offering numerous advantages over traditional fossil fuels, including abundance and low cost. However, because solar and wind power are intermittent, RE faces production and capacity constraints. In particular, RE currently has a higher electricity cost than fossil fuels. Furthermore, one of the key challenges in RE development is investment management (IM). Uncertainty and conflicting criteria often characterize IM decisions, making them more complex to solve [23].

Selecting RE alternatives is a multi-criteria problem involving numerous, often conflicting, factors. It is necessary to weigh the benefits and drawbacks of each option when determining the selection criteria. Therefore, a multi-criteria approach should be employed to address this issue. Such criteria add complexity and ambiguity to the evaluation process. In many decision-making situations, the assessments of decision-makers are imprecise, and providing exact values for features or criteria may be difficult. Decision-makers often use linguistic terms to express the suitability of alternative energy policies across various subjective criteria and their weights, as most evaluation parameters cannot be precisely measured. This ambiguity in human judgment can be effectively modeled using fuzzy logic. A subfield of fuzzy set (FS) theory, fuzzy logic employs membership functions to handle imprecise data. Zadeh formalized fuzzy set theory in 1965 [52].

Multi-criteria decision-making (MCDM) techniques have been applied in several studies to rank RESs. These techniques are particularly suitable when selecting the best RES, as they involve balancing numerous complex factors. The authors of [17] focused on ranking RE options in India using fuzzy analytic hierarchy process (AHP) and an energy index metric. Eleven techno-economic and environmental criteria were used to evaluate different alternatives, and the results were combined to create this index. Using data envelopment analysis (DEA), the authors of [33] conducted a thorough analysis of the efficiency of seven key RESs, providing valuable insights into their relative effectiveness. In order to prioritize RES in Iran, Kolagar et al. presented a hybrid approach combining DEA and the fuzzy best-worst method [23]. This strategy encompasses the economic, technological, political, social, and environmental dimensions of sustainability.

Furthermore, compared to alternative decision-making strategies, a three-stage sensitivity analysis confirmed the importance of the DEA-fuzzy best-worst method approach. In [35], the authors examined the feasibility of several RESs for hydrogen production in Pakistan using a combination of fuzzy Delphi, fuzzy AHP, and environmental DEA techniques. The researchers focused on six RES options: municipal solid waste, solar, wind, biomass, geothermal, and micro-hydro. In [39], the authors used AHP and fuzzy Vlekriterijumsko KOmpromisno Rangiranje (VIKOR) techniques to evaluate several locations for solar power facilities. Each criterion and sub-criterion was weighted via AHP, and feasible options were ranked using fuzzy VIKOR. In [44], the authors developed an integrated system combining fuzzy technique for order of preference by similarity to ideal solution (TOPSIS), fuzzy AHP, and slack-based DEA to rank and prioritize Pakistan's most efficient and sustainable hydrogen generation methods. A literature-based SWOT analysis was performed in [26], where 12 strategies derived from SWOT factors were weighted using fuzzy analytic hierarchy process (FAHP) based on feedback from 15 energy industry experts and ranked using Grey-TOPSIS. The available research demonstrates the effectiveness of DEA and fuzzy AHP techniques in selecting and ranking RESs.

A multi-criteria decision support framework was developed by [34] to rank RE supply systems in Turkey. The framework integrates fuzzy TOPSIS and interval Shannon entropy algorithms. An MCDM approach based on axiomatic design and AHP methodologies was presented by [19] to assist countries in selecting renewable energy investments. In [4], intuitionistic fuzzy TOPSIS was used to evaluate renewable energy technologies for electricity generation in Turkey. To assess Turkey's energy landscape and rank various RESs, [18] proposed a hybrid model based on benefit, opportunity, cost, and risk (BOCR) and ANP. Lee and Chang introduced a novel MCDM technique integrating weighted sum model (WSM), VIKOR, TOPSIS, and ELimination Et Choix Traduisant la REalite (ELECTRE) [25], using Shannon entropy weighting to evaluate the significance of each factor. Using a hybrid decision-making model combining TOPSIS-I and best worst method (BWM), [1] identified the most significant options for sustainable RE. AHP and VIKOR were used in [32] to help decision-makers choose RE options, illustrating how the priority and performance metrics of Chicago's RE options vary across criteria. The intuitionistic fuzzy evaluation based on distance from average solution (EDAS) approach was used in [10] to assess Turkey's best RE options. To support sustainable energy decision-making, [14] proposed a comprehensive integrated methodology combining entropy, proximity indexed value (PIV), multi-attributive border approximation area comparison (MABAC), measurement alternatives and ranking according to compromise solution (MARCOS), weighted aggregates sum product assessment (WASPAS), multi-objective optimization by ratio analysis (MOORA), EDAS, grey relational analysis (GRA), and TOPSIS. In [9], the authors introduced a decision framework integrating BWM and interval-valued Pythagorean fuzzy WASPAS techniques to assess RESs in developing nations. Using an integrated DEA and fuzzy AHP approach, [27] examined various RESs in India, including solar, geothermal, hydro, biomass, wave, onshore, and offshore wind energy. In [36], the authors assessed five RE technologies in Isfahan—solar photovoltaic, biomass, wind, concentrated solar power, and geothermal energy—using a combination of fuzzy Decision-Making Trial and Evaluation Laboratory (DEMATEL)-ANP and fuzzy multi-objective optimization by ratio analysis plus full multiplicative form (MULTI-MOORA) algorithms. The TOPSIS approach was employed by [29] to select suitable locations for RE resources.

To analyze energy policy and investment in RESs, [43] proposed an MCDM framework that first uses a correlation coefficient based on interval type-2 FSs to analyze criteria via a balanced scorecard technique, followed by fuzzy TOPSIS and fuzzy DEMATEL under interval type-2 FSs. A novel weighted correlation coefficient was introduced by [21] to evaluate relationships between two Fermatean FSs. In [13], the authors developed a new approach to the Fermatean fuzzy correlation coefficient with potential practical applications. In [22], the authors explored four categories of correlation coefficients for interval-valued Fermatean hesitant FSs. Several applications of fuzzy extensions in real-world problems have also been discussed [2, 3, 11, 12, 46].

Although interval-valued intuitionistic fuzzy sets (IVIFSs) and intuitionistic fuzzy sets (IFSs) can handle partial information, they cannot effectively manage uncertain and inconsistent decision information frequently encountered in real-world scenarios. Therefore, the concept of neutrosophic sets (NSs) was first introduced by [37], primarily from a philosophical standpoint. NSs independently and simultaneously consider truth, falsity, and indeterminacy memberships. In [31], the authors provided a comprehensive review of the theory and applications of NSs in literature, spanning pattern classification, segmentation, clustering, and decision-making. In [28], the concept of NSs was applied to topological spaces. In [5], the authors offered a detailed overview of various types of NSs, highlighting their differences, significance, and applications. In [53], the authors discussed interval NSs in multi-criteria decision problems using aggregation operations.

Applying NSs in science and engineering remains challenging. Hence, the concept of single-valued neutrosophic sets (SVNSs), a subclass of NSs, was introduced by [42]. When handling highly ambiguous and uncertain data, SVNSs are particularly useful. Like NSs, SVNSs consist of truth, indeterminacy, and falsity membership values; however, in SVNSs, these parameters take values within the unit interval $[0, 1]$. In [8], the authors used SVNSs to propose a hybrid set called single-valued neutrosophic fuzzy sets (SVNFSs), addressing certain practical challenges of nonstandard intervals in neutrosophic fuzzy sets (NFSs). The concept was further used for decision-making in investment analysis based on similarity measures. Discrimination information is primarily quantified using cross-entropy and similarity metrics. To determine the ranking order of all alternatives in MADM problems, [49] proposed three vector similarity measures between SVNSs. In [50], the authors developed modified cosine similarity measures for SVNSs based on the cosine function to overcome the limitations of existing similarity measures.

Building on the correlation coefficient of IFSs, [47] introduced the correlation coefficient of SVNSs and applied it to practical problems. The method in [47] was later extended to interval NSs [7]. Similarly, a correlation coefficient between neutrosophic refined sets (NRSs) was discussed in [6] and applied to diagnostic processes. In [48], the authors proposed a distinct correlation coefficient between SVNSs that was proved more efficient than the method in [47] and was applied to multi-attribute decision-making. In [15], the authors introduced a modified correlation coefficient between SVNSs (based on [47]) for fault diagnosis. In [54], the authors presented a statistical correlation coefficient for SVNSs based on covariance and variance, applying it to medical diagnosis. A distinct correlation measure for Pythagorean neutrosophic sets was discussed in [16]. In [20], the authors reviewed existing correlation coefficient methods under SVNSs with applications in multi-criteria decision-making. In [51], the authors extended the method used in [47] to dynamic SVN multisets for multi-attribute decision processes. Likewise, in [40], the authors defined a Pearson-based correlation coefficient in probabilistic SVN hesitant FSs with applications in supply chain management. In [45], the authors addressed practical problems by extending the method in [47] under SVNSs to neutrosophic cubic sets, while in [55], the authors extended it to linguistic NSs. By combining m-polar FSs with interval-valued neutrosophic soft sets (IVNSSs), the authors in [30] proposed a correlation coefficient method under m-polar IVNSSs.

The repetition of correlation coefficient models under SVNSs [15, 47, 48, 54] indicates that they either lack precision or are not expressed from a statistical perspective. To this end, this chapter presents a novel technique for the correlation coefficient under SVNSs, which possesses strong discriminative ability, produces precise results, and satisfies the axioms of a correlation coefficient. Furthermore, the proposed correlation coefficient under SVNSs is applied to analyze RES selection preferences, offering an eco-friendly alternative evaluation framework. The work presents the following contributions:

- Formulation of a novel correlation coefficient under SVNSs that enhances precision, reliability, and satisfies the axioms of a correlation coefficient.
- Demonstration of the theoretical properties of the proposed correlation coefficient to validate its formulation.
- Comparative analysis between the proposed and existing correlation coefficients under SVNSs, showcasing its superior precision.
- Application of the proposed method to analyze RES selection preferences for promoting eco-friendly energy utilization.

The remainder of this work is structured as follows: Section 6.2 reviews SVNSs and existing correlation coefficients. Section 6.3 introduces the novel correlation coefficient under SVNSs, along with its numerical validation and theoretical properties. Section 6.4 presents the decision-making algorithm for

RES selection based on the proposed method. Section 6.5 provides a comparative analysis demonstrating the superiority of the proposed approach. Finally Section 6.6 concludes the work and outlines future research directions.

---

## 6.2 Preliminaries

This section presents NSs, SVNSs, and enlists the existing correlation coeffient's methods. In this section, we take $\mathfrak{Y} = \{\mathfrak{y}_1, \mathfrak{y}_2, \cdots, \mathfrak{y}_k\}$ as the universal set.

**Definition 6.1:** [37]: An NS designated as $\mathfrak{M}$ is of the form:

$$\mathfrak{M} = \{(\mathfrak{y}_p, T_{\mathfrak{M}}(\mathfrak{y}_\mathfrak{p}), I_{\mathfrak{M}}(\mathfrak{y}_\mathfrak{p}), F_{\mathfrak{M}}(\mathfrak{y}_\mathfrak{p}))|\mathfrak{y} \in \mathfrak{Y}\}, \tag{6.1}$$

where $p = 1, 2, \cdots, k$, and $T_{\mathfrak{M}}$, $I_{\mathfrak{M}}$, $F_{\mathfrak{M}}\colon \mathfrak{Y} \to ]^{-}0, 1^{+}[$ are functions that define the truth function, indeterminancy function, and falsity function, respectively. It is expedient to state that $T_{\mathfrak{M}}$ and $F_{\mathfrak{M}}$ can also be called membership function and non-membership function, respectively. The codomain of these functions is a non-standard unit interval with the lower endpoint, $^{-}0 = 0 - \epsilon$ and the upper endpoint, $^{+}1 = 1 + \epsilon$.

The neutroscphic components must satisfy the inequality,

$$^{-}0 \leq \sup T_{\mathfrak{M}}(\mathfrak{y}_\mathfrak{p}) + \sup I_{\mathfrak{M}}(\mathfrak{y}_\mathfrak{p}) + \sup F_{\mathfrak{M}}(\mathfrak{y}_\mathfrak{p}) \leq 3^{+}, \tag{6.2}$$

meaning the three functions are independent, unlike the cases in FS and its generalizations. For $^{-}0 \leq \sup T_{\mathfrak{M}}(\mathfrak{y}_\mathfrak{p}) + \sup I_{\mathfrak{M}}(\mathfrak{y}_\mathfrak{p}) + \sup F_{\mathfrak{M}}(\mathfrak{y}_\mathfrak{p}) \leq 2^{+}$, it means only two of the functions are independent. Also, for $^{-}0 \leq \sup T_{\mathfrak{M}}(\mathfrak{y}_\mathfrak{p}) + \sup I_{\mathfrak{M}}(\mathfrak{y}_\mathfrak{p}) + \sup F_{\mathfrak{M}}(\mathfrak{y}_\mathfrak{p}) \leq 1^{+}$, it means only one of the functions is independent.

**Definition 6.2:** [38]: Suppose $\mathfrak{M}$, $\mathfrak{M}_1$, and $\mathfrak{M}_2$ are NSs in $\mathfrak{Y}$. Then

1. Complement: $\mathfrak{M}^c = \{(\mathfrak{y}_p, 1^{+} - T_{\mathfrak{M}}(\mathfrak{y}_\mathfrak{p}), 1^{+} - I_{\mathfrak{M}}(\mathfrak{y}_\mathfrak{p}), 1^{+} - F_{\mathfrak{M}}(\mathfrak{y}_\mathfrak{p}))|\mathfrak{y} \in \mathfrak{Y}\}$.

2. Null NS: $\inf T_{\mathfrak{M}}(\mathfrak{y}_\mathfrak{p}) = \sup T_{\mathfrak{M}}(\mathfrak{y}_\mathfrak{p}) = 0$, $\inf I_{\mathfrak{M}}(\mathfrak{y}_\mathfrak{p}) = \sup I_{\mathfrak{M}}(\mathfrak{y}_\mathfrak{p}) = 1$, and $\inf F_{\mathfrak{M}}(\mathfrak{y}_\mathfrak{p}) = \sup F_{\mathfrak{M}}(\mathfrak{y}_\mathfrak{p}) = 0$, $\forall \mathfrak{y} \in \mathfrak{Y}$.

3. Inclusion: $\mathfrak{M}_1 \subseteq \mathfrak{M}_2$ implies that

$$\begin{aligned}
&\inf T_{\mathfrak{M}_1}(\mathfrak{y}_\mathfrak{p}) \leq \text{Inf } T_{\mathfrak{M}_2}(\mathfrak{y}_\mathfrak{p}), \sup T_{\mathfrak{M}_1}(\mathfrak{y}_\mathfrak{p}) \leq \sup T_{\mathfrak{M}_2}(\mathfrak{y}_\mathfrak{p}),\\
&\inf I_{\mathfrak{M}_1}(\mathfrak{y}_\mathfrak{p}) \geq \inf I_{\mathfrak{M}_2}(\mathfrak{y}_\mathfrak{p}), \sup I_{\mathfrak{M}_1}(\mathfrak{y}_\mathfrak{p}) \geq \sup I_{\mathfrak{M}_2}(\mathfrak{y}_\mathfrak{p}),\\
&\inf F_{\mathfrak{M}_1}(\mathfrak{y}_\mathfrak{p}) \geq \inf F_{\mathfrak{M}_2}(\mathfrak{y}_\mathfrak{p}), \sup F_{\mathfrak{M}_1}(\mathfrak{y}_\mathfrak{p}) \geq \sup F_{\mathfrak{M}_2}(\mathfrak{y}_\mathfrak{p}),
\end{aligned}$$

$\forall \mathfrak{y} \in \mathfrak{Y}$, where inf and sup denote infimum and supremum operations, respectively.

4. Intersection:

$$\mathfrak{M}_1 \cap \mathfrak{M}_2 = \Big\{\big(\mathfrak{y}_{p}, \min\{T_{\mathfrak{M}_1}(\mathfrak{y}_\mathfrak{p}), T_{\mathfrak{M}_2}(\mathfrak{y}_\mathfrak{p})\}, \max\{I_{\mathfrak{M}_1}(\mathfrak{y}_\mathfrak{p}), I_{\mathfrak{M}_2}(\mathfrak{y}_\mathfrak{p})\}, \\ \max\{F_{\mathfrak{M}_1}(\mathfrak{y}_\mathfrak{p}), F_{\mathfrak{M}_2}(\mathfrak{y}_\mathfrak{p})\}\big)|\mathfrak{y} \in \mathfrak{Y}\Big\}.$$

5. Union:

$$\mathfrak{M}_1 \cap \mathfrak{M}_2 = \Big\{\big(\mathfrak{y}_{p}, \max\{T_{\mathfrak{M}_1}, (\mathfrak{y}_\mathfrak{p}), T_{\mathfrak{M}_2}(\mathfrak{y}_\mathfrak{p})\}, \min\{I_{\mathfrak{M}_1}(\mathfrak{y}_\mathfrak{p}), I_{\mathfrak{M}_2}(\mathfrak{y}_\mathfrak{p})\}, \\ \min\{F_{\mathfrak{M}_1}(\mathfrak{y}_\mathfrak{p}), F_{\mathfrak{M}_2}(\mathfrak{y}_\mathfrak{p})\}\big)|\mathfrak{y} \in \mathfrak{Y}\Big\}.$$

Because of the difficulty in assigning values to the neutrosophic components in practical application sense, the non-standard interval, $]^{-}0, 1^{+}[$ was replaced by a standard interval, $[0, 1]$ to introduce a flexible kind of NS called SVNS.

**Definition 6.3:** [41]: An SVNS designated as $\mathfrak{N}$ is of the form:

$$\mathfrak{N} = \{(\mathfrak{y}_p, T_\mathfrak{N}(\mathfrak{y}_\mathfrak{p}), I_\mathfrak{N}(\mathfrak{y}_\mathfrak{p}), F_\mathfrak{N}(\mathfrak{y}_\mathfrak{p}))|\mathfrak{y} \in \mathfrak{Y}\}, \tag{6.3}$$

where $p = 1, 2, \cdots, k$, and $T_\mathfrak{N}$, $I_\mathfrak{N}$, $F_\mathfrak{N}\colon \mathfrak{Y} \to [0, 1]$ are functions that define the membership function, indeterminancy function, and non-membership function, respectively, in such a way that

$$0 \le T_\mathfrak{N}(\mathfrak{y}_\mathfrak{p}) + I_\mathfrak{N}(\mathfrak{y}_\mathfrak{p}) + F_\mathfrak{N}(\mathfrak{y}_\mathfrak{p}) \le 3, \tag{6.4}$$

where all the functions are independent. For two independent functions and one independent function, we have

$$\left.\begin{aligned} 0 \le T_\mathfrak{N}(\mathfrak{y}_\mathfrak{p}) + I_\mathfrak{N}(\mathfrak{y}_\mathfrak{p}) + F_\mathfrak{N}(\mathfrak{y}_\mathfrak{p}) \le 2 \\ 0 \le T_\mathfrak{N}(\mathfrak{y}_\mathfrak{p}) + I_\mathfrak{N}(\mathfrak{y}_\mathfrak{p}) + F_\mathfrak{N}(\mathfrak{y}_\mathfrak{p}) \le 1 \end{aligned}\right\}. \tag{6.5}$$

In case, $T_\mathfrak{N}(\mathfrak{y}_\mathfrak{p}) + I_\mathfrak{N}(\mathfrak{y}_\mathfrak{p}) + F_\mathfrak{N}(\mathfrak{y}_\mathfrak{p}) = 1$, it means $\mathfrak{N}$ is an IFS in $\mathfrak{Y}$, and $I_\mathfrak{N}(\mathfrak{y}_\mathfrak{p})$ is the hesitation margin. The pair $\big(T_\mathfrak{N}(\mathfrak{y}_\mathfrak{p}), I_\mathfrak{N}(\mathfrak{y}_\mathfrak{p}), F_\mathfrak{N}(\mathfrak{y}_\mathfrak{p})\big)$ is known as single-valued neutrosophic number (SVNN), simply presented as $\big(T_\mathfrak{N}, I_\mathfrak{N}, F_\mathfrak{N}\big)$.

**Definition 6.4:** [41]: Suppose $\mathfrak{N} = \big(T_\mathfrak{N}, I_\mathfrak{N}, F_\mathfrak{N}\big)$, $\mathfrak{N}_1 = \big(T_{\mathfrak{N}_1}, I_{\mathfrak{N}_1}, F_{\mathfrak{N}_1}\big)$, and $\mathfrak{N}_2 = \big(T_{\mathfrak{N}_2}, I_{\mathfrak{N}_2}, F_{\mathfrak{N}_2}\big)$ are SVNNs in $\mathfrak{Y}$. Then

1. Complement: $\mathfrak{N}^c = \big(F_\mathfrak{N}, 1 - I_\mathfrak{N}, T_\mathfrak{N}\big)$.
2. Null NS: $T_\mathfrak{N} = 0$, $I_\mathfrak{N} = 1$, and $F_\mathfrak{N} = 0$. Thus, $\mathfrak{N} = (0, 1, 0)$.
3. Inclusion: $\mathfrak{N}_1 \subseteq \mathfrak{N}_2$ implies that

$$T_{\mathfrak{M}_1} \le T_{\mathfrak{M}_2}, I_{\mathfrak{M}_1} \ge I_{\mathfrak{M}_2}, F_{\mathfrak{M}_1} \ge F_{\mathfrak{M}_2}.$$

4. Intersection:

$$\mathfrak{M}_1 \cap \mathfrak{M}_2 = \big(\min\{T_{\mathfrak{M}_1}, T_{\mathfrak{M}_2}\}, \max\{I_{\mathfrak{M}_1}, I_{\mathfrak{M}_2}\}, \max\{F_{\mathfrak{M}_1}, F_{\mathfrak{M}_2}\}\big).$$

5. Union:

$$\mathfrak{M}_1 \cup \mathfrak{M}_2 = \big(\max\{T_{\mathfrak{M}_1}, T_{\mathfrak{M}_2}\}, \min\{I_{\mathfrak{M}_1}, I_{\mathfrak{M}_2}\}, \min\{F_{\mathfrak{M}_1}, F_{\mathfrak{M}_2}\}\big).$$

6. Sum:

$$\begin{aligned}\mathfrak{M}_1 \oplus \mathfrak{M}_2 &= (T_{\mathfrak{M}_1} + T_{\mathfrak{M}_2} - T_{\mathfrak{M}_1}T_{\mathfrak{M}_2}, I_{\mathfrak{M}_1} + I_{\mathfrak{M}_2} - I_{\mathfrak{M}_1}I_{\mathfrak{M}_2}, F_{\mathfrak{M}_1} \\ &\quad + F_{\mathfrak{M}_2} - F_{\mathfrak{M}_1}F_{\mathfrak{M}_2}\big).\end{aligned}$$

7. Product:

$$\mathfrak{M}_1 \odot \mathfrak{M}_2 = (T_{\mathfrak{M}_1}T_{\mathfrak{M}_2}, I_{\mathfrak{M}_1}I_{\mathfrak{M}_2}, F_{\mathfrak{M}_1}F_{\mathfrak{M}_2}\big).$$

To effectively estimate the relation between two arbitrary SVNSs, the idea of correlation coefficient was deployed. The correlation coefficient under SVNSs is a function defined as follows:

**Definition 6.5:** [47]: Let $\mathfrak{N}$, $\mathfrak{N}_1$, and $\mathfrak{N}_2$ be SVNSs in $\mathfrak{Y}$. Then, the correlation coefficient between $\mathfrak{N}_1$ and $\mathfrak{N}_2$ designated by $\Delta(\mathfrak{N}_1, \mathfrak{N}_2)$ is a function, $\Delta$: SVNS $\times$ SVNS $\rightarrow [-1, 1]$ which satisfies:

1. $\Delta(\mathfrak{N}_1, \mathfrak{N}_2) \in [-1, 1]$,
2. $\Delta(\mathfrak{N}_1, \mathfrak{N}_2) = 1 \Leftrightarrow \mathfrak{N}_1 = \mathfrak{N}_2$,
3. $\Delta(\mathfrak{N}_1, \mathfrak{N}_2) = \Delta(\mathfrak{N}_2, \mathfrak{N}_1)$.

### 6.2.1 Existing methods of correlation coefficient for SVNSs

The practical applications of SVNSs have been extensively explored by leveraging the flexibility of the correlation coefficient. The known methods for computing the correlation coefficient between SVNSs are enumerated as follows. Let $\mathfrak{N}_1$ and $\mathfrak{N}_2$ be SVNSs defined on $\mathfrak{Y} = \{\mathfrak{y}_1, \mathfrak{y}_2, \ldots, \mathfrak{y}_k\}$, where $p = 1, 2, \ldots, k$. Then, the existing methods of correlation coefficient between $\mathfrak{N}_1$ and $\mathfrak{N}_2$ in $\mathfrak{Y}$ are given by (6.6), (6.8), (6.9), and (6.13), respectively.
Ye [47]

$$\Delta_1(\mathfrak{N}_1, \mathfrak{N}_2) = \frac{C(\mathfrak{N}_1, \mathfrak{N}_2)}{\sqrt{E(\mathfrak{N}_1)}\sqrt{E(\mathfrak{N}_2)}}, \tag{6.6}$$

where

$$\left.\begin{aligned} C(\mathfrak{N}_1, \mathfrak{N}_2) &= \sum_{p=1}^{k} \Big(T_{\mathfrak{N}_1}(\mathfrak{y}_p)T_{\mathfrak{N}_2}(\mathfrak{y}_p) + I_{\mathfrak{N}_1}(\mathfrak{y}_p)I_{\mathfrak{N}_2}(\mathfrak{y}_p) + F_{\mathfrak{N}_1}(\mathfrak{y}_p)F_{\mathfrak{N}_2}(\mathfrak{y}_p)\Big) \\ E(\mathfrak{N}_1) &= \sum_{p=1}^{k} \Big(T^2_{\mathfrak{N}_1}(\mathfrak{y}_p) + I^2_{\mathfrak{N}_1}(\mathfrak{y}_p) + F^2_{\mathfrak{N}_1}(\mathfrak{y}_p)\Big) \\ E(\mathfrak{N}_2) &= \sum_{p=1}^{k} \Big(T^2_{\mathfrak{N}_2}(\mathfrak{y}_p) + I^2_{\mathfrak{N}_2}(\mathfrak{y}_p) + F^2_{\mathfrak{N}_2}(\mathfrak{y}_p)\Big) \end{aligned}\right\}. \tag{6.7}$$

Iryna et al. [15]

$$\Delta_2(\mathfrak{N}_1, \mathfrak{N}_2) = \frac{2C(\mathfrak{N}_1, \mathfrak{N}_2)}{E(\mathfrak{N}_1) + E(\mathfrak{N}_2)}, \tag{6.8}$$

where $C(\mathfrak{N}_1, \mathfrak{N}_2)$, $E(\mathfrak{N}_1)$, and $E(\mathfrak{N}_2)$ are as stated in (6.7).
Ye [48]

$$\Delta_3(\mathfrak{N}_1, \mathfrak{N}_2) = \frac{1}{3k}\sum_{p=1}^{k}\Big(A_p(1-\Delta T_p) + B_p(1-\Delta I_p) + C_p(1-\Delta F_p)\Big), \tag{6.9}$$

where

$$\left.\begin{aligned} \Delta T_p &= |T_{\mathfrak{N}_1}(\mathfrak{y}_p) - T_{\mathfrak{N}_2}(\mathfrak{y}_p)| \\ \Delta I_p &= |I_{\mathfrak{N}_1}(\mathfrak{y}_p) - I_{\mathfrak{N}_2}(\mathfrak{y}_p)| \\ \Delta F_p &= |F_{\mathfrak{N}_1}(\mathfrak{y}_p) - F_{\mathfrak{N}_2}(\mathfrak{y}_p)| \end{aligned}\right\}, \tag{6.10}$$

$$\left.\begin{aligned} A_p &= \frac{3 - \Delta T_p - \Delta T_{\max}}{3 - \Delta T_{\min} - \Delta T_{\max}} \\ B_p &= \frac{3 - \Delta I_p - \Delta I_{\max}}{3 - \Delta I_{\min} - \Delta I_{\max}} \\ C_p &= \frac{3 - \Delta F_p - \Delta F_{\max}}{3 - \Delta F_{\min} - \Delta F_{\max}} \end{aligned}\right\}, \tag{6.11}$$

$$\left.\begin{aligned} \Delta T_{\min} &= \min_{1\le p\le k}\{\Delta T_p\}, \Delta I_{\min} = \min_{1\le p\le k}\{\Delta I_p\} \\ \Delta F_{\min} &= \min_{1\le p\le k}\{\Delta F_p\}, \Delta T_{\max} = \max_{1\le p\le k}\{\Delta T_p\} \\ \Delta I_{\max} &= \max_{1\le p\le k}\{\Delta I_p\}, \Delta F_{\max} = \max_{1\le p\le k}\{\Delta F_p\} \end{aligned}\right\}. \tag{6.12}$$

Zhou et al. [54]

$$\Delta_4(\mathfrak{N}_1, \mathfrak{N}_2) = \frac{\mathrm{Cov}(\mathfrak{N}_1, \mathfrak{N}_2)}{\sqrt{\mathrm{Dev}(\mathfrak{N}_1)\mathrm{Dev}(\mathfrak{N}_2)}} = \frac{\mathrm{Dev}(\mathfrak{N}_1)\mathrm{Dev}(\mathfrak{N}_2)}{\sqrt{\mathrm{Dev}(\mathfrak{N}_1)\mathrm{Dev}(\mathfrak{N}_2)}}, \tag{6.13}$$

where

$$\left.\begin{aligned} \mathrm{Dev}(\mathfrak{N}_1) &= \Sigma_{p=1}^{k}\Big(\big(T^2_{\mathfrak{N}_1}(\mathfrak{y}_p) - \overline{T}^2_{\mathfrak{N}_1}\big) - \big(I^2_{\mathfrak{N}_1}(\mathfrak{y}_p) - \overline{I}^2_{\mathfrak{N}_1}\big) - \big(F^2_{\mathfrak{N}_1}(\mathfrak{y}_p) - \overline{F}^2_{\mathfrak{N}_1}\big)\Big) \\ \mathrm{Dev}(\mathfrak{N}_2) &= \Sigma_{p=1}^{k}\Big(\big(T^2_{\mathfrak{N}_2}(\mathfrak{y}_p) - \overline{T}^2_{\mathfrak{N}_2}\big) - \big(I^2_{\mathfrak{N}_2}(\mathfrak{y}_p) - \overline{I}^2_{\mathfrak{N}_2}\big) - \big(F^2_{\mathfrak{N}_2}(\mathfrak{y}_p) - \overline{F}^2_{\mathfrak{N}_2}\big)\Big) \end{aligned}\right\}, \tag{6.14}$$

in which

$$\left.\begin{aligned} \overline{T}_{\mathfrak{N}_1} &= \frac{\Sigma_{p=1}^{k} T_{\mathfrak{N}_1}(\mathfrak{y}_p)}{k}, \overline{T}_{\mathfrak{N}_2} = \frac{\Sigma_{p=1}^{k} T_{\mathfrak{N}_2}(\mathfrak{y}_p)}{k} \\ \overline{I}_{\mathfrak{N}_1} &= \frac{\Sigma_{p=1}^{k} I_{\mathfrak{N}_1}(\mathfrak{y}_p)}{k}, \overline{I}_{\mathfrak{N}_2} = \frac{\Sigma_{p=1}^{k} I_{\mathfrak{N}_2}(\mathfrak{y}_p)}{k} \\ \overline{F}_{\mathfrak{N}_1} &= \frac{\Sigma_{p=1}^{k} F_{\mathfrak{N}_1}(\mathfrak{y}_p)}{k}, \overline{F}_{\mathfrak{N}_2} = \frac{\Sigma_{p=1}^{k} F_{\mathfrak{N}_2}(\mathfrak{y}_p)}{k} \end{aligned}\right\}. \tag{6.15}$$

Although these correlation coefficients are well formulated, we observe that: (i) the methods in [15, 47, 48] are defined within $[0, 1]$ and indicate only the degree of correlation between the SVNSs without revealing the direction of the relationship; (ii) the method in [54], although defined within $[-1, 1]$, considers only the deviation of the SVNSs. Since none of the existing correlation coefficient methods under SVNSs are based on the statistical viewpoint of Pearson's model, it is therefore reasonable to formulate a novel correlation coefficient from a statistical perspective that demonstrates reliability, correlation strength, and the direction of the relationship.

---

## 6.3 A Novel Correlation Coefficient between Single-Valued Neutrosophic Sets

In this section, a novel approach of correlation coefficient for SVNSs is presented based on Pearson's correlation perspective.

**Definition 6.6:** Let $\mathfrak{N}_1$ and $\mathfrak{N}_2$ be SVNSs in $\mathfrak{Y} = \{\mathfrak{y}_1, \mathfrak{y}_2, \cdots, \mathfrak{y}_k\}$, for $p = 1, 2, \cdots, k$. Then, the novel approach of correlation coefficient for $\mathfrak{N}_1$ and $\mathfrak{N}_2$ in $\mathfrak{Y}$ is defined as (6.16).

$$\Delta(\mathfrak{N}_1, \mathfrak{N}_2) = \frac{T(\mathfrak{N}_1, \mathfrak{N}_2) + I(\mathfrak{N}_1, \mathfrak{N}_2) + F(\mathfrak{N}_1, \mathfrak{N}_2)}{3}, \tag{6.16}$$

where the correlation compartments are:

$$\left.\begin{aligned}
T(\mathfrak{N}_1, \mathfrak{N}_2) &= \frac{\Sigma_{p=1}^{k}(T_{\mathfrak{N}_1}(\mathfrak{y}_p) - \overline{T}_{\mathfrak{N}_1})(T_{\mathfrak{N}_2}(\mathfrak{y}_p) - \overline{T}_{\mathfrak{N}_2})}{\sqrt{\Sigma_{p=1}^{k}(T_{\mathfrak{N}_1}(\mathfrak{y}_p) - \overline{T}_{\mathfrak{N}_1})^2}\sqrt{\Sigma_{p=1}^{k}(T_{\mathfrak{N}_2}(\mathfrak{y}_p) - \overline{T}_{\mathfrak{N}_2})^2}} \\
I(\mathfrak{N}_1, \mathfrak{N}_2) &= \frac{\Sigma_{p=1}^{k}(I_{\mathfrak{N}_1}(\mathfrak{y}_p) - \overline{I}_{\mathfrak{N}_1})(I_{\mathfrak{N}_2}(\mathfrak{y}_p) - \overline{I}_{\mathfrak{N}_2})}{\sqrt{\Sigma_{p=1}^{k}(I_{\mathfrak{N}_1}(\mathfrak{y}_p) - \overline{I}_{\mathfrak{N}_1})^2}\sqrt{\Sigma_{p=1}^{k}(I_{\mathfrak{N}_2}(\mathfrak{y}_p) - \overline{I}_{\mathfrak{N}_2})^2}} \\
F(\mathfrak{N}_1, \mathfrak{N}_2) &= \frac{\Sigma_{p=1}^{k}(F_{\mathfrak{N}_1}(\mathfrak{y}_p) - \overline{F}_{\mathfrak{N}_1})(F_{\mathfrak{N}_2}(\mathfrak{y}_p) - \overline{F}_{\mathfrak{N}_2})}{\sqrt{\Sigma_{p=1}^{k}(F_{\mathfrak{N}_1}(\mathfrak{y}_p) - \overline{F}_{\mathfrak{N}_1})^2}\sqrt{\Sigma_{p=1}^{k}(F_{\mathfrak{N}_2}(\mathfrak{y}_p) - \overline{F}_{\mathfrak{N}_2})^2}}
\end{aligned}\right\}, \tag{6.17}$$

and the mean values of the SVNSs are:

$$\left.\begin{aligned}
\overline{T}_{\mathfrak{N}_1} &= \frac{\Sigma_{p=1}^{k} T_{\mathfrak{N}_1}(\mathfrak{y}_p)}{k}, \; \overline{T}_{\mathfrak{N}_2} = \frac{\Sigma_{p=1}^{k} T_{\mathfrak{N}_2}(\mathfrak{y}_p)}{k} \\
\overline{I}_{\mathfrak{N}_1} &= \frac{\Sigma_{p=1}^{k} I_{\mathfrak{N}_1}(\mathfrak{y}_p)}{k}, \; \overline{I}_{\mathfrak{N}_2} = \frac{\Sigma_{p=1}^{k} I_{\mathfrak{N}_2}(\mathfrak{y}_p)}{k} \\
\overline{F}_{\mathfrak{N}_1} &= \frac{\Sigma_{p=1}^{k} F_{\mathfrak{N}_1}(\mathfrak{y}_p)}{k}, \; \overline{F}_{\mathfrak{N}_2} = \frac{\Sigma_{p=1}^{k} F_{\mathfrak{N}_2}(\mathfrak{y}_p)}{k}
\end{aligned}\right\}. \tag{6.18}$$

Now, we describe the properties of the novel method of correlation coefficient between SVNSs as follows:

**Theorem 6.1:** *The novel correlation coefficient for SVNSs $\mathfrak{N}_1$ and $\mathfrak{N}_2$ in $\mathfrak{Y}$ satisfies the conditions of correlation coefficient.*

**Proof:** *In this proof, the authenticity of the new correlation coefficient between $\mathfrak{N}_1$ and $\mathfrak{N}_2$ is established. To do this, we need to show that $\Delta(\mathfrak{N}_1, \mathfrak{N}_2)$ satisfies the conditions in Definition 6.5. To prove that $\Delta(\mathfrak{N}_1, \mathfrak{N}_2) \in [-1, 1]$, we need to verify $-1 \leq \Delta(\mathfrak{N}_1, \mathfrak{N}_2) \leq 1$, meaning $\Delta(\mathfrak{N}_1, \mathfrak{N}_2) \geq -1$ and $\Delta(\mathfrak{N}_1, \mathfrak{N}_2) \leq 1$. Since*

$$\left.\begin{aligned}
T(\mathfrak{N}_1, \mathfrak{N}_2) &= \frac{\Sigma_{p=1}^{k}(T_{\mathfrak{N}_1}(\mathfrak{y}_p) - \overline{T}_{\mathfrak{N}_1})(T_{\mathfrak{N}_2}(\mathfrak{y}_p) - \overline{T}_{\mathfrak{N}_2})}{\sqrt{\Sigma_{p=1}^{k}(T_{\mathfrak{N}_1}(\mathfrak{y}_p) - \overline{T}_{\mathfrak{N}_1})^2}\sqrt{\Sigma_{p=1}^{k}(T_{\mathfrak{N}_2}(\mathfrak{y}_p) - \overline{T}_{\mathfrak{N}_2})^2}} \geq -1 \\
I(\mathfrak{N}_1, \mathfrak{N}_2) &= \frac{\Sigma_{p=1}^{k}(I_{\mathfrak{N}_1}(\mathfrak{y}_p) - \overline{I}_{\mathfrak{N}_1})(I_{\mathfrak{N}_2}(\mathfrak{y}_p) - \overline{I}_{\mathfrak{N}_2})}{\sqrt{\Sigma_{p=1}^{k}(I_{\mathfrak{N}_1}(\mathfrak{y}_p) - \overline{I}_{\mathfrak{N}_1})^2}\sqrt{\Sigma_{p=1}^{k}(I_{\mathfrak{N}_2}(\mathfrak{y}_p) - \overline{I}_{\mathfrak{N}_2})^2}} \geq -1 \\
F(\mathfrak{N}_1, \mathfrak{N}_2) &= \frac{\Sigma_{p=1}^{k}(F_{\mathfrak{N}_1}(\mathfrak{y}_p) - \overline{F}_{\mathfrak{N}_1})(F_{\mathfrak{N}_2}(\mathfrak{y}_p) - \overline{F}_{\mathfrak{N}_2})}{\sqrt{\Sigma_{p=1}^{k}(F_{\mathfrak{N}_1}(\mathfrak{y}_p) - \overline{F}_{\mathfrak{N}_1})^2}\sqrt{\Sigma_{p=1}^{k}(F_{\mathfrak{N}_2}(\mathfrak{y}_p) - \overline{F}_{\mathfrak{N}_2})^2}} \geq -1
\end{aligned}\right\},$$

*it is certain that $\Delta(\mathfrak{N}_1, \mathfrak{N}_2) \geq -1$. In continuation, assume*

$$T(\mathfrak{N}_1, \mathfrak{N}_2) = \kappa, I(\mathfrak{N}_1, \mathfrak{N}_2) = \xi,$$
$$F(\mathfrak{N}_1, \mathfrak{N}_2) = \varrho,$$

*then $\Delta(\mathfrak{N}_1, \mathfrak{N}_2) = \dfrac{\kappa + \xi + \varrho}{3}$. Hence,*

$$\begin{aligned}
\Delta(\mathfrak{N}_1, \mathfrak{N}_2) - 1 &= \frac{\kappa + \xi + \varrho}{3} - 1 = \frac{(\kappa + \xi + \varrho) - 3}{3} \\
&= \frac{-\big(-(\kappa + \xi + \varrho) + 3\big)}{3} \\
&= -\frac{\big(-(\kappa + \xi + \varrho) + 3\big)}{3}.
\end{aligned}$$

*Thus, $\Delta(\mathfrak{N}_1, \mathfrak{N}_2) - 1 \leq 0 \implies \Delta(\mathfrak{N}_1, \mathfrak{N}_2) \leq 1$. Hence, $\Delta(\mathfrak{N}_1, \mathfrak{N}_2) \in [-1, 1]$, which satisfies condition (1).*

*Next, we prove condition (2). Assume $\Delta(\mathfrak{N}_1, \mathfrak{N}_2) = 1$, then $T(\mathfrak{N}_1, \mathfrak{N}_2) = 1$, $I(\mathfrak{N}_1, \mathfrak{N}_2) = 1$, and $F(\mathfrak{N}_1, \mathfrak{N}_2) = 1$. Hence, $\mathfrak{N}_1 = \mathfrak{N}_2$.*

*Conversely, if* $\mathfrak{N}_1 = \mathfrak{N}_2$*, then*

$$\left.\begin{aligned} T(\mathfrak{N}_1,\mathfrak{N}_2) &= \frac{\Sigma_{p=1}^{k}(T_{\mathfrak{N}_1}(\mathfrak{y}_p)-\overline{T}_{\mathfrak{N}_1})(T_{\mathfrak{N}_1}(\mathfrak{y}_p)-\overline{T}_{\mathfrak{N}_1})}{\sqrt{\Sigma_{p=1}^{k}(T_{\mathfrak{N}_1}(\mathfrak{y}_p)-\overline{T}_{\mathfrak{N}_1})^2}\sqrt{\Sigma_{p=1}^{k}(T_{\mathfrak{N}_1}(\mathfrak{y}_p)-\overline{T}_{\mathfrak{N}_1})^2}}=1\\ I(\mathfrak{N}_1,\mathfrak{N}_2) &= \frac{\Sigma_{p=1}^{k}(I_{\mathfrak{N}_1}(\mathfrak{y}_p)-\overline{I}_{\mathfrak{N}_1})(I_{\mathfrak{N}_1}(\mathfrak{y}_p)-\overline{I}_{\mathfrak{N}_1})}{\sqrt{\Sigma_{p=1}^{k}(I_{\mathfrak{N}_1}(\mathfrak{y}_p)-\overline{I}_{\mathfrak{N}_1})^2}\sqrt{\Sigma_{p=1}^{k}(I_{\mathfrak{N}_1}(\mathfrak{y}_p)-\overline{I}_{\mathfrak{N}_1})^2}}=1\\ F(\mathfrak{N}_1,\mathfrak{N}_2) &= \frac{\Sigma_{p=1}^{k}(F_{\mathfrak{N}_1}(\mathfrak{y}_p)-\overline{F}_{\mathfrak{N}_1})(F_{\mathfrak{N}_1}(\mathfrak{y}_p)-\overline{F}_{\mathfrak{N}_1})}{\sqrt{\Sigma_{p=1}^{k}(F_{\mathfrak{N}_1}(\mathfrak{y}_p)-\overline{F}_{\mathfrak{N}_1})^2}\sqrt{\Sigma_{p=1}^{k}(F_{\mathfrak{N}_1}(\mathfrak{y}_p)-\overline{F}_{\mathfrak{N}_1})^2}}=1 \end{aligned}\right\}.$$

*Thus,* $\Delta(\mathfrak{N}_1,\mathfrak{N}_2) = 1$ *since* $\Delta(\mathfrak{N}_1,\mathfrak{N}_2) = \dfrac{T(\mathfrak{N}_1,\mathfrak{N}_2)+I(\mathfrak{N}_1,\mathfrak{N}_2)+F(\mathfrak{N}_1,\mathfrak{N}_2)}{3}$.
*These complete the proof of condition (2).*

*Finally, the proof of condition (3) follows since*

$$\begin{aligned} T(\mathfrak{N}_1,\mathfrak{N}_2) &= \frac{\Sigma_{p=1}^{k}(T_{\mathfrak{N}_1}(\mathfrak{y}_p)-\overline{T}_{\mathfrak{N}_1})(T_{\mathfrak{N}_2}(\mathfrak{y}_p)-\overline{T}_{\mathfrak{N}_2})}{\sqrt{\Sigma_{p=1}^{k}(T_{\mathfrak{N}_1}(\mathfrak{y}_p)-\overline{T}_{\mathfrak{N}_1})^2}\sqrt{\Sigma_{p=1}^{k}(T_{\mathfrak{N}_2}(\mathfrak{y}_p)-\overline{T}_{\mathfrak{N}_2})^2}}\\ &= \frac{\Sigma_{p=1}^{k}(T_{\mathfrak{N}_2}(\mathfrak{y}_p)-\overline{T}_{\mathfrak{N}_2})(T_{\mathfrak{N}_1}(\mathfrak{y}_p)-\overline{T}_{\mathfrak{N}_1})}{\sqrt{\Sigma_{p=1}^{k}(T_{\mathfrak{N}_2}(\mathfrak{y}_p)-\overline{T}_{\mathfrak{N}_2})^2}\sqrt{\Sigma_{p=1}^{k}(T_{\mathfrak{N}_1}(\mathfrak{y}_p)-\overline{T}_{\mathfrak{N}_1})^2}}\\ &= T(\mathfrak{N}_2,\mathfrak{N}_1). \end{aligned}$$

*Similarly,* $I(\mathfrak{N}_1,\mathfrak{N}_2) = I(\mathfrak{N}_2,\mathfrak{N}_1)$ *and* $F(\mathfrak{N}_1,\mathfrak{N}_2) = F(\mathfrak{N}_2,\mathfrak{N}_1)$*. Hence,* $\Delta(\mathfrak{N}_1,\mathfrak{N}_2) = \Delta(\mathfrak{N}_2,\mathfrak{N}_1)$.

**Remark 6.1:** If $\mathfrak{N}_1$ and $\mathfrak{N}_2$ are SVNSs in $\mathfrak{Y}$, then $T(\mathfrak{N}_1,\mathfrak{N}_2)$, $I(\mathfrak{N}_1,\mathfrak{N}_2)$, and $F(\mathfrak{N}_1,\mathfrak{N}_2)$ also satisfy Definition 6.5.

### 6.3.1 Numerical validity

Here, the validity of the new approach of correlation coefficient between SVNSs is shown in comparison to the existing ones [15, 47, 48, 54], using the SVNNs in Table 6.1.

From the first case in Table 6.1, it is observed that the methods $\Delta_1$ and $\Delta_2$ [15, 47] yield the same correlation coefficient, although they are not identical. The method $\Delta_4$ [54] produces a zero correlation coefficient between the SVNSs, which is unreliable. Interestingly, the method $\Delta_3$ [48] indicates a somewhat positive relationship between the SVNSs, which is inconsistent with direct observation. In contrast, the proposed method reveals a negative relationship between the SVNSs, which aligns with the observed relationship.

In the second case, the proposed approach indicates a positive relationship between the SVNSs. Although the approaches in [15, 47, 48] yield comparatively higher correlation values, their results are less reliable due to their

**TABLE 6.1**: Correlation coefficient's results.

| Cases | Ye [47] | Iryna et al. [15] | Ye [48] | Zhou et al. [54] | New method |
|---|---|---|---|---|---|
| $\mathfrak{N}_1 = (1, \frac{1}{2}, 0), (0, \frac{2}{5}, 1), (1, \frac{1}{2}, 0)$, $\mathfrak{N}_2 = (0, \frac{1}{2}, 1), (1, \frac{2}{5}, 0), (0, \frac{1}{2}, 1)$ | 0.1803 | 0.1803 | 0.3333 | 0.0000 | −0.3333 |
| $\mathfrak{N}_1 = (\frac{1}{2}, \frac{3}{10}, \frac{3}{10}), (\frac{3}{5}, \frac{2}{5}, \frac{1}{2}), (\frac{4}{5}, \frac{1}{10}, \frac{3}{10})$, $\mathfrak{N}_2 = (\frac{11}{20}, \frac{3}{10}, \frac{2}{5}), (\frac{2}{5}, \frac{9}{20}, \frac{3}{5}), (\frac{3}{5}, \frac{1}{5}, \frac{1}{5})$ | 0.9676 | 0.9658 | 0.9119 | 0.3672 | 0.5979 |
| $\mathfrak{N}_1 = (1, \frac{1}{2}, 0), (0, \frac{3}{5}, 1), (\frac{1}{2}, \frac{3}{10}, \frac{2}{5})$, $\mathfrak{N}_2 = (0, \frac{3}{5}, 1), (1, \frac{11}{20}, 0), (\frac{7}{10}, \frac{1}{5}, \frac{3}{10})$ | 0.3631 | 0.3629 | 0.6574 | 0.3539 | −0.4105 |

inability to capture both the magnitude and direction of the correlation. In the final case, the proposed method again identifies a negative relationship between the SVNSs, consistent with the observed relationship, whereas the existing methods yield contradictory results and interpretations. Overall, the proposed approach proves more robust, consistent, and reliable than the existing methods [15, 47, 48, 54]. Furthermore, the proposed method is developed based on Pearson's correlation framework, unlike the existing approaches.

---

## 6.4 Application Example of Renewable Energy Source Selection Analysis

The choice of a suitable RES for effective power and energy supply is essential for achieving sustainable development. Selecting an appropriate RES is a challenging decision-making problem due to the complexity of the selection criteria and the potential environmental impacts of different energy sources.

This application example demonstrates a decision-making approach based on computational intelligence, in which ten selection criteria are considered: land-use requirements, energy storage capacity, aesthetic effect, noise pollution, environmental impact, adaptability to climate variability, technological know-how, maintenance cost, economic impact, and supply chain resilience. These selection criteria are represented by the set $\mathfrak{S} = \{\mathfrak{S}_1, \mathfrak{S}_2, \ldots, \mathfrak{S}_{10}\}$ in the order of their presentation.

The candidate RESs are represented by the set $\mathfrak{R} = \{\mathfrak{R}_1, \mathfrak{R}_2, \ldots, \mathfrak{R}_{10}\}$, whose elements correspond to photovoltaic energy, hydroelectric energy, hy-

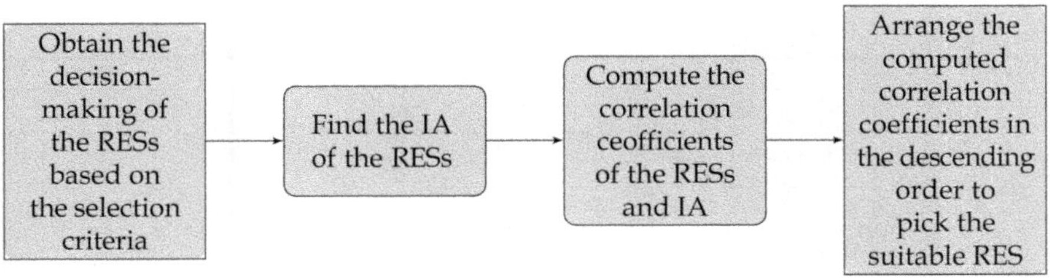


**FIGURE 6.1**: Decision-making algorithm representation.

drogen fuel cell power plant, solar energy, bioenergy, hydropower energy, wind energy, geothermal energy, tidal energy, and nuclear energy, respectively. The objective of this application example is to determine the most appropriate RES using the proposed decision-making method based on the newly developed correlation coefficient.

### 6.4.1 Decision-making algorithm for the selection process

**Step 1:** Obtain the decision matrix of the RESs based on the selection criteria, represented by (6.19).

$$\mathfrak{K}_p = \{\mathfrak{S}_p(\mathfrak{R}_p)\}_{(k\times k)}(p = 1, 2, \cdots, k). \tag{6.19}$$

**Step 2:** Find the ideal alternative (IA), $\tilde{\mathfrak{R}}$, which is defined by (6.20).

$$\tilde{\mathfrak{R}} = \left( \max_{1\leq p\leq k}\{T_{\mathfrak{R}_p}(\mathfrak{S}_p)\}, \min_{1\leq p\leq k}\{I_{\mathfrak{R}_p}(\mathfrak{S}_p)\}, \min_{1\leq p\leq k}\{F_{\mathfrak{R}_p}(\mathfrak{S}_p)\} \right). \tag{6.20}$$

**Step 3:** Compute the correlation ceofficients of the RESs and IA, using $\Delta(\tilde{\mathfrak{R}}, \mathfrak{R}_1)$, $\Delta(\tilde{\mathfrak{R}}, \mathfrak{R}_2)$, $\cdots$, $\Delta(\tilde{\mathfrak{R}}, \mathfrak{R}_{10})$.
**Step 4:** Arrange the computed correlation coefficients in the descending order to pick the suitable RES. The decision-making algorithm is shown by Figure 6.1.

### 6.4.2 Algorithm application

By modifying the data in [2] in terms of SVNSs, we obtain the information in Table 6.2, which is the decision matrix.

Next, the IA, $\tilde{\mathfrak{R}}$, is given as follows:

$$\tilde{\mathfrak{R}} = \begin{bmatrix} (0.72, 0.2, 0.2) & (0.75, 0.2, 0.2) & (0.6, 0.2, 0.16) & (0.7, 0.2, 0.12) & (0.72, 0.2, 0.2) \\ (0.58, 0.2, 0.3) & (0.68, 0.2, 0.18) & (0.54, 0.1, 0.28) & (0.78, 0.2, 0.12) & (0.72, 0.2, 0.12) \end{bmatrix},$$

where the entries of the first row are for $\mathfrak{S}_1$, $\mathfrak{S}_2$, $\mathfrak{S}_3$, $\mathfrak{S}_4$, and $\mathfrak{S}_5$, and the entries of the second row are for $\mathfrak{S}_6$, $\mathfrak{S}_7$, $\mathfrak{S}_8$, $\mathfrak{S}_9$, and $\mathfrak{S}_{10}$. The computed

**TABLE 6.2**: RESs and their selection criteria.

| $\mathfrak{R}/\mathfrak{S}$ | $\mathfrak{S}_1$ | $\mathfrak{S}_2$ | $\mathfrak{S}_3$ | $\mathfrak{S}_4$ | $\mathfrak{S}_5$ |
|---|---|---|---|---|---|
| $\mathfrak{R}_1$ | $(0.42, 0.3, 0.6)$ | $(0.34, 0.4, 0.23)$ | $(0.6, 0.2, 0.16)$ | $(0.7, 0.3, 0.23)$ | $(0.32, 0.5, 0.42)$ |
| $\mathfrak{R}_2$ | $(0.72, 0.4, 0.52)$ | $(0.12, 0.4, 0.35)$ | $(0.33, 0.6, 0.22)$ | $(0.55, 0.2, 0.51)$ | $(0.36, 0.2, 0.54)$ |
| $\mathfrak{R}_3$ | $(0.56, 0.2, 0.54)$ | $(0.54, 0.5, 0.12)$ | $(0.4, 0.4, 0.59)$ | $(0.4, 0.4, 0.45)$ | $(0.7, 0.2, 0.2)$ |
| $\mathfrak{R}_4$ | $(0.4, 0.3, 0.45)$ | $(0.3, 0.4, 0.7)$ | $(0.4, 0.4, 0.21)$ | $(0.5, 0.4, 0.54)$ | $(0.35, 0.2, 0.6)$ |
| $\mathfrak{R}_5$ | $(0.3, 0.4, 0.45)$ | $(0.64, 0.4, 0.45)$ | $(0.6, 0.3, 0.3)$ | $(0.25, 0.3, 0.23)$ | $(0.12, 0.3, 0.35)$ |
| $\mathfrak{R}_6$ | $(0.52, 0.3, 0.2)$ | $(0.56, 0.4, 0.3)$ | $(0.45, 0.5, 0.3)$ | $(0.5, 0.3, 0.25)$ | $(0.5, 0.3, 0.7)$ |
| $\mathfrak{R}_7$ | $(0.7, 0.3, 0.3)$ | $(0.26, 0.4, 0.36)$ | $(0.18, 0.2, 0.36)$ | $(0.6, 0.3, 0.12)$ | $(0.24, 0.3, 0.7)$ |
| $\mathfrak{R}_8$ | $(0.68, 0.4, 0.25)$ | $(0.75, 0.3, 0.3)$ | $(0.42, 0.4, 0.56)$ | $(0.52, 0.3, 0.4)$ | $(0.72, 0.2, 0.6)$ |
| $\mathfrak{R}_9$ | $(0.28, 0.3, 0.68)$ | $(0.6, 0.2, 0.62)$ | $(0.34, 0.4, 0.6)$ | $(0.46, 0.3, 0.32)$ | $(0.56, 0.2, 0.3)$ |
| $\mathfrak{R}_{10}$ | $(0.18, 0.3, 0.47)$ | $(0.18, 0.4, 0.33)$ | $(0.28, 0.3, 0.7)$ | $(0.26, 0.4, 0.31)$ | $(0.3, 0.4, 0.4)$ |

| $\mathfrak{R}/\mathfrak{S}$ | $\mathfrak{S}_6$ | $\mathfrak{S}_7$ | $\mathfrak{S}_8$ | $\mathfrak{S}_9$ | $\mathfrak{S}_1 0$ |
|---|---|---|---|---|---|
| $\mathfrak{R}_1$ | $(0.38, 0.3, 0.54)$ | $(0.58, 0.3, 0.7)$ | $(0.45, 0.3, 0.3)$ | $(0.26, 0.2, 0.35)$ | $(0.5, 0.3, 0.45)$ |
| $\mathfrak{R}_2$ | $(0.18, 0.3, 0.52)$ | $(0.31, 0.2, 0.7)$ | $(0.25, 0.2, 0.5)$ | $(0.78, 0.3, 0.45)$ | $(0.28, 0.3, 0.41)$ |
| $\mathfrak{R}_3$ | $(0.58, 0.3, 0.46)$ | $(0.62, 0.3, 0.45)$ | $(0.4, 0.3, 0.36)$ | $(0.62, 0.3, 0.12)$ | $(0.23, 0.5, 0.12)$ |
| $\mathfrak{R}_4$ | $(0.28, 0.3, 0.4)$ | $(0.26, 0.2, 0.36)$ | $(0.42, 0.3, 0.28)$ | $(0.29, 0.2, 0.35)$ | $(0.18, 0.2, 0.4)$ |
| $\mathfrak{R}_5$ | $(0.48, 0.3, 0.54)$ | $(0.24, 0.2, 0.54)$ | $(0.44, 0.3, 0.54)$ | $(0.26, 0.3, 0.38)$ | $(0.34, 0.3, 0.6)$ |
| $\mathfrak{R}_6$ | $(0.58, 0.3, 0.34)$ | $(0.37, 0.2, 0.75)$ | $(0.48, 0.2, 0.5)$ | $(0.4, 0.3, 0.55)$ | $(0.62, 0.2, 0.3)$ |
| $\mathfrak{R}_7$ | $(0.52, 0.3, 0.35)$ | $(0.55, 0.2, 0.7)$ | $(0.52, 0.1, 0.4)$ | $(0.58, 0.2, 0.6)$ | $(0.33, 0.2, 0.54)$ |
| $\mathfrak{R}_8$ | $(0.54, 0.2, 0.45)$ | $(0.68, 0.2, 0.18)$ | $(0.54, 0.2, 0.38)$ | $(0.54, 0.2, 0.23)$ | $(0.72, 0.2, 0.45)$ |
| $\mathfrak{R}_9$ | $(0.36, 0.2, 0.3)$ | $(0.54, 0.3, 0.25)$ | $(0.3, 0.2, 0.4)$ | $(0.32, 0.2, 0.55)$ | $(0.18, 0.2, 0.36)$ |
| $\mathfrak{R}_{10}$ | $(0.26, 0.2, 0.5)$ | $(0.32, 0.2, 0.6)$ | $(0.2, 0.2, 0.5)$ | $(0.38, 0.2, 0.6)$ | $(0.3, 0.2, 0.35)$ |

correlation ceofficients between the RESs and IA are as follows:

$$\Delta(\tilde{\mathfrak{R}}, \mathfrak{R}_1) = 0.1668, \Delta(\tilde{\mathfrak{R}}, \mathfrak{R}_2) = 0.6146, \Delta(\tilde{\mathfrak{R}}, \mathfrak{R}_3) = 0.6233,$$
$$\Delta(\tilde{\mathfrak{R}}, \mathfrak{R}_4) = -0.0598, \Delta(\tilde{\mathfrak{R}}, \mathfrak{R}_5) = 0.4249, \Delta(\tilde{\mathfrak{R}}, \mathfrak{R}_6) = 0.6219,$$
$$\Delta(\tilde{\mathfrak{R}}, \mathfrak{R}_7) = 0.3267, \Delta(\tilde{\mathfrak{R}}, \mathfrak{R}_8) = 0.7, \Delta(\tilde{\mathfrak{R}}, \mathfrak{R}_9) = 0.066,$$
$$\Delta(\tilde{\mathfrak{R}}, \mathfrak{R}_{10}) = 0.2609.$$

The correlation coefficients are presented in Figure 6.2 for easy interpretation. The preference ordering of the RESs obtained by ranking the correlation coefficients are as follows:

$$\mathfrak{R}_8 \succ \mathfrak{R}_3 \succ \mathfrak{R}_6 \succ \mathfrak{R}_2 \succ \mathfrak{R}_5 \succ \mathfrak{R}_7 \succ \mathfrak{R}_{10} \succ \mathfrak{R}_1 \succ \mathfrak{R}_9 \succ \mathfrak{R}_4.$$

From this preference ordering, the prefered RES to deploy for an effective RE supply is $\mathfrak{R}_8$, which represents geothermal energy source.

## 6.5 Comparative Analysis Discussion

This section showcases the compliance of the new correlation method with the existing correlation methods between SVNSs [15, 47, 48, 54]. The comparative results using the correlation methods via a decision-making technique based on the RESs' data are presented in Table 6.3.

**TABLE 6.3**: Correlation coefficient values between RESs and IA.

| **FFSMs** | $(\mathfrak{R}_1, \tilde{\mathfrak{R}})$ | $(\mathfrak{R}_2, \tilde{\mathfrak{R}})$ | $(\mathfrak{R}_3, \tilde{\mathfrak{R}})$ | $(\mathfrak{R}_4, \tilde{\mathfrak{R}})$ | $(\mathfrak{R}_5, \tilde{\mathfrak{R}})$ | $(\mathfrak{R}_6, \tilde{\mathfrak{R}})$ | $(\mathfrak{R}_7, \tilde{\mathfrak{R}})$ | $(\mathfrak{R}_8, \tilde{\mathfrak{R}})$ | $(\mathfrak{R}_9, \tilde{\mathfrak{R}})$ | $(\mathfrak{R}_{10}, \tilde{\mathfrak{R}})$ |
|---|---|---|---|---|---|---|---|---|---|---|
| Ye [47] | 0.8354 | 0.7727 | 0.8773 | 0.7528 | 0.7694 | 0.8601 | 0.8123 | 0.9314 | 0.7956 | 0.6938 |
| Iryna et al. [15] | 0.8351 | 0.7727 | 0.8773 | 0.7465 | 0.7672 | 0.8599 | 0.8123 | 0.9291 | 0.7926 | 0.6864 |
| Ye [48] | 0.9241 | 0.8875 | 0.9428 | 0.8586 | 0.8525 | 0.925 | 0.9095 | 0.9529 | 0.8865 | 0.7976 |
| Zhou et al. [54] | 0.0 | 0.159 | 0.0 | 0.0 | 0.1144 | 0.1965 | 0.1815 | 0.22 | 0.0 | 0.0 |
| New method | 0.1668 | 0.6146 | 0.6233 | −0.0598 | 0.4249 | 0.6219 | 0.3267 | 0.7 | 0.066 | 0.2609 |

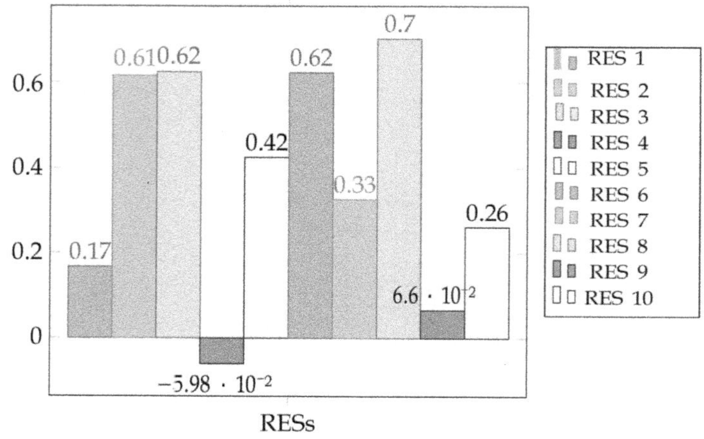


**FIGURE 6.2**: Correlation coefficient values.

**TABLE 6.4**: Preference ordering of the RESs.

| **Methods** | **Preference ordering** | **Suitable RES** |
|---|---|---|
| Ye [47] | $\mathfrak{R}_8 \succ \mathfrak{R}_3 \succ \mathfrak{R}_6 \succ \mathfrak{R}_1 \succ \mathfrak{R}_7 \succ \mathfrak{R}_9 \succ \mathfrak{R}_2 \succ \mathfrak{R}_5 \succ \mathfrak{R}_4 \succ \mathfrak{R}_{10}$ | $\mathfrak{R}_8$ |
| Iryna et al. [15] | $\mathfrak{R}_8 \succ \mathfrak{R}_3 \succ \mathfrak{R}_6 \succ \mathfrak{R}_1 \succ \mathfrak{R}_7 \succ \mathfrak{R}_9 \succ \mathfrak{R}_2 \succ \mathfrak{R}_5 \succ \mathfrak{R}_4 \succ \mathfrak{R}_{10}$ | $\mathfrak{R}_8$ |
| Ye [48] | $\mathfrak{R}_8 \succ \mathfrak{R}_3 \succ \mathfrak{R}_6 \succ \mathfrak{R}_1 \succ \mathfrak{R}_7 \succ \mathfrak{R}_2 \succ \mathfrak{R}_9 \succ \mathfrak{R}_4 \succ \mathfrak{R}_5 \succ \mathfrak{R}_{10}$ | $\mathfrak{R}_8$ |
| Zhou et al. [54] | $\mathfrak{R}_8 \succ \mathfrak{R}_6 \succ \mathfrak{R}_7 \succ \mathfrak{R}_2 \succ \mathfrak{R}_5 \succ \mathfrak{R}_1 = \mathfrak{R}_3 = \mathfrak{R}_4 = \mathfrak{R}_9 = \mathfrak{R}_{10}$ | $\mathfrak{R}_8$ |
| New method | $\mathfrak{R}_8 \succ \mathfrak{R}_3 \succ \mathfrak{R}_6 \succ \mathfrak{R}_2 \succ \mathfrak{R}_5 \succ \mathfrak{R}_7 \succ \mathfrak{R}_{10} \succ \mathfrak{R}_1 \succ \mathfrak{R}_9 \succ \mathfrak{R}_4$ | $\mathfrak{R}_8$ |

From the outcomes presented in Table 6.3, it is observed that although the methods in [15, 47, 48, 54] yield comparatively large correlation coefficient values, they cannot be relied upon for accurate results, as they fail to indicate the direction of the correlation between the RESs and the IA. This shortcoming limits the robustness and interpretability of the aforementioned methods. In contrast, the proposed method produces results that clearly reflect both the strength and direction of the correlation between the RESs and the IA.

Using the correlation values of the RESs and the IA in Table 6.3, the corresponding preference ordering is obtained and presented in Table 6.4.

From the preference ordering in Table 6.4, we observe that the RES, $\mathfrak{R}_8$, which stands for geothermal energy is the suitable renewable source of energy to be selected. It is interesting to note that all the methods yield the same final decision, although their preference orderings are not exactly identical. Again, it is noticed that the method in [54] could not classify $\mathfrak{R}_1$, $\mathfrak{R}_3$, $\mathfrak{R}_4$, $\mathfrak{R}_9$, and $\mathfrak{R}_{10}$, which shows the limitation of the method.

## 6.6 Conclusion

In conclusion, this chapter presents an effective framework for addressing the inherent complexities in the selection process of RESs by employing a novel correlation coefficient under SVNSs within a decision-making context. By utilizing the proposed correlation coefficient, this chapter provides a more precise and flexible representation of the uncertainty inherent in selection analysis. A comprehensive comparative analysis has been conducted to validate the effectiveness of the new correlation measure, which is grounded in a statistical perspective. The application of the proposed method to the selection of suitable RESs demonstrates its practicality and ability to efficiently prioritize the evaluation criteria. The results indicate that the new correlation coefficient under SVNSs outperforms existing approaches by capturing both the magnitude and direction of the correlation, thereby offering a more reliable interpretation of the decision benchmarks for RES selection. Furthermore, this approach can be extended to other domains of neutrosophic sets to address various decision-making problems.

## Conflict of interest

The authors declare that they have no competing interests.

### Data availability statement

This chapter has no associated data.

The abbreviations used in this chapter and their meaning are given below:

| Abbreviations | Meanings |
|---|---|
| AHP | Analytic Hierarchy Process |
| ANP | Analytic Network Process |
| BOCR | Benefit, Opportunity, Cost, and Risk |
| BWM | Best Worst Method |
| DEA | Data Envelopment Analysis |
| DEMATEL | Decision-Making Trial and Evaluation Laboratory |
| EDAS | Evaluation based on Distance from Average Solution |
| ELECTRE | ELimination Et Choix Traduisant la REalite |
| FAHP | Fuzzy Analytic Hierarchy Process |

| | |
|---|---|
| FS | Fuzzy Set |
| GRA | grey relational analysis |
| GW | Gigawatt |
| IA | Ideal Alternative |
| IFS | Intuitionistic Fuzzy Set |
| IM | Infrastructural Management |
| IRENA | International Renewable Energy Agency |
| IVIFS | Interval-Valued Intuitionistic Fuzzy Set |
| IVNSS | Interval-Valued Neutrosophic Soft Set |
| MABAC | Multi-Attributive Border Approximation Area Comparison |
| MARCOS | Measurement Alternatives and Ranking according to Compromise Solution |
| MCDM | Multi-Criteria Decision-Making |
| MOORA | Multi-Objective Optimization by Ratio Analysis |
| MULTIMOORA | multi-objective optimization by ratio analysis plus full multiplicative form |
| NFS | Neutrosophic Fuzzy Set |
| NRS | Neutrosophic Refined Set |
| NS | Neutrosophic Set |
| PIV | Proximity Indexed Value |
| RE | Renewable Energy |
| REN21 | Renewable Energy Policy Network for the 21st Century |
| RES | Renewable Energy Source |
| SVNFS | Single-Valued Neutrosophic Fuzzy Set |
| SVNS | Single-Valued Neutrosophic Set |
| TOPSIS | Technique for Order of Preference by Similarity to Ideal Solution |
| VIKOR | Vlekriterijumsko KOmpromisno Rangiranje |
| WASPAS | Weighted Aggregates Sum Product Assessment |
| WSM | Weighted Sum Model |

## Bibliography

[1] Alshamrani, A., Majumder, P., Das, A., Hezam, I. M., and Bozanic, D. (2023). An integrated BWM-TOPSIS-I approach to determine the ranking of alternatives and application of sustainability analysis of renewable energy. *Axioms*, 12:159.

[2] Anum, M. T., Ejegwa, P. A., and Feng, Y. (2024a). *An enhanced Pythagorean fuzzy distance method with application in the selection of renewable energy source via MCDA.* 2024 International Conference on New Trends in Computational Intelligence (NTCI), Qingdao, China.

[3] Anum, M. T., Zhang, H., Ejegwa, P. A., and Feng, Y. (2024b). *Tendency coefficient-based weighted distance measure for intuitionistic fuzzy sets*

*with applications.* Proceedings of the 12th International Conference on Intelligent Control and Information Processing (ICICIP 2024), Nanjing, China.

[4] Boran, F., Boran, K., and Menlik, T. (2012). The evaluation of renewable energy technologies for electricity generation in turkey using intuitionistic fuzzy TOPSIS. *Energy Sources B: Econ. Plan. Policy*, 7:81–90.

[5] Broumi, S., Bakali, A., Talea, M., Smarandache, F., Ulucay, V., Sahin, M., Dey, A., Dhar, M., Tan, R. P., Bahnasse, A., and Pramanik, S. (2018). *Neutrosophic sets: an overview.* In: Florentin Smarandache, Surapati Pramanik (eds.), *New Trends in Neutrosophic Theory and Applications, Volume II*, Infinite Study.

[6] Broumi, S. and Deli, I. (2016). Correlation measure for neutrosophic refined sets and its application in medical diagnosis. *Palestine J. Math.*, 5(1):135–143.

[7] Broumi, S. and Smarandache, F. (2013). Correlation coefficient of interval neutrosophic set. *Applied Mech. Mat.*, 436:511–517.

[8] Das, S., Roy, B. K., Kar, M. B., Kar, S., and Pamucar, D. (2020). Neutrosophic fuzzy set and its application in decision making. *J. Amb. Intell. Human. Comput.*, DOI:10.1007/s12652-020-01808-3.

[9] Dehshiri, H., Amiri, S. J. M., Mostafaeipour, A., and Le, T. (2024). Evaluation of renewable energy projects based on sustainability goals using a hybrid Pythagorean fuzzy-based decision approach. *Energy*, 297:131272.

[10] Dumrul, C., Bilgili, F., Zarali, F., Dumrul, Y., and Kilicarslan, Z. (2024). The evaluation of renewable energy alternatives in Turkey using intuitionistic-fuzzy EDAS methodology. *Environ. Sci. Pollut. Res.*, 31:15503–15524.

[11] Ejegwa, P. A., Kausar, N., Aydin, N., Feng, Y., and Olanrewaju, O. A. (2024a). A new Fermatean fuzzy spearman-like correlation coefficient and its application in evaluating insecurity problem via multi-criteria decision-making approach. *Heliyon*, 10(22):e40403.

[12] Ejegwa, P. A. and Sarkar, A. (2023). Novel correlation measure for generalized orthopair fuzzy sets and its decision-making applications. *Operat. Res. Forum*, 4(32):23.

[13] Ejegwa, P. A., Wanzenke, T. D., Ogwuche, I. O., and Isife, K. I. (2024b). A robust correlation coefficient for Fermatean fuzzy sets based on Spearman's correlation measure with application to clustering and selection process. *J. Appl. Math. Comput.*, 70:1747–1770.

[14] Husain, A. M., Hasan, M. M., Khan, Z. A., and Asjad, M. (2024). A robust decision-making approach for the selection of an optimal renewable energy source in India. *Energy Convers. Manage.*, 301:117989.

[15] Iryna, S., Zhong, Y., Jiang, W., Deng, X., and Geng, J. (2020). Single-valued neutrosophic set correlation coefficient and its application in fault diagnosis. *Symmetry*, 12:1371.

[16] Jansi, R., Mohana, K., and Smarandache, F. (2019). Correlation measure for Pythagorean neutrosophic sets with t and f as dependent neutrosophic components. *Neut. Set Syst.*, 30:202–212.

[17] Jha, S. K. and Puppala, H. (2017). Prospects of renewable energy sources in india: Prioritization of alternative sources in terms of energy index. *Energy*, 127:116–127.

[18] Kabak, M. and Dagdeviren, M. (2014). Prioritization of renewable energy sources for turkey by using a hybrid MCDM methodology. *Energy Convers. Manage.*, 79:25–33.

[19] Kahraman, C., Kaya, I., and Cebi, S. (2009). A comparative analysis for multiattribute selection among renewable energy alternatives using fuzzy axiomatic design and fuzzy analytic hierarchy process. *Energy*, 34:1603–1616.

[20] Karaaslan, F. (2019). *Correlation coefficient of neutrosophic sets and its applications in decision-making,* In: C. Kahraman and I. Otay (eds.), *Fuzzy Multi-criteria Decision-Making using Neutrosophic Sets*, volume 369. Studies in Fuzziness and Soft Computing, Springer.

[21] Kirisci, M. (2022). Correlation coefficients of fermatean fuzzy sets with a medical application. *J. Mathemat. Sci. Model.*, 5:16–23.

[22] Kirisci, M. (2023). Data analysis for panoramic X-ray selection: Fermatean fuzzy type correlation coefficients approach. *Eng. Appl. Artif. Intell.*, 126:106824.

[23] Kolagar, M., Hosseini, S. M. H., Felegari, R., and Fattahi, P. (2020). Policy-making for renewable energy sources in search of sustainable development: A hybrid DEA-FBWM approach. *Environ. Syst. Dec.*, 40:485–509.

[24] Lee, H. C. and Chang, C. T. (2018a). Comparative analysis of MCDM methods for ranking renewable energy sources in Taiwan. *Ren. Sustain. Energy Rev.*, 92:883–896.

[25] Lee, H. C. and Chang, C. T. (2018b). Comparative analysis of MCDM methods for ranking renewable energy sources in Taiwan. *Renew. Sustain. Energy Rev.*, 92:883–896.

[26] Longsheng, C., Shah, S. A. A., Solangi, Y. A., Ahmad, M., and Ali, S. (2022). An integrated SWOT-multi-criteria analysis of implementing sustainable waste-to-energy in Pakistan. *Renew. Energy*, 195:1438–1453.

[27] Luhanniwal, J., Agarwal, S., and Mathur, T. (2025). An integrated DEA-fuzzy AHP method for prioritization of renewable energy sources in India. *Scient. Rep.*, 15:838.

[28] Lupianez, F. G. (2017). On neutrosophic sets and topology. *Procedia Comput. Sci.*, 120:975–982.

[29] Murtaja, M., Bajpai, S., and Sharma, M. K. (2025). Optimal renewable energy sources selection using multi-criteria decision making- TOPSIS approach in Indian condition. *Indian Chem. Eng.*, DOI:10.1080/00194506.2025.2454254.

[30] Naveed, H. and Ali, S. (2024). Multi-criteria decision-making approach based on correlation coefficient for multi-polar interval-valued neutrosophic soft set. *Neut. Syst. Appl.*, 24(1):2.

[31] Otay, I. and Karaaslan, F. (2019). *A state-of-the-art review of neutrosophic sets and theory,* In: C. Kahraman and I. Otay (eds.), *Fuzzy Multi-criteria Decision-Making using Neutrosophic Sets*, volume 369. Studies in Fuzziness and Soft Computing, Springer.

[32] Ozturk, B., Chen, X., and Budak, G. (2024). Ranking renewable energy alternatives for Chicago by integrated analytic hierarchy process and Vlekriterijumsko KOmpromisno Rangiranje. *Int. J. Environ. Sci. Technol.*, 21:5411–5420.

[33] Saglam, U. (2018). *The efficiency assessment of renewable energy sources with data envelopment analysis.* In the 2018 Annual Meeting of the Decision Sciences Institute Proceedings, Chicago, IL.

[34] Şengül, Ü., Eren, M., Shiraz, S.E., Gezder, V. and Şengül, A.B. (2015). Fuzzy TOPSIS method for ranking renewable energy supply systems in Turkey. *Renew. Energy*, 75:617–625.

[35] Shah, S. A. A. (2020). Feasibility study of renewable energy sources for developing the hydrogen economy in Pakistan. *Int. J. Hydrogen Energy*, 45:15841–15854.

[36] Shamsipour, R., Nohegar, A., Ehsani, A. H., and Zand, A. D. (2025). A comprehensive assessing of renewable energy sources ranking by fuzzy DANP and fuzzy MULTIMOORA (Case study Isfahan Province, Iran). *Int. J. Environ. Res.*, 19:137.

[37] Smarandache, F. (1998). *A Unifying Field in Logics. Neutrosophy: Neutrosophic Probability, Set and Logic.* American Research Press.

[38] Smarandache, F. (2005). Neutrosophic set-a generalization of the intuitionistic fuzzy set. *Int. J. Pure Applied Math.*, 24(3):287–297.

[39] Solangi, Y. A., Shah, S. A. A., Zameer, H., Ikram, M., and Saracoglu, B. O. (2019). Assessing the solar PV power project site selection in Pakistan: based on AHP-fuzzy VIKOR approach. *Environ. Sci. Pollut. Res.*, 26:30286–30302.

[40] Thakur, R., Raj, M., and Malik, S. C. (2025). Correlation coefficient measures for probabilistic single valued neutrosophic hesitant fuzzy sets and its application in supply chain management. *J. Fuzzy Exten. Appl.*, 6(2):391–409.

[41] Wang, H., Smarandache, F., Zhang, Y., and Sunderraman, R. (2010a). *Single valued neutrosophic sets,* In: *F. Smarandache, Multispace and Multistructure*, vol. IV, pp. 410–413, North-European Scientific Publishers.

[42] Wang, H., Smarandache, F., Zhang, Y. Q., and Sunderraman, R. (2010b). Single valued neutrosophic sets. *Multistep Multistruct.*, 4:410–413.

[43] Wang, S., Li, W., Dincer, H., and Yuksel, S. (2019). Recognitive approach to the energy policies and investments in renewable energy resources via the fuzzy hybrid models. *Energies*, 12(23):4536.

[44] Xu, L., Wang, Y., Shah, S. A. A., Zameer, H., Solangi, Y. A., Walasai, G. D., and Siyal, Z. A. (2019). Economic viability and environmental efficiency analysis of hydrogen production processes for the decarbonization of energy systems. *Processes*, 7:494.

[45] Xue, H., Yu, M., and Chen, C. (2019). Research on novel correlation coefficient of neutrosophic cubic sets and its applications. *Mathemat. Probl. Eng.*, 2019:Article ID 7453025, 10.

[46] Yan, D., Wu, K., Ejegwa, P. A., Xie, X., and Feng, Y. (2023). Pythagorean fuzzy partial correlation measure and its application. *Symmetry*, 15:216.

[47] Ye, J. (2013). Multicriteria decision-making method using the correlation coefficient under single-valued neutrosophic environment. *Int. J. Gen. Syst.*, 42:386–394.

[48] Ye, J. (2014a). Improved correlation coefficients of single valued neutrosophic sets and interval neutrosophic sets for multiple attribute decision making. *J. Intell. Fuzzy Syst.*, 27:2453–2462.

[49] Ye, J. (2014b). Vector similarity measures of simplified neutrosophic sets and their application in multicriteria decision-making. *Int. J. Fuzzy Syst.*, 16(2):204–215.

[50] Ye, J. (2015). Improved cosine similarity measures of simplified neutrosophic sets for medical diagnoses. *Artif. Intell. Med.*, 63(3).

[51] Ye, J. (2017). Correlation coefficient between dynamic single valued neutrosophic multisets and its multiple attribute decision-making method. *Inf.*, 8:41.

[52] Zadeh, L. A. (1965). Fuzzy sets. *Inf. Cont.*, 8:338–353.

[53] Zhang, H. Y., Wang, J. Q., and Chen, X. H. (2014). Interval neutrosophic sets and their application in multicriteria decision making problems. *Scient. World J.*, 2014:Article ID 645953, 15.

[54] Zhou, X., Lin, M., and Wang, W. (2023). Statistical correlation coefficients for single-valued neutrosophic sets and their applications in medical diagnosis. *AIMS Math.*, 8(7):16340–16359.

[55] Zhu, Y., Cui, W., and Ye, J. (2019). New correlation coefficients between linguistic neutrosophic numbers and their group decision making method. *J. New Theory*, 28:74–83.

# Chapter 7

## Several Similarity Measures of Trigonometric Functions for Refined Neutrosophic Sets and Their Application in Medical Diagnosis

**Norzieha Mustapha, Fathiah Hasnizul, Suriana Alias, and Roliza Md Yasin**

### 7.1 Introduction

Zadeh introduced fuzzy set theory in 1965. It has since become a popular framework for dealing with uncertainty in real-world situations. However, fuzzy set theory and its extensions can only handle limited types of information. These theories are not enough for uncertain, inconsistent, or unreliable data [1]. For example, take a medical test where a patient has 40% evidence indicating the presence of a disease, 25% evidence suggesting otherwise, 15% results that are inconclusive, and 20% data that is missing or unreliable. This scenario exceeds what intuitionistic fuzzy sets (IFSs) can represent, as they cannot fully capture the uncertainty and incomplete information present [2].

To overcome these limitations, the concept of the neutrosophic set was introduced in [3]. Neutrosophic sets extend IFSs by adding an independent measure of indeterminacy. This framework defines decision-making through three components: membership, indeterminacy, and non-membership degrees. Together, these offer a more thorough evaluation process. Unlike IFS, the neutrosophic approach directly addresses ambiguity, imprecision, incompleteness, and vagueness, all of which are common in real-world information. This ability makes neutrosophic theory very useful in many areas, as it offers a clearer measurement of indeterminacy and a better representation of the complex, uncertain nature of real-world challenges.

In [4], the authors developed the idea of neutrosophic multi-set (NmS), also known as refined neutrosophic set (RNS), and examined its distinctive characteristics. In [5], the authors studied RNSs and explored their core

DOI: 10.1201/9781003683339-7

characteristics, including intersection, relationship, convexity, and extremely convexity, making them more flexible than intuitionistic fuzzy multisets. RNS is a generalization of fuzzy multisets and intuitionistic fuzzy multisets.

Similarity measurements play an important part in data clustering. The link between two neutrosophic sets is studied by calculating their corresponding distances [6]. It's important to point out that similarity and distance (dissimilarity) measurements are complimentary, meaning that when one increases, the other decreases. Using a similarity measure can aid in speedier and more accurate diagnosis by identifying patterns and similarities among cases [7].

Similarity measures for RNS can be highly useful in multicriteria decision-making (MCDM). It is a useful method for dealing with multiple uncertainties, ambiguities, and vague data that are frequently seen in actual decision-making processes. When diagnosing an illness with multiple symptoms that do not clearly indicate a specific disease, refined neutrosophic values can be used, and similarity measurements can aid in determining the most likely diagnosis.

The diagnosis of illnesses can be tough because symptoms are often unclear, and there is a lot of information from modern medical technologies. In many situations, it is hard to place symptoms under one illness since each symptom can have different degrees of truth, uncertainty, and falsity. To tackle this challenge, the authors of [8] and [9] introduced the concept of the RNS for medical diagnosis. This model allows for multiple truth, uncertainty, and non-membership functions. This research has explored correlation measures for RNS and their use in decision-making. The RNS framework is essential in medical diagnostics because it offers a clearer and more flexible way to deal with uncertain information.

This chapter examines several trigonometric similarity measures, including cosine, tangent, cotangent, and cosecant. It evaluates how effective these measures are at identifying patients' diseases. The RNS framework helps collect and represent various medical data points, while trigonometric similarity measures are effective tools for comparing patient symptoms with disease profiles. This ultimately supports more accurate diagnostic choices.

## 7.2 Preliminaries

In this section, we present the preliminary definitions of RNS.

**Definition 7.1:** *RNS [10]*
*Let $X$ be a universe, a RNS on $X$ can be defined as follows:*
$A = \{\langle x, (T_A^1(x), T_A^2(x), \ldots T_A^p(x)), (I_A^1(x), I_A^2(x), \ldots I_A^p(x)), (F_A^1(x), F_A^2(x), \ldots F_A^p(x))\rangle : x \in X\}$ *where* $T_A^1(x), T_A^2(x), \ldots T_A^p(x) : X \to [0,1]$ *such that* $0 \leq T_A^j(x) + I_A^j(x) + F_A^j \leq 3$ *for* $j = 1, 2, \ldots, p$ *for any*

*$x \in X$, $(T_A^1(x), T_A^2(x), \ldots T_A^p(x))$, $(I_A^1(x), I_A^2(x), \ldots I_A^p(x))$ and $(F_A^1(x), F_A^2(x), \ldots F_A^p(x))$ is the truth-membership sequence, indeterminate-membership sequence, and falsity-membership sequence of the element $x$, respectively. Also $p$ is called the dimension of RNS $A$.*

**Definition 7.2:** *RNS[5]*
*Let $A = \{\langle x, (T_A^1(x), T_A^2(x), \ldots T_A^p(x)), (I_A^1(x), I_A^2(x), \ldots I_A^p(x)), (F_A^1(x), F_A^2(x), \ldots F_A^p(x))\rangle : x \in X\}$ and*
*$B = \{\langle x, (T_B^1(x), T_B^2(x), \ldots T_B^p(x)), (I_B^1(x), I_B^2(x), \ldots I_B^p(x)), (F_B^1(x), F_B^2(x), \ldots F_B^p(x))\rangle : x \in X\}$*
*be two RNS. Then*
*$A \subseteq B \Rightarrow T_A^j(x) \leq T_B^j(x), I_A^j(x) \geq I_B^j(x) \& \; F_A^j(x) \leq F_B^j(x)$ for all $x \in X$.*

---

## 7.3 Trigonometric Similarity Measures for Refined Neutrosophic Set

Several trigonometric similarity measures for RNS are presented in this section.

**Definition 7.3:** *Trigonometric similarity measure [1]*
*Let $A = \{\langle x, (T_A^1(x), T_A^2(x), \ldots T_A^p(x)), (I_A^1(x), I_A^2(x), \ldots I_A^p(x)), (F_A^1(x), F_A^2(x), \ldots F_A^p(x))\rangle : x \in X\}$ and*
*$B = \{\langle x, (T_B^1(x), T_B^2(x), \ldots T_B^p(x)), (I_B^1(x), I_B^2(x), \ldots I_B^p(x)), (F_B^1(x), F_B^2(x), \ldots F_B^p(x))\rangle : x \in X\}$*
*be two RNS. Then the trigonometric similarity measure is defined as*

**(a) Cosecant similarity measure, $T_{RNS}^1(A, B)$** [1]

$$T_{RNS}^1(A, B) = \frac{1}{6np} \sum_{j=1}^{p} \left[ \sum_{i=1}^{n} \csc \right. \tag{7.1}$$

$$\left. \left\langle \frac{\pi(1 + |T_A^j(x_i) - T_B^j(x_i)| + |I_A^j(x_i) - I_B^j(x_i)| + |F_A^j(x_i) - F_B^j(x_i)|)}{4n} \right\rangle \right]$$

**(b) Cosine similarity measure, $T_{RNS}^2(A, B)$** [11]

$$T_{RNS}^2(A, B) = \frac{1}{p} \sum_{j=1}^{p} \tag{7.2}$$

$$\times \left\{ \frac{1}{n} \sum_{i=1}^{n} \cos \left[ \frac{\pi |T_A^j(x_i) - T_B^j(x_i)| + |I_A^j(x_i) - I_B^j(x_i)| + |F_A^j(x_i) F_B^j(x_i)|}{6} \right] \right\}$$

**(c) Tangent similarity measure, $T^3_{RNS}(A,B)$ [12]**

$$T^3_{RNS}(A,B) = \frac{1}{p}\sum_{j=1}^{P} \tag{7.3}$$

$$\left[\frac{1}{n}\sum_{i=1}^{n}\left\langle 1 - \tan\left(\frac{\pi}{12}\left(|T^j_A(x_i) - T^j_B(x_i)| + |I^j_A(x_i) - I^j_B(x_i)| + |F^j_A(x_i) - F^j_B(x_i)|\right)\right)\right\rangle\right]$$

**(d) Cotangent similarity measure, $T^4_{RNS}(A,B)$ [13]**

$$T^4_{RNS}(A,B) = \frac{1}{p}\sum_{j=1}^{P} \tag{7.4}$$

$$\left[\frac{1}{n}\sum_{i=1}^{n}\left\langle \cot\left(\frac{\pi}{12}\left(3 + |T^j_A(x_i) - T^j_B(x_i)| + |I^j_A(x_i) - I^j_B(x_i)| + |F^j_A(x_i) - F^j_B(x_i)|\right)\right)\right\rangle\right]$$

The trigonometric similarity measures, $T^k_{RNS}(A,B)$ for $k = 1,2,3,4$ between RNS($A$) and RNS($B$) satisfied the properties as follows:

**Proposition 7.1:**

*(i)* $T^k_{RNS}(A,B) > 0$.

*(ii)* $T^k_{RNS}(A,B) = T^k_{RNS}(B,A)$.

*(iii) If* $A \subseteq B \subseteq C$, *then* $T^k_{RNS}(A,C) \leq T^k_{RNS}(A,B)$ *and* $T^k_{RNS}(A,C) \leq T^k_{RNS}(B,C)$.

These properties of trigonometric similarity measures (7.1) – (7.4) have been proved in [1], [11], [13], and [12], respectively.

---

## 7.4 Application of Trigonometric Similarity Measures in Medical Diagnosis

This chapter uses four trigonometric similarity measures, cosecant, cosine, tangent, and cotangent, within the framework of RNS to analyze medical data. Medical diagnosis often involves uncertainty, vagueness, and imprecise information. RNS provides a flexible mathematical structure to capture these aspects through its truth, indeterminacy, and falsity components. By integrating trigonometric similarity measures, the analysis gains additional perspectives that improve the comparison between patient profiles and disease patterns, thus supporting more accurate and reliable diagnostic decisions.

To show their practical use, two case studies are considered. Each case involves patients with symptoms that may relate to multiple possible diseases. This requires careful evaluation to avoid misclassification. The refined

**TABLE 7.1**: The relation between patients and symptoms.

| $Q$ | **Temperature** | **Headache** | **Stomach pain** | **Cough** | **Chest pain** |
|---|---|---|---|---|---|
| $P_1$ | (0.8,0.1,0.1)<br>(0.6,0.3,0.3)<br>(0.6,0.3,0.1) | (0.6,0.1,0.3)<br>(0.5,0.2,0.4)<br>(0.5,0.1,0.2) | (0.2,0.8,0.0)<br>(0.3,0.5,0.2)<br>(0.2,0.3,0.4) | (0.6,0.1,0.3)<br>(0.4,0.4,0.4)<br>(0.4,0.3,0.3) | (0.1,0.6,0.3)<br>(0.3,0.4,0.5)<br>(0.2,0.5,0.4) |
| $P_2$ | (0.0,0.8,0.2)<br>(0.2,0.6,0.4)<br>(0.1,0.6,0.4) | (0.4,0.4,0.2)<br>(0.5,0.4,0.1)<br>(0.4,0.6,0.3) | (0.6,0.1,0.3)<br>(0.4,0.2,0.5)<br>(0.3,0.2,0.4) | (0.1,0.7,0.2)<br>(0.2,0.7,0.5)<br>(0.3,0.5,0.4) | (0.1,0.8,0.1)<br>(0.3,0.6,0.4)<br>(0.3,0.6,0.3) |
| $P_3$ | (0.8,0.1,0.1)<br>(0.6,0.4,0.1)<br>(0.5,0.3,0.3) | (0.8,0.1,0.1)<br>(0.6,0.2,0.4)<br>(0.6,0.1,0.3) | (0.0,0.6,0.4)<br>(0.2,0.5,0.5)<br>(0.3,0.4,0.6) | (0.2,0.7,0.1)<br>(0.2,0.5,0.5)<br>(0.1,0.6,0.3) | (0.0,0.5,0.5)<br>(0.2,0.5,0.3)<br>(0.3,0.3,0.4) |
| $P_4$ | (0.6,0.1,0.3)<br>(0.4,0.3,0.2)<br>(0.5,0.2,0.3) | (0.5,0.4,0.1)<br>(0.4,0.4,0.4)<br>(0.5,0.2,0.4) | (0.3,0.4,0.3)<br>(0.2,0.4,0.5)<br>(0.1,0.5,0.4) | (0.7,0.2,0.1)<br>(0.5,0.2,0.4)<br>(0.6,0.4,0.1) | (0.3,0.4,0.3)<br>(0.4,0.3,0.4)<br>(0.3,0.5,0.5) |

neutrosophic framework effectively models uncertainty in patient data, while the four similarity measures offer different but complementary ways to identify the most likely diagnosis. The results of these case studies, presented in Sections 7.4.1 and 7.4.2, highlight the consistency and variations among the trigonometric approaches, demonstrating their potential for real-world medical decision-making.

### 7.4.1 Case study 1

Medical data collected from [1] are presented in Tables 7.1 and 1.2. Consider a group of four patients, $P = \{P_1, P_2, P_3, P_4\}$ and five symptoms, $S=$ {Temperature, Headache, Stomach pain, Cough, Chest pain}. The set of diseases being examined is defined as $D =$ {Viral fever, Malaria, Typhoid, Stomach problem, Chest problem}. The refined neutrosophic relation $Q(P \to S)$ is given in Table 7.1. The neutrosophic relation $R(S \to D)$ is given as in Table 7.2. Tables 7.3–7.6 present the results of the trigonometric similarity measures for the four patients $(P_1, P_2, P_3, P_4)$ across the five diseases. These measures quantify how closely a patient's symptoms match the related disease profiles. A higher similarity score suggests a greater chance of matching with the disease.

**TABLE 7.2**: The relation between symptoms and diseases.

| $R$ | **Viral fever** | **Malaria** | **Typhoid** | **Stomach problem** | **Chest problem** |
|---|---|---|---|---|---|
| Temperature | (0.6,0.3,0.3) | (0.2,0.5,0.3) | (0.2,0.6,0.4) | (0.1,0.6,0.6) | (0.1,0.6,0.4) |
| Headache | (0.4,0.5,0.3) | (0.6,0.4,0.1) | (0.1,0.5,0.4) | (0.2,0.4,0.6) | (0.1,0.6,0.4) |
| Stomach pain | (0.1,0.6,0.3) | (0.0,0.6,0.4) | (0.2,0.5,0.5) | (0.8,0.2,0.2) | (0.1,0.7,0.1) |
| Cough | (0.4,0.4,0.4) | (0.4,0.1,0.5) | (0.2,0.5,0.5) | (0.1,0.7,0.4) | (0.4,0.5,0.4) |
| Chest pain | (0.1,0.7,0.4) | (0.1,0.6,0.3) | (0.1,0.6,0.4) | (0.1,0.7,0.4) | (0.8,0.2,0.2) |

**TABLE 7.3**: Cosecant similarity measure's result for case study 1 [1].

| $T$ | Viral fever | Malaria | Typhoid | Stomach problem | Chest problem |
|---|---|---|---|---|---|
| $P_1$ | **0.7889** | 0.6867 | 0.6686 | 0.5906 | 0.6164 |
| $P_2$ | 0.7016 | 0.6931 | **0.7685** | 0.7441 | 0.6690 |
| $P_3$ | **0.7237** | 0.6609 | 0.7166 | 0.6067 | 0.5821 |
| $P_4$ | **0.7656** | 0.6896 | 0.6815 | 0.5781 | 0.6163 |

**TABLE 7.4**: Cosine similarity measure's result for case study 1.

| $T$ | Viral fever | Malaria | Typhoid | Stomach problem | Chest problem |
|---|---|---|---|---|---|
| $P_1$ | **0.9725** | 0.9362 | 0.9215 | 0.8710 | 0.8886 |
| $P_2$ | 0.9386 | 0.9369 | 0.9644 | **0.9647** | 0.9044 |
| $P_3$ | **0.9575** | 0.9129 | 0.9285 | 0.8717 | 0.8733 |
| $P_4$ | **0.9714** | 0.9491 | 0.9359 | 0.8794 | 0.9155 |

**TABLE 7.5**: Tangent similarity measure's result for case study 1.

| $T$ | Viral fever | Malaria | Typhoid | Stomach problem | Chest problem |
|---|---|---|---|---|---|
| $P_1$ | **0.8964** | 0.8352 | 0.8184 | 0.7553 | 0.7762 |
| $P_2$ | 0.8422 | 0.8386 | **0.8837** | 0.8768 | 0.8049 |
| $P_3$ | **0.8643** | 0.8091 | 0.8393 | 0.7620 | 0.7540 |
| $P_4$ | **0.8893** | 0.8466 | 0.8335 | 0.7566 | 0.7960 |

In Table 7.3, we see that using the cosecant similarity measure [1], patients $P_1, P_3$ and $P_4$ have the highest similarity to viral fever. This suggests that viral fever is the most likely condition for these patients. Meanwhile, $P_2$ shows the strongest similarity to typhoid (0.7685) but also has a high similarity to stomach problem (0.7411). This indicates possible co-occurrence or shared symptoms. Tables 1.4–1.6 show the similarity measures based on the cosine, tangent, and cotangent functions, which generally reveal patterns similar to those found in cosecant in Table 7.3.

In Table 7.4, the cosine similarity measure gives values above 0.87 across all diseases, indicating very strong similarities overall. Notably, $P_2$ is most closely linked to stomach problem (0.9647) and shows a similarity to typhoid (0.9644), again suggesting possible overlap between the two conditions.

The results in Tables 7.5 and 7.6, which use tangent and cotangent similarity measures, show some differences. Tangent similarity values range from 0.75 to 0.88, while cotangent values are generally lower, ranging from 0.61 to 0.81. This shows more distinction across disease profiles.

Overall, the trigonometric similarity measures for the RNS in case study 1 reveal consistent but detailed patterns across all four approaches. Viral

**TABLE 7.6**: Cotangent similarity measure's result for case study 1.

| $T$ | Viral fever | Malaria | Typhoid | Stomach problem | Chest problem |
|---|---|---|---|---|---|
| $P_1$ | **0.8173** | 0.7254 | 0.7038 | 0.6193 | 0.6464 |
| $P_2$ | 0.7372 | 0.7314 | **0.7987** | 0.7849 | 0.6906 |
| $P_3$ | **0.7660** | 0.6922 | 0.7395 | 0.6318 | 0.6149 |
| $P_4$ | **0.8041** | 0.7376 | 0.7222 | 0.6152 | 0.6645 |

**TABLE 7.7**: The relation between patients and test values.

| | ELISA test | Blood test | Viral culture |
|---|---|---|---|
| $Q_1$ | (0.6, 0.2, 0.2)(0.5, 0.2, 0.2) | (0.8, 0.1, 0.1)(0.8, 0.1, 0.1) | (0.4, 0.2, 0.4)(0.3, 0.2, 0.5) |
| $Q_2$ | (0.8, 0.1, 0.1)(0.7, 0.2, 0.1) | (0.4, 0.2, 0.4)(0.3, 0.3, 0.4) | (0.6, 0.2, 0.3)(0.5, 0.1, 0.3) |
| $Q_3$ | (0.4, 0.1, 0.5)(0.4, 0.1, 0.5) | (0.3, 0.1, 0.6)(0.3, 0.2, 0.5) | (0.8, 0.1, 0.1)(0.8, 0.1, 0.1) |

fever consistently appears as the most likely diagnosis for patients $P_1, P_3$, and $P_4$. However, the diagnosis for $P_2$ varies based on the chosen similarity measure. Cosecant, tangent, and cotangent identify typhoid as the closest match, while cosine points to stomach problem. Among the four measures, cosine similarity produces the highest numerical values, mostly above 0.90. This reflects its strong ability to differentiate between diagnoses. Cosecant and cotangent yield more moderate values but show similar diagnostic trends, while tangent closely matches both. This comparison shows that while viral fever is commonly identified for most patients, the diagnosis for $P_2$ depends on the selected similarity measure. In general, cosine similarity appears to be the most stable and reliable among the trigonometric measures. It provides stronger links between patient symptoms and disease profiles, while the other measures offer more conservative estimates.

### 7.4.2 Case study 2

The data shown in Tables 7.7 and 7.8 come from [9]. This case study includes three patients who show symptoms linked to three viruses: herpes simplex, hepatitis, and influenza. For diagnostics, three tests have been used: the enzyme-linked immunosorbent assay test (ELISA test), blood test, and viral culture.

Let $Q = \{Q_1, Q_2, Q_3\}$ denote the set of patients, where $Q_1, Q_2, Q_3$ represent individual patients. Likewise, let $V = \{V_1, V_2, V_3\}$ represent the set of viruses, and $T = \{T_1, T_2, T_3\}$ represent the set of diagnostic tests.

The results of case study 2, as shown in Tables 7.9–7.12 using trigonometric similarity measures for RNS, show consistent diagnostic patterns across patients $Q_1, Q_2$, and $Q_3$. However, there is some variation based on the chosen similarity function.

**TABLE 7.8**: The relation between test values and viruses

| | **Herpes simplex** | **Hepatitis** | **Influenza** |
|---|---|---|---|
| ELISA test | (0.8, 0.1, 0.1)(0.8, 0.1, 0.1) | (0.5, 0.1, 0.4)(0.5, 0.1, 0.4) | (0.4, 0.1, 0.5)(0.4, 0.1, 0.5) |
| Blood test | (0.4, 0.1, 0.5)(0.4, 0.1, 0.5) | (0.8, 0.1, 0.1)(0.8, 0.1, 0.1) | (0.4, 0.1, 0.5)(0.4, 0.1, 0.5) |
| Viral culture | (0.5, 0.1, 0.4)(0.5, 0.1, 0.4) | (0.4, 0.1, 0.5)(0.4, 0.1, 0.5) | (0.8, 0.1, 0.1)(0.8, 0.1, 0.1) |

**TABLE 7.9**: Cosecant similarity measure's result for case study 2.

| $T$ | **Herpes simplex** | **Hepatitis** | **Influenza** |
|---|---|---|---|
| $Q_1$ | 0.4399 | **0.5552** | 0.3800 |
| $Q_2$ | **0.5456** | 0.4091 | 0.4349 |
| $Q_3$ | 0.4387 | 0.4133 | **0.6091** |

**TABLE 7.10**: Cosine similarity measure's result for case study 2.

| $T$ | **Herpes simplex** | **Hepatitis** | **Influenza** |
|---|---|---|---|
| $Q_1$ | 0.9573 | **0.9925** | 0.9206 |
| $Q_2$ | **0.9923** | 0.9392 | 0.9553 |
| $Q_3$ | 0.9530 | 0.9247 | **0.9982** |

**TABLE 7.11**: Tangent similarity measure's result for case study 2.

| $T$ | **Herpes simplex** | **Hepatitis** | **Influenza** |
|---|---|---|---|
| $Q_1$ | 0.8634 | **0.9519** | 0.8007 |
| $Q_2$ | **0.9475** | 0.8321 | 0.8590 |
| $Q_3$ | 0.8589 | 0.8224 | **0.9825** |

**TABLE 7.12**: Cotangent similarity measure's result for case study 2.

| $T$ | **Herpes simplex** | **Hepatitis** | **Influenza** |
|---|---|---|---|
| $Q_1$ | 0.7644 | **0.9108** | 0.6699 |
| $Q_2$ | **0.9023** | 0.7169 | 0.7573 |
| $Q_3$ | 0.7588 | 0.7091 | **0.9668** |

From Table 7.9, it can be seen that the cosecant similarity measure gives relatively lower values than the other measures, identifying hepatitis as the most likely diagnosis for $Q_1$, herpes simplex for $Q_2$, and influenza for $Q_3$.

In contrast, as presented in Table 7.10, the cosine similarity measure produces very high similarity values, all above 0.90. It clearly suggests hepatitis for $Q_1$, herpes simplex for $Q_2$, and influenza for $Q_3$.

In Table 7.11, it can be seen that the tangent similarity measure follows a similar pattern, also indicating hepatitis for $Q_1$, herpes simplex for $Q_2$, and influenza for $Q_3$. Its values are slightly lower but still strong.

Meanwhile, Table 7.12 show that the cotangent similarity measure supports the same conclusions, although its values are lower than those from cosine and tangent.

Overall, all four measures consistently agree on the most likely disease for each patient: hepatitis for $Q_1$, herpes simplex for $Q_2$, and influenza for $Q_3$. Among them, cosine similarity offers the strongest distinction, followed closely by tangent, while cosecant and cotangent provide more moderate yet steady confirmation of the diagnostic results.

## 7.5 Conclusion and Recommendation

This chapter discusses several trigonometric similarity measures, including cosine, tangent, cosecant, and cotangent, for RNSs. It demonstrates their effectiveness in medical diagnosis through two case studies.

The findings from case study 1 show that viral fever is the most common diagnosis for most patients $(P_1, P_3, P_4)$. This indicates that the similarity measures effectively capture consistent diagnostic trends. However, for $P_2$, the results vary depending on the similarity measure. Cosine suggests a stomach problem, while cosecant, tangent, and cotangent consistently point to typhoid. This shows that while most trigonometric measures are similar, there can be slight differences, especially in borderline cases. Notably, cosine similarity produces the highest numerical values (generally over 0.90), reinforcing its reliability and stability in distinguishing disease profiles.

In case study 2, all four measures identify hepatitis $(Q_1)$, herpes simplex $(Q_2)$, and influenza $(Q_3)$ consistently. This demonstrates strong agreement across different trigonometric approaches. Cosine similarity again produces the highest values, followed closely by tangent, while cosecant and cotangent offer moderate but supportive outcomes. This overall consistency strengthens the argument for trigonometric similarity measures as useful tools for diagnostics within neutrosophic settings.

Based on the findings from the two case studies, it is recommended to prioritize cosine similarity as the most reliable trigonometric measure for RNS in medical diagnosis. Its stability, high accuracy, and strong discriminative power make it a great choice. Tangent similarity can also be a useful method to provide additional validation. In cases where diagnoses are unclear, using multiple similarity measures at the same time is advisable. This approach can improve decision-making by showing consistent results across different methods. Additionally, integrating these measures into clinical decision support systems would help practitioners manage uncertainty and overlapping symptoms better. Future research should look into applying trigonometric similarity measures to larger datasets and a broader range of diseases. It should also explore using hybrid models with machine learning to improve predictive ability and expand their application in various areas.

---

## Acknowledgements

Thank you to Universiti Teknologi MARA, Cawangan Kelantan, for the support. Thank you to our group members for sharing information and giving full cooperation to complete the study.

---

## Bibliography

[1] AE. Samuel and R. Narmadhagnanam. Neutrosophic refined sets in medical diagnosis. *International Journal of Fuzzy Mathematical Archive*, 14 (1):117–123, 2017.

[2] Y. Zeng, H. Ren, T. Yang, S. Xiao, and N. Xiong. A novel similarity measure of single-valued neutrosophic sets based on modified manhattan distance and its applications. *Electronics (2079-9292)*, 11(6):941–956, 2022.

[3] F. Smarandache. *Neutrosophy: Neutrosophic Probability, Set, and Logic: Analytic Synthesis & Synthetic Analysis.* American Research Press, Rehoboth, NM, 1998.

[4] S. Ye and J. Ye. Dice similarity measure between single valued neutrosophic multisets and its application in medical diagnosis. *Neutrosophic sets and systems*, 6(1):49–55, 2014.

[5] I. Deli, S. Broumi, and F. Smarandache. On neutrosophic refined sets and their applications in medical diagnosis. *Journal of new theory*, (6): 88–98, 2015.

[6] MA. Dasan, E. Bementa, M. Aslam, and VFL. Flower. Multi-attribute decision-making problem in career determination using single-valued neutrosophic distance measure. *Complex & Intelligent Systems*, pages 1–15, 2024.

[7] M. Saeed, MU. Nisa, MH. Saeed, T. Alballa, and HAE. Khalifa. Detecting patterns of infection-induced fertility using fermatean neutrosophic set with similarity analysis. *IEEE Access*, 11:112320–112333, 2023.

[8] S. Broumi and I. Deli. Correlation measure for neutrosophic refined sets and its application in medical diagnosis. *Palestine Journal of Mathematics*, 5(1):135–143, 2015.

[9] F. Smarandache, K Hemabala, and BS. Kumar. Medical diagnosis via refined neutrosophic fuzzy logic: Detection of illness using neutrosophic sets. *Journal of Advanced Zoology*, 44(1):15–24, 2023.

[10] S. Broumi and F. Smarandache. Extended hausdorff distance and similarity measures for neutrosophic refined sets and their application in medical diagnosis. *Journal of New Theory*, (7):64–78, 2015.

[11] S. Broumi and F. Smarandache. Neutrosophic refined similarity measure based on cosine function. *Neutrosophic Sets and Systems*, 6:42–49, 2014.

[12] K. Mondal and S. Pramanik. Neutrosophic refined similarity measure based on tangent function and its application to multi attribute decision making. *Journal of New theory*, (8):41–50, 2015.

[13] K. Mondal and S. Pramanik. Neutrosophic refined similarity measure based on cotangent function and its application to multi-attribute decision making. *Global Journal of Advanced Researc*, 2(2):486–496, 2015.

# Chapter 8

## *Enhancing Antiviral Drug Repurposing for Future Pandemics with a Linguistic Single-Valued Neutrosophic MAGDM Framework*

**Sumera Naz, Sabahat Zafar, and Shariq Aziz Butt**

### 8.1 Introduction

Redirecting, repurposing, repositioning, and re-profiling are other terms for the process of discovering new applications for current medications that go beyond their original medical use. The issue with productivity, global pressure for prices to rise, and the increasing number of regulations obstacles one must overcome have forced several drug researchers to come out with more innovative versions of their old medications with new targets and uses. The conventional drug discovery method is a laborious, costly, and intricate one. The time and expense involved in finding new drugs for infectious diseases are greatly decreased by the medication repurposing strategy. The effectiveness of new medications that target viral proteins and host components is limited by the advent of resistant viruses and undesirable side effects. Among the worst pandemics in history is HIV/AIDS. The World Health Organization estimates that 26 million individuals have passed away since 1981, with 1.6 million of those deaths occurring in 2012 alone. Chloroquine and its hydroxyl counterpart were reported to decrease HIV-1 replication in multiple investigations [1]. One of the best illustrations of host adaptation and a virus's capacity to totally disrupt cellular physiological functions in the infected cell are shown by chloroquine.

Global pandemics like COVID-19, SARS, and MERS have highlighted the pressing need for quick and efficient treatment approaches to fight new viral illnesses. Antiviral drug repurposing, or finding and using existing medications for novel viral targets, has drawn a lot of attention among the available response mechanisms since it can avoid early-stage development, lower costs, and

DOI: 10.1201/9781003683339-8 

shorten treatment times [2–4]. Naz et al. [5] proposed Maclaurin symmetric mean operators under fuzzy two-tuple linguistic environment for the selection of appropriate treatment to control COVID-19 outbreaks. However, choosing appropriate repurposed medications is a difficult process that involves a number of competing factors, including availability, regulatory status, side effects, therapeutic efficacy, resistance potential, and production scalability. In addition, making decisions in situations with such high stakes frequently involves a great deal of uncertainty, inaccurate expert opinions, and insufficient information. A systematic, transparent, and adaptable decision-support framework that can manage both quantitative and qualitative evaluations in unpredictable circumstances is necessary for decision-makers to handle these complexity. A useful strategy in this respect is to use multi-attribute group decision-making (MAGDM) techniques, which enable the methodical assessment of several options based on professional opinions on a variety of criteria. However, when used in settings with ambiguous, inconsistent, or insufficient information conditions common in the early phases of a pandemic, conventional MAGDM [6] procedures frequently fail.

In modern decision-making domains, MAGDM is a significant and popular issue. Numerous advancements have been made in a variety of domains since Zadeh [7] introduced the fuzzy set (FS) theory, including intelligent fuzzy control, decision support system, and so forth. One of FS's drawbacks, though, is that it solely accounts for the level of membership; it ignores the level of non-membership. Consequently, Atanassov [8] proposed the intuitionistic fuzzy set (IFS) as an extension of FS, which can successfully compensate for FS's drawbacks by including a non-membership. Despite their successful use and advancement, FS and IFS are not equipped to handle all types of unusual circumstances in actual situations involving fuzzy conditions. Linguistic term sets (LTSs) are useful when handling complex or ill-defined circumstances because, for example, individuals prefer to use terms like "very bad," "somewhat bad," "good," and so forth to express their evaluations in daily life rather than numerical values or FS and IFS.

The three primary approaches for computing with words (CWW) that are most commonly employed in relation to representation models of LTSs are as follows: (1) transformation function-based models, like triangular models, in which fuzzy numbers and LTSs match identically fuzzy number [9], trapezoidal fuzzy number, and so forth; (2) linguistic symbolic representation-based models that directly compute words using the index of linguistic labels [10]: (3) two-tuple symbolic representation-based models that convert linguistic data into consecutive two-tuple linguistic terms (2TLTs) [11]. Since the evaluated LTSs for the first two approaches are discrete, the computation outputs might not match the original LTSs, potentially leading to information and detail loss. Herrera et al.'s [12–14] (2TL) representation has been widely used in many domains and has performed well in recent years because it has the exact characteristic of expressing any counting of information in the discourse universe while also resolving information distortion and loss in the linguistic

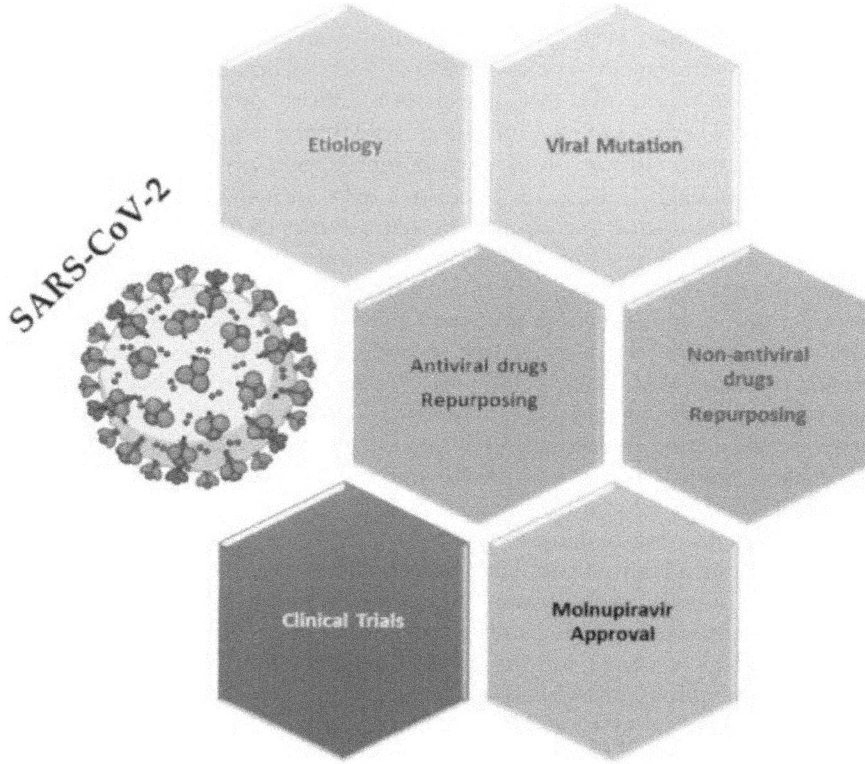


**FIGURE 8.1**: Antiviral drug repurposing.

information processing. While the faulty-membership (FM) connected with LTs or 2TLTs is uncertain, it is impliedly acknowledged that the truth-membership (TM) associated with them is 1 when we utilize them to express our preferences. To overcome this shortcoming, an increasing number of researchers shift their focus to the coupling of certain FSs with linguistic representation frameworks. For example, LTSs and IFSs can be combined to create intuitionistic linguistic sets (ILSs), which were proposed by Wang and Li [15]. Naz et al. [16] established the new strategy 2TL bipolar FS with (HM) operator and numerical illustration was provided for the selection of the best photovoltaic cell to demonstrate the flexibility of the proposed operators. Linguistic intuitionistic fuzzy numbers (LIFNs) were proposed by Chen et al. [17] through the integration of IFSs and linguistic models. The point view of IFS and linguistic truth valued lattice implication algebra serve as the foundation for linguistic truth-valued intuitionistic fuzzy sets (LTVIFSs) [18]. The inability of ILS to resolve certain conflicting or missing information in real-world issues remains a flaw. For example, when evaluating emergency response solutions for sustainable community development, let's say that one decision-maker (DM) declares that his or her evaluation of the attribute of rescuing capacity is "good." In addition, he or she estimates that there is a 70% chance that the judge is correct, a 40% chance that the judge is incorrect, and a 30% chance that the judge is unsure. FSs and IFSs cannot be used to immediately resolve this issue. Thus, it is necessary to propose fresh theory and methodology. Finch [19] introduced every single-valued point in the domain that has a distinct value within the range, making it either many-to-one or one-to-one. A complex variable that is single-valued in a complex function has the same value throughout. Monsalve and Arena [20] extended a local version of a bounded operator's single-valued extension property along with a few related quantities that are crucial to Fredholm's theory. In [21], Broumi and Smarandache presented the idea of (NSs). Smarandache [22] first offered the NS as a solution to this kind of issue by introducing an independent indeterminacy-membership (IM) based on IFS, which can be thought of as an extension of FS, IFS, and so forth. The three membership components of the NS are the TM, the FM, and the IM, where all three variables are totally unrelated to one another. Since the NS originates from a philosophical perspective, it is challenging to apply it to real-world scientific and technical processes. Over time, a number of NS subclasses have been proposed to address the aforementioned issue, and this has grown to be an intriguing and well-liked area of study. A single-valued neutrosophic linguistic set (SVNLS), its mathematical properties, and its operations were presented by Wang et al. [9]. Wang [24] introduced set-theoretic operators and proposed a single-valued neutrosophic set (SVNS) approach to address MAGDM problems, while also exploring the concept of fuzzy graphs, specifically SVNS fuzzy graphs, to investigate effective solutions for challenging MAGDM scenarios [25, 26]. A neutrosophic subsethood was described by Sahin and Kucuk [27] measure for SVNS, and to illustrate the efficacy of the planned subsethood meter, a case study of an application to a multi-attribute decision-making problem was provided. We can see from the SVNS that it provides us with an extra way to depict inaccurate, partial, and inconsistent data present in the real world. In this manner, it would make more sense in dealing with conflicting and ambiguous information. Numerous scholars have since looked at further NS subclasses, including SVNLS, neutrosophic soft sets [28], interval neutrosophic sets (INSs) [29], and other NSs [30]. Naz et al. [31] use 2T-SVNS for software quality assessment. It is used in this study's unique MAGDM framework to express expert judgments with three independent components: truth-membership, indeterminacy-membership, and falsity-membership. This helps to overcome these restrictions. Compared to more conventional fuzzy or intuitionistic methods, this sophisticated model helps decision-makers better reflect the hesitancy and ambiguity present in language evaluations. In order to preserve the semantic interpretability of expert inputs and improve computational accuracy throughout the aggregation process, the 2TL model is further incorporated. To ensure objective prioritization of evaluation criteria, the entropy method is adopted for weight determination, allowing the model to reflect the informational value of each criterion without subjective bias. The aggregation of expert judgments is performed using the two-tuple single-valued neutrosophic weighted average (2T-SVNWA) operator, which consolidates uncertain linguistic inputs into a comprehensive group assessment. Finally, the ELECTRE-I [32] method is used to conduct pairwise comparisons and establish an outranking relationship among candidate antiviral drugs, resulting in a robust and justifiable ranking of alternatives. This integrated approach addresses the limitations of existing models and offers a powerful tool for public health authorities, researchers, and pharmaceutical decision-makers to prioritize antiviral candidates in preparation for future pandemics.

### 8.1.1 Motivation

Novel viral pandemics like COVID-19, SARS, and MERS have frequently challenged the global healthcare landscape. These incidents have brought attention to the pressing necessity for prompt and efficient therapeutic interventions. Because they need a lot of time and resources, traditional drug research pipelines are not appropriate for urgent public health emergencies. Antiviral drug repurposing, or the process of finding new therapeutic applications for already-approved medications, has therefore become a key tactic to hasten treatment readiness. Effectiveness, safety, availability, affordability, and scalability are just a few of the competing factors that must be considered while making repurposing decisions, all of which are subject to considerable subjectivity and ambiguity. This ambiguous and hesitant information is frequently difficult for traditional decision-making techniques to capture. In order to facilitate well-founded antiviral medication selection, a sophisticated and reliable MAGDM methodology that can handle ambiguous, linguistic, and uncertain expert viewpoints is urgently needed.

### 8.1.2 Objective and contribution

Developing and implementing a novel MAGDM framework specifically designed for the assessment and selection of repurposed antiviral medications in the context of future pandemics is the main goal of this chapter. The suggested framework successfully models the linguistic, ambiguous, and uncertain character of expert assessments by utilizing 2T-SVNS. The framework reduces reliance on subjective judgments by using the entropy weighting method to determine the relative importance of evaluation criteria, ensuring impartiality in decision-making. Finally, the ELECTRE-I outranking technique is applied to rank the antiviral drug alternatives, offering a robust decision-making mechanism that accommodates non-compensatory criteria and conflicting expert preferences under uncertain conditions. The key contributions of this work are as follows:

- **Novel MAGDM Framework:** To provide a more precise and adaptable decision-making environment under uncertainty, a novel integrated MAGDM model is proposed by merging 2T-SVNS, entropy weighting, 2T-SVNWA aggregation, and the ELECTRE-I approach.
- **Application to a High-Impact Domain:** To illustrate the framework's usefulness in crucial healthcare decision-making, it is applied to the urgent and pertinent real-world domain of antiviral medication repurposing for upcoming pandemics.
- **Enhanced Modeling of Expert opinions:** Experts can convey qualitative evaluations in natural language while taking truth, indeterminacy, and falsity into consideration by using two-tuple linguistic neutrosophic sets. This preserves nuanced opinions.
- **Evidence-Based and Scalable Decision Support:** The method makes it possible to choose antiviral candidates in a transparent, reliable, and scalable manner, which may help international health organizations and health authorities plan for future outbreaks.

### 8.1.3 Key learnings

In light of upcoming pandemics, this chapter offers a number of crucial insights into the creation and use of sophisticated decision-making tools for antiviral medication repurposing.

- First of all, the incorporation of 2TL-SVNS shows how well it captures the linguistic, and imprecise character of expert assessments, which are typical in decision-making contexts connected to healthcare.
- Second, using the entropy method to determine attribute weights is a dependable and impartial way to reduce subjective bias, particularly when working with a variety of ambiguous expert opinions.
- Third, the 2T-SVNWA aggregation operator preserves the depth of individual evaluations while guaranteeing a fair consolidation of group viewpoints.
- Last but not least, the ELECTRE-I outranking approach provides a clear and rational ranking of options, demonstrating its effectiveness in handling conflicting criteria and non-compensatory choice circumstances.

  All things considered, the suggested MAGDM framework shows great adaptability to additional multi-criteria situations under uncertainty in addition to improving the quality and dependability of complex healthcare decision-making.

### 8.1.4 Research gaps

Even though repurposing antiviral drugs is becoming more popular, the complicated, unpredictable, and multi-criteria character of this field is frequently not adequately addressed by current decision-making frameworks. The literature identifies a number of significant research gaps:

- **Gap 1:** The ambiguous and imprecise expert opinions that are frequently included in healthcare evaluations are difficult to capture using traditional MAGDM approaches, especially when judgments must be made quickly during pandemics.
- **Gap 2:** Many current models are unable to handle linguistic data in a way that is consistent with the natural way that experts communicate their opinions, which might result in oversimplification or meaning loss.
- **Gap 3:** Most decision-making frameworks use arbitrary or subjective weighting of factors, which can create bias and undermine the process's legitimacy.
- **Gap 4:** Common ranking and aggregation techniques often assume that poor performance in one criterion can be compensated for by good performance in another, which is not always acceptable in critical domains, such as drug selection.

**How will it be overcome:** The above limitations are addressed through the development of an integrated decision-making framework.

- 2T-SVNS provides a more accurate representation of expert input in imprecise situations by allowing a more complete modeling of uncertainty, indeterminacy, and linguistic words.
- In order to improve the trustworthiness of group consensus, the 2T-SVNWA operator is used to aggregate expert opinions while preserving the qualitative and quantitative components of the linguistic evaluations.
- Instead of depending on personal preferences, the entropy-based weighting approach guaranties that criterion weights are determined objectively, taking into account the inherent information and unpredictability in the data.
- In order to address non-compensatory interactions between criteria, the ELECTRE-I outranking approach is finally presented. This makes it especially appropriate for healthcare decisions when trade-offs are not always acceptable. By addressing these key gaps, the proposed MAGDM approach enhances both the robustness and credibility of antiviral drug selection decisions, paving the way for its application in other high-stakes, uncertainty-driven domains.

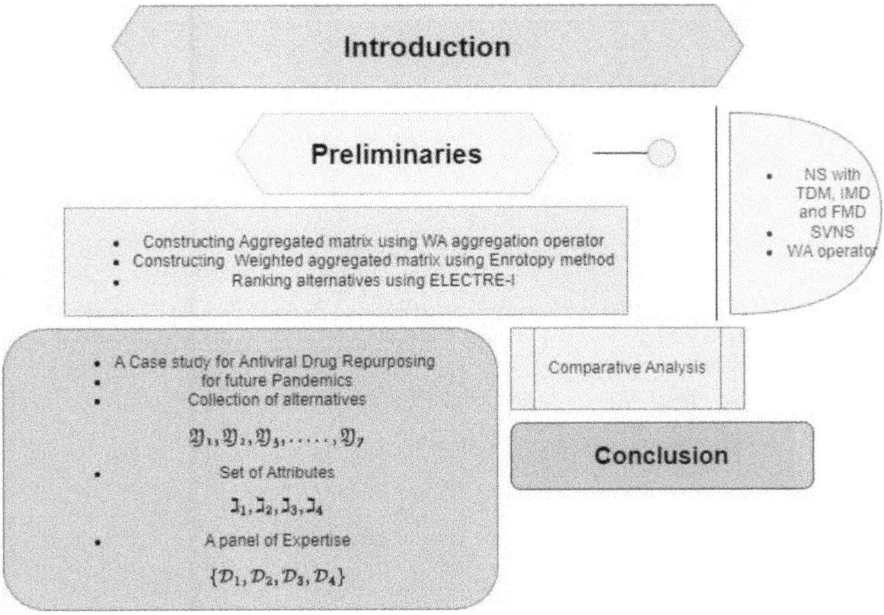


**FIGURE 8.2**: Structure of the chapter.

### 8.1.5 Outline

To fully address the purpose of the study, this chapter is divided into multiple sections. An interpretation of earlier research is provided in Section 15.1. The background details and definitions of important terms related to the analysis of the set 2T-LSVNS are given in Section 8.2. The formulation, underlying ideas, characteristics, and uses of these operators in various DM scenarios are also covered. The study methodology, which includes the approach, techniques, and processes utilized to develop the 2T-LSVNS-entropy-ELECTRE-I model, is presented in Section 8.3. A case study that illustrates the practical application of the 2T-LSVNS-entropy-ELECTRE-I model in the context of antiviral drug repurposing for future pandemics is presented in Section 8.4. The model's DM results, a list of alternatives and attributes, and a comparison with current methods are shown in Section 8.5. Section 8.6 summarizes major findings, identifies limitations, and suggests future research directions, further investigation, and improvement in the field. The structure of the chapter is clearly presented in Figure 8.2.

## 8.2 Preliminaries

In this section, we review some fundamental ideas that will be used throughout the chapter in order to better understand the established research work. These ideas include NS, SVNS, 2T-LSVNS, and Weighted average operator.

**Definition 8.1:** *[12]: Let there exists a linguistic term set $S = \{\flat_t | t = 0, 1, \ldots, \tau\}$ with odd cardinality, where $\flat_t$ indicates a possible linguistic term for a linguistic variable. For instance, an LTS S having seven terms can be described as follows: $S^7 = \{\flat_0^7$: none, $\flat_1^7$ : very low, $\flat_2^7$ : low, $\flat_3^7$ : medium, $\flat_4^7$ : high, $\flat_5^7$ : very high, $\flat_6^7$ : perfect$\}$.*

*If $\flat_t, \flat_\jmath \in S$, then the LTS meets the following characteristics:*

**(i)** *The set is ordered: $\flat_t > \flat_\jmath$, if and only if $t > \jmath$.*

**(ii)** *Max operator: $\max(\flat_t, \flat_\jmath) = \flat_t$, if and only if $t \geq \jmath$.*

**(iii)** *Min operator: $\min(\flat_t, \flat_\jmath) = \flat_t$, if and only if $t \leq \jmath$.*

**(iv)** *Negative operator: $Neg(\flat_t) = \flat_\jmath$ such that $\jmath = \tau - t$.*

The 2TL representation model is based on the idea of symbolic translation and is useful for representing the linguistic assessment information by means of a two-tuple $(\flat_t, \upsilon_t)$, where $\flat_t$ is a linguistic label from predefined LTS $S$ and $\upsilon_t$ is the value of symbolic translation, and $\upsilon_t \in [-0.5, 0.5)$.

**Definition 8.2:** *[14]: Let $\varrho$ be the result of an aggregation of the indices of a set of labels assessed in an LTS S, i.e., the result of a symbolic aggregation operation, $\varrho \in [0, \tau]$, where $\tau$ is the cardinality of S. Let $t = round(\varrho)$ and $\upsilon = \varrho - t$ be two values, such that, $t \in [0, \tau]$ and $\upsilon \in [-0.5, 0.5)$. Then $\upsilon$ is called a symbolic translation.*

**Definition 8.3:** *[14]: Let $S = \{\flat_t | t = 1, \ldots, \tau\}$ be a LTS and $\varrho \in [0, \tau]$ is a number value representing the aggregation result of linguistic symbolic. Then the function $\Delta$ used to obtain the 2TL information equivalent to $\varrho$ is defined as:*

$$\Delta : [0, \tau] \to S \times [-0.5, 0.5),$$

$$\Delta(\varrho) = \begin{cases} \flat_t, t = round(\varrho) \\ \upsilon = \varrho - t, \ \upsilon \in [-0.5, 0.5). \end{cases} \tag{8.1}$$

**Definition 8.4:** *[14]: Let $S = \{\flat_t | t = 1, \ldots, \tau\}$ be an LTS and $(\flat_t, \upsilon_t)$ be a two-tuple. There exists a function $\Delta^{-1}$ that restores the two-tuple to its equivalent numerical value $\varrho \in [0, \tau] \subset R$, where*

$$\Delta^{-1} : S \times [-0.5, 0.5) \to [0, \tau],$$
$$\Delta^{-1}(\flat_t, \upsilon) = t + \upsilon = \varrho. \tag{8.2}$$

**Definition 8.5:** *[33, 34]: An NS $X$, made up of three functions including the TMD $P_X(\vartheta)$, the IMD $L_X(\vartheta)$, and the FMD $M_X(\vartheta)$, can be expressed as follows in a fix set $N$:*

$$X = \{\vartheta, P_X(\vartheta), L_X(\vartheta), M_X(\vartheta)|\vartheta \in N\}, \quad (8.3)$$

*where $P_X(\vartheta)$, $L_X(\vartheta)$, $M_X(\vartheta)$ are lying in $]0^-, 1^+[$ and $0^- \leq sup(P_X(\vartheta)) + sup(L_X(\vartheta)) + sup(M_X(\vartheta)) \leq 3^+$.*
*A new function named as IMD is introduced in NS, although it is challenging to use in actual applications. Hence SVNS is utilized.*

**Definition 8.6:** *[35]: An SVNS $Y$ in a fix set $N$ can be expressed in the following form:*

$$Y = \{\vartheta, P_Y(\vartheta), L_Y(\vartheta), M_Y(\vartheta)|\vartheta \in N\}, \quad (8.4)$$

*where TMD $P_Y(\vartheta)$, IMD $L_Y(\vartheta)$ and FMD $M_Y(\vartheta)$ all belong to $[0, 1]$ and satisfy $0 \leq P_Y(\vartheta) + L_Y(\vartheta) + M_Y(\vartheta) \leq 3$.*

**Definition 8.7:** *[36]: Let $S = \{\mathbf{s}_t|t = 0, 1, \ldots, \tau\}$ be a linguistic term set with odd cardinality. If $((\mathbf{s}_a(\vartheta), \alpha(\vartheta)), (\mathbf{s}_b(\vartheta), \beta(\vartheta)),$ $(\mathbf{s}_c(\vartheta), \delta(\vartheta))$ is defined for $\mathbf{s}_a(\vartheta), \mathbf{s}_b(\vartheta), \mathbf{s}_c(\vartheta) \in S$, $\alpha(\vartheta), \beta(\vartheta), \delta(\vartheta)) \in [-0.5, 0.5)$, where $(\mathbf{s}_a(\vartheta), \alpha(\vartheta))$,$(\mathbf{s}_b(\vartheta), \beta(\vartheta))$, and $(\mathbf{s}_c(\vartheta), \delta(\vartheta))$ represent the TMD, IMD, and FMD by 2TL sets, respectively. The 2T-LSVNS is defined as:*

$$\aleph = \{(\vartheta, ((\mathbf{s}_a(\vartheta), \alpha(\vartheta)), (\mathbf{s}_b(\vartheta), \beta(\vartheta)), (\mathbf{s}_c(\vartheta), \delta(\vartheta))))|\vartheta \in N\}, \quad (8.5)$$

*where $0 \leq \Delta^{-1}(\mathbf{s}_a(\vartheta), \alpha(\vartheta)) \leq \tau, 0 \leq \Delta^{-1}(\mathbf{s}_b(\vartheta), \beta(\vartheta)) \leq \tau, 0 \leq \Delta^{-1}(\mathbf{s}_c(\vartheta), \delta(\vartheta)) \leq \tau$ and $0 \leq (\Delta^{-1}(\mathbf{s}_a(\vartheta), \alpha(\vartheta))) + (\Delta^{-1}(\mathbf{s}_b(\vartheta), \beta(\vartheta))) + (\Delta^{-1}(\mathbf{s}_c(\vartheta), \delta(\vartheta))) \leq 3\tau$.*

To compare any two 2T-LSVN numbers (2T-LSVNNs), their score value and accuracy value are defined as follows:

**Definition 8.8:** *[37]: Let $\mathcal{Y} = ((\mathbf{s}_a, \alpha), (\mathbf{s}_b, \beta), (\mathbf{s}_c, \delta))$ be the 2T-LSVNN. Then the score function $\Theta$ of a 2T-LSVNN can be represented as:*

$$\Theta(\mathcal{Y}) = \Delta\left(\frac{(2\tau + \Delta^{-1}(\mathbf{s}_a, \alpha) - \Delta^{-1}(\mathbf{s}_b, \beta) - \Delta^{-1}(\mathbf{s}_c, \delta))}{3}\right), \ \Delta^{-1}(\Theta(\mathcal{Y})) \in [0, \tau], \quad (8.6)$$

*and its accuracy function $\beth$ is defined as:*

$$\beth(\mathcal{Y}) = \Delta\left(\Delta^{-1}(\mathbf{s}_a, \alpha) - \Delta^{-1}(\mathbf{s}_c, \delta)\right), \ \Delta^{-1}(\beth(\mathcal{Y})) \in [-\tau, \tau]. \quad (8.7)$$

**Definition 8.9:** *[37]: Let $\mathcal{Y}_1 = ((\mathbf{s}_{a_1}, \alpha_1), (\mathbf{s}_{b_1}, \beta_1), (\mathbf{s}_{c_1}, \delta_1))$ and $\mathcal{Y}_2 = ((\mathbf{s}_{a_2}, \alpha_2), (\mathbf{s}_{b_2}, \beta_2), (\mathbf{s}_{c_2}, \delta_2))$ be two 2T-LSVNNs. Then these two 2T-LSVNNs can be compared according to the following rules:*

**(1)** *If $\Theta(\mathcal{Y}_1) > \Theta(\mathcal{Y}_2)$, then $\mathcal{Y}_1 > \mathcal{Y}_2$;*

**(2)** *If* $\Theta(\mathcal{Y}_1) = \Theta(\mathcal{Y}_2)$, *then*

- *If* $\beth(\mathcal{Y}_1) > \beth(\mathcal{Y}_2)$, *then* $\mathcal{Y}_1 > \mathcal{Y}_2$;
- *If* $\beth(\mathcal{Y}_1) = \beth(\mathcal{Y}_2)$, *then* $\mathcal{Y}_1 = \mathcal{Y}_2$.

The novel operational laws for addition, multiplication, scalar multiplication, and power rules of the 2T-LSVNNs can be described as follows:

**Definition 8.10:** *[37]: Let* $\mathcal{Y} = ((\mathbf{s}_a, \alpha), (\mathbf{s}_b, \beta), (\mathbf{s}_c, \delta))$, $\mathcal{Y}_1 = ((\mathbf{s}_{a_1}, \alpha_1), (\mathbf{s}_{b_1}, \beta_1), (\mathbf{s}_{c_1}, \delta_1))$, *and* $\mathcal{Y}_2 = ((\mathbf{s}_{a_2}, \alpha_2), (\mathbf{s}_{b_2}, \beta_2), (\mathbf{s}_{c_2}, \delta_2))$ *be three 2T-LSVNNs. Then*

1. $$\mathcal{Y}_1 \oplus \mathcal{Y}_2 = \begin{pmatrix} \Delta\left(\tau\left(\frac{\Delta^{-1}(\mathbf{s}_{a_1},\alpha_1)}{\tau} + \frac{\Delta^{-1}(\mathbf{s}_{a_2},\alpha_2)}{\tau} - \frac{\Delta^{-1}(\mathbf{s}_{a_1},\alpha_1)}{\tau}.\frac{\Delta^{-1}(\mathbf{s}_{a_2},\alpha_2)}{\tau}\right)\right), \\ \Delta\left(\tau\left(\frac{\Delta^{-1}(\mathbf{s}_{b_1},\beta_1)}{\tau}.\frac{\Delta^{-1}(\mathbf{s}_{b_2},\beta_2)}{\tau}\right)\right), \Delta\left(\tau\left(\frac{\Delta^{-1}(\mathbf{s}_{c_1},\delta_1)}{\tau}.\frac{\Delta^{-1}(\mathbf{s}_{c_2},\delta_2)}{\tau}\right)\right) \end{pmatrix};$$
2. $$\mathcal{Y}_1 \otimes \mathcal{Y}_2 = \begin{pmatrix} \Delta\left(\tau\left(\frac{\Delta^{-1}(\mathbf{s}_{a_1},\alpha_1)}{\tau}.\frac{\Delta^{-1}(\mathbf{s}_{a_2},\alpha_2)}{\tau}\right)\right), \\ \Delta\left(\tau\left(\frac{\Delta^{-1}(\mathbf{s}_{b_1},\beta_1)}{\tau} + \frac{\Delta^{-1}(\mathbf{s}_{b_2},\beta_2)}{\tau} - \frac{\Delta^{-1}(\mathbf{s}_{b_1},\beta_1)}{\tau}.\frac{\Delta^{-1}(\mathbf{s}_{b_2},\beta_2)}{\tau}\right)\right), \\ \Delta\left(\tau\left(\frac{\Delta^{-1}(\mathbf{s}_{c_1},\delta_1)}{\tau} + \frac{\Delta^{-1}(\mathbf{s}_{c_2},\delta_2)}{\tau} - \frac{\Delta^{-1}(\mathbf{s}_{c_1},\delta_1)}{\tau}.\frac{\Delta^{-1}(\mathbf{s}_{c_2},\delta_2)}{\tau}\right)\right) \end{pmatrix};$$
3. $$\lambda\mathcal{Y} = \left( \Delta\left(\tau\left(1 - \left(1 - \frac{\Delta^{-1}(\mathbf{s}_a,\alpha)}{\tau}\right)^{\lambda}\right)\right), \Delta\left(\tau\left(\frac{\Delta^{-1}(\mathbf{s}_b,\beta)}{\tau}\right)^{\lambda}\right), \Delta\left(\tau\left(\frac{\Delta^{-1}(\mathbf{s}_c,\delta)}{\tau}\right)^{\lambda}\right) \right),$$ $\lambda > 0$;
4. $$\mathcal{Y}^{\lambda} = \left(\Delta\left(\tau\left(\frac{\Delta^{-1}(\mathbf{s}_a,\alpha)}{\tau}\right)^{\lambda}\right), \Delta\left(\tau\left(1 - \left(1 - \frac{\Delta^{-1}(\mathbf{s}_b,\beta)}{\tau}\right)^{\lambda}\right)\right), \Delta\left(\tau\left(1 - \left(1 - \frac{\Delta^{-1}(\mathbf{s}_c,\delta)}{\tau}\right)^{\lambda}\right)\right)\right),$$ $\lambda > 0$.

**Definition 8.11:** *[38]: Let* $\delta_j = \{(\int_{t_j}, \alpha_j), (\int_{i_j}, \beta_j), (\int_{f_j}, \chi_j)\}$, $(j = 1, 2, \ldots, n)$ *be a set of 2T-LNNs. The 2T-LSVNWA operators can be presented as:*

$$2T\text{-}LSWNWA(\delta_1, \delta_2, \ldots, \delta_n) = w_1\delta_1 \oplus w_2\delta_2 \otimes \cdots \otimes w_n\delta_n = \bigoplus_{j=1}^{n} w_j\delta_j$$

$$= \left\langle \Delta\left(k\left(1 - \prod_{j=1}^{n}\left(1 - \frac{\Delta^{-1}(\int_{t_j}, \alpha_j)}{k}\right)^{w_j}\right)\right), \Delta\left(k\prod_{j=1}^{n}\left(\frac{\Delta^{-1}(\int_{i_j}, \beta_j)}{k}\right)^{w_j}\right), \Delta\left(k\prod_{j=1}^{n}\left(\frac{\Delta^{-1}(\int_{f_j}, \chi_j)}{k}\right)^{w_j}\right)\right\rangle$$

---

## 8.3 Decision Analysis with Entropy-ELECTRE-I Approach in MAGDM Environment

Our investigation yields an original decision-support synthesis combining ELECTRE-I's ranking strengths with ENTROPY's objective weighting in

2T-LSVNS space. This novel MAGDM approach employs 2T-LSVNS-WA aggregation to tackle complex decision problems in real-world applications, demonstrating that the integration of the AO operator enhances MAGDM reliability and achieves measurable performance improvements compared to non-operator methods. The proposed 2T-LSVNS-ELECTRE-I algorithm is detailed here, featuring ENTROPY-based weight calibration to ensure balanced criterion evaluation.

### 8.3.1 Entropy weighting approach

The German physicist and mathematician Rudolf Julius Emanuel Clausius (1822-1888) introduced the concept of entropy in 1865 [39, 40]. This approach calculates the weights of each criterion based on the fundamental information. It can eliminate the unpredictability and inconsistency in human opinions, which can lead to errors in determining criteria weights. As a result, the weights gained by using the entropy method are more likely to be accurate because they are based on actual alternative performance data.

**Step 1:** The 2T-LSVNS decision engine processes a constrained solution space containing all permissible alternative selections, $\mathfrak{Y} = \{\mathfrak{Y}_1, \mathfrak{Y}_2, \ldots, \mathfrak{Y}_m\}$ that undergo comprehensive analysis across multiple defined criteria $\beth = \{\beth_1, \beth_2, \ldots, \beth_n\}$, whose weight vector is represented by $\mathfrak{w} = \{\mathfrak{w}_1, \mathfrak{w}_2, \ldots, \mathfrak{w}_n\}$ and set of expert is represented by $\mathfrak{E} = \{\mathfrak{E}_1, \mathfrak{E}_2, \ldots, \mathfrak{E}_e\}$.
Let $\mathfrak{D}^k = [\mathbf{X}^{\mathbf{k}}_{ij}]_{m \times n} = \langle\{\flat_{\vartheta^{(k)}}(\hbar^{(k)})\}, \{\flat_{\phi^{(k)}}(\hbar^{(k)})\}\rangle$ denote the expert assessment matrix composed of heterogeneous 2T-LSVNS judgments covering the solution space and evaluation dimensions of alternative $\mathfrak{Y}_m$ with respect to attribute $\beth_n$.

$$\mathfrak{D}^k = [\mathbf{X}_{ij}]_{m \times n}$$
$$= \begin{bmatrix} \langle\{\flat_{(11)}(\hbar^{(k)})\}, \{\flat_{(11)}(\hbar^{(k)})\}\rangle, & \langle\{\flat_{(12)}(\hbar^{(k)})\}, \{\flat_{(12)}(\hbar^{(k)})\}\rangle, & \ldots & \langle\{\flat_{(1j)}(\hbar^{(k)})\}, \{\flat_{(1j)}(\hbar^{(k)})\}\rangle, \\ \langle\{\flat_{(21)}(\hbar^{(k)})\}, \{\flat_{(21)}(\hbar^{(k)})\}\rangle, & \langle\{\flat_{(22)}(\hbar^{(k)})\}, \{\flat_{(22)}(\hbar^{(k)})\}\rangle, & \ldots & \langle\{\flat_{(2j)}(\hbar^{(k)})\}, \{\flat_{(2j)}(\hbar^{(k)})\}\rangle, \\ \vdots & \vdots & \ddots & \vdots \\ \langle\{\flat_{(i1)}(\hbar^{(k)})\}, \{\flat_{(i1)}(\hbar^{(k)})\}\rangle, & \langle\{\flat_{(i2)}(\hbar^{(k)})\}, \{\flat_{(i2)}(\hbar^{(k)})\}\rangle, & \ldots & \langle\{\flat_{(ij)}(\hbar^{(k)})\}, \{\flat_{(ij)}(\hbar^{(k)})\}\rangle \end{bmatrix}_{m \times n}$$
$$i = 1, \ldots, m, j = 1, \ldots, n. \tag{8.8}$$

**Step 2:** To construct the aggregated 2T-LSVNs decision matrix, the individual evaluations provided by the DMs must be integrated into a unified assessment. This integration is achieved using the 2T-LSVNWA operator presented in Eq. 8.11, which synthesizes the individual judgments of the DMs into a collective decision representation. So matrix is represented as

$$\mathfrak{D} = [\mathbf{X}_{\imath\jmath}]_{\mathfrak{m}\times\mathfrak{n}} = \langle\{\flat_{\vartheta}(\hbar)\}, \{\flat_{\phi}(\hbar)\}\rangle$$
$$= \begin{bmatrix} \langle\{\flat_{(11)}(\hbar)\}, \{\flat_{(11)}(\hbar)\}\rangle, & \langle\{\flat_{(12)}(\hbar)\}, \{\flat_{(12)}(\hbar)\}\rangle, & \cdots & \langle\{\flat_{(1j)}(\hbar)\}, \{\flat_{(1j)}(\hbar)\}\rangle, \\ \langle\{\flat_{(21)}(\hbar)\}, \{\flat_{(21)}(\hbar)\}\rangle, & \langle\{\flat_{(22)}(\hbar)\}, \{\flat_{(22)}(\hbar)\}\rangle, & \cdots & \langle\{\flat_{(2j)}(\hbar)\}, \{\flat_{(2j)}(\hbar)\}\rangle, \\ \vdots & \vdots & \ddots & \vdots \\ \langle\{\flat_{(i1)}(\hbar)\}, \{\flat_{(i1)}(\hbar)\}\rangle, & \langle\{\flat_{(i2)}(\hbar)\}, \{\flat_{(i2)}(\hbar)\}\rangle, & \cdots & \langle\{\flat_{(ij)}(\hbar)\}, \{\flat_{(ij)}(\hbar)\}\rangle \end{bmatrix}_{\mathfrak{m}\times\mathfrak{n}}$$
$$\imath = 1, \ldots, \mathfrak{m}, \jmath = 1, \ldots, \mathfrak{n}. \tag{8.9}$$

**Step 3:** We establish uniformity across the evaluation matrix through criteria normalization, removing dimensional inconsistencies that could distort comparisons. The normalized criteria values are calculated using Eq. 8.10.

$$\mathfrak{r}_{\imath\jmath} = \frac{\chi_{\imath\jmath}}{\sum\limits_{\imath=1}^{\mathfrak{n}} \chi_{\imath\jmath}}, \tag{8.10}$$

in which $\mathfrak{r}_{\imath\jmath}$ represents the normalized values of $\jmath$th criteria for $\imath$th alternative.

**Step 4:** We have attained the desired degree of entropy "$\mathcal{E}_\jmath$" for ($\jmath = 1, 2, \ldots, \mathfrak{n}$; $0 \leq \mathcal{E}_\jmath \leq 1$). Equation 8.11 is executed to determine measurement values throughout the attribute set.

$$\mathcal{E}_\jmath = -\frac{1}{ln\mathfrak{n}} \sum_{\imath=1}^{\mathfrak{n}} \mathfrak{r}_{\imath\jmath} ln(\mathfrak{r}_{\imath\jmath}). \tag{8.11}$$

**Step 5:** Calculate the extent of variations ($\mathfrak{d}_\jmath$) for all criteria as:

$$\mathfrak{d}_\jmath = 1 - \mathcal{E}_\jmath. \tag{8.12}$$

**Step 6:** The uncertainty measures obtained from entropy calculations are denoted by "$wj$". These are normalized values $dj$, derived by transforming the $\jmath$th criterion using the following formula:

$$\mathfrak{w}_\jmath = \frac{\mathfrak{d}_\jmath}{\sum\limits_{\jmath=1}^{\mathfrak{n}} \mathfrak{d}_\jmath}, \tag{8.13}$$

Observed that $\sum\limits_{\jmath=1}^{\mathfrak{n}} \mathfrak{w}_\jmath = 1, \jmath = 1, 2, \ldots, \mathfrak{n}$.

### 8.3.2 ELECTRE-I method

**Step 7:** Evaluate the combined weighted normalized 2T-SVNS decision matrix in which $t_{\theta\vartheta} = \mathfrak{P}_{\theta\vartheta} + \mathfrak{F}_{\theta\vartheta}, \quad \theta = 1, 2, \ldots, \mathfrak{P}, \quad \vartheta = 1, 2, \ldots, \mathfrak{F}$.

**Step 8:** The notion of concordance and discordance relies on the consideration of both negative and positive aspects, as the evaluations of alternatives are expressed using 2T-SVNS. In the following, we define the concordance sets $C_{\Lambda'\Lambda}$ and discordance sets $D_{\Lambda'\Lambda}$, respectively:

$$C_{\Lambda'\Lambda} = \{1 \leq \vartheta \leq \mathfrak{F} | t_{\Lambda'\vartheta} \geq t_{\Lambda\vartheta}, \quad \Lambda' \neq \Lambda, \quad \Lambda', \Lambda = 1, 2, \ldots, \mathfrak{P}\},$$
$$D_{\Lambda'\Lambda} = \{1 \leq \vartheta \leq \mathfrak{F} | t_{\Lambda'\vartheta} < t_{\Lambda\vartheta}, \quad \Lambda' \neq \Lambda, \quad \Lambda', \Lambda = 1, 2, \ldots, \mathfrak{P}\}$$

where $t_{\theta\vartheta} = \mathfrak{P}_{\theta\vartheta} + \mathfrak{F}_{\theta\vartheta}, \quad \theta = 1, 2, \ldots, \mathfrak{P}, \quad \vartheta = 1, 2, \ldots, \mathfrak{F}$.

**Step 9:** The concordance indices are determined as follows:

$$C_{\Lambda'\Lambda} = \sum_{\vartheta \in C_{\Lambda'\Lambda}} \omega_{\vartheta}.$$

Consequently, the following can be used to establish the concordance matrix:

$$C = \begin{pmatrix} - & c_{12} & \cdots & c_{1\mathfrak{P}} \\ c_{21} & - & \cdots & c_{2\mathfrak{P}} \\ \vdots & \vdots & \ddots & \vdots \\ c_{\mathfrak{P}1} & c_{\mathfrak{P}2} & \cdots & - \end{pmatrix}.$$

**Step 10:** The given formula is employed to compute the discordance index values.

$$d_{\Lambda'\Lambda} = \frac{\max_{\vartheta} \in D_{\Lambda'\Lambda} \sqrt{\frac{1}{2}[(\mathfrak{P}_{\Lambda'\vartheta} - \mathfrak{P}_{\Lambda\vartheta})^2 + (\mathfrak{F}_{\Lambda'\vartheta} - \mathfrak{F}_{\Lambda\vartheta})^2]}}{\max_{\vartheta} \sqrt{\frac{1}{2}[(\mathfrak{P}_{\Lambda'\vartheta} - \mathfrak{P}_{\Lambda\vartheta})^2 + (\mathfrak{F}_{\Lambda'\vartheta} - \mathfrak{F}_{\Lambda\vartheta})^2]}}.$$

Thus, the discordance matrix can be constructed as follows:

$$D = \begin{pmatrix} - & d_{12} & \cdots & d_{1\mathfrak{P}} \\ d_{21} & - & \cdots & d_{2\mathfrak{P}} \\ \vdots & \vdots & \ddots & \vdots \\ d_{\mathfrak{P}1} & d_{\mathfrak{P}2} & \cdots & - \end{pmatrix}.$$

**Step 11:** The determination of concordance and discordance levels is essential for evaluating the alternatives. The concordance and discordance indices are averaged to obtain the respective concordance level $\bar{c}$ and discordance level $\bar{d}$. So, that

$$\bar{c} = \frac{1}{r(r-1)} \sum_{\substack{\Lambda'=1, \\ \Lambda' \neq \Lambda}}^{\mathfrak{P}} \sum_{\substack{\Lambda=1, \\ \Lambda \neq \Lambda'}}^{\mathfrak{P}} C_{\Lambda'\Lambda},$$

$$\bar{d} = \frac{1}{r(r-1)} \sum_{\substack{\Lambda'=1, \\ \Lambda' \neq \Lambda}}^{\mathfrak{P}} \sum_{\substack{\Lambda=1, \\ \Lambda \neq \Lambda'}}^{\mathfrak{P}} d_{\Lambda'\Lambda}.$$

**Step 12:** The concordance dominance matrix is defined as:

$$L = \begin{pmatrix} - & l_{12} & \cdots & l_{1\mathfrak{P}} \\ l_{21} & - & \cdots & l_{2\mathfrak{P}} \\ \vdots & \vdots & \ddots & \vdots \\ l_{\mathfrak{P}1} & l_{\mathfrak{P}2} & \cdots & - \end{pmatrix}$$

where the values $l_{\theta\vartheta}$ are defined as:

$$l_{\theta\vartheta} = \begin{cases} 1, & if \quad c_{\theta\vartheta} \geq c, \\ 0, & if \quad c_{\theta\vartheta} < c. \end{cases}$$

**Step 13:** The discordance dominance matrix is defined as:

$$P = \begin{pmatrix} - & p_{12} & \cdots & p_{1\mathfrak{P}} \\ p_{21} & - & \cdots & p_{2\mathfrak{P}} \\ \vdots & \vdots & \ddots & \vdots \\ p_{\mathfrak{P}1} & p_{\mathfrak{P}2} & \cdots & - \end{pmatrix}$$

where the values $p_{\theta\vartheta}$ are defined as:

$$p_{\theta\vartheta} = \begin{cases} 1, & if \quad c_{\theta\vartheta} \geq d, \\ 0, & if \quad c_{\theta\vartheta} < d. \end{cases}$$

**Step 14:** The definition of the aggregated dominance matrix is:

$$Z = \begin{pmatrix} - & z_{12} & \cdots & z_{1\mathfrak{P}} \\ z_{21} & - & \cdots & z_{2\mathfrak{P}} \\ \vdots & \vdots & \ddots & \vdots \\ z_{\mathfrak{P}1} & z_{\mathfrak{P}2} & \cdots & - \end{pmatrix}$$

where the values $z_{\theta\vartheta}$ are calculated as:

$$z_{\theta\vartheta} = l_{\theta\vartheta} p_{\theta\vartheta}.$$

**Step 15:** The final step of the process involves ranking the alternatives based on the aforementioned outranking values $z_{\theta\vartheta}$. In other words, an arrow from $D_\theta$ into $D_\vartheta$ exists for any pair of alternatives $D_\theta$ and $D_\vartheta$ if and only if $z_{\theta\vartheta} = 1$. As a result, there are three scenarios that may occur:

**Case 1:** A distinctive arrow leads from $D_\theta$ into $D_\vartheta$.

**Case 2:** Two primary paths connect $D_\theta$ and $D_\vartheta$.

**Case 3:** Neither $D_\theta$ nor $D_\vartheta$ is connected by an arrow.

In the first case, we determine that $D_\theta$ is preferred to $D_\vartheta$. In the second case, $D_\theta$ and $D_\vartheta$ are considered to be equal. In the third case, $D_\theta$ and $D_\vartheta$ are deemed incomparable.

The graphical representation of this method is presented in Figure 8.3.

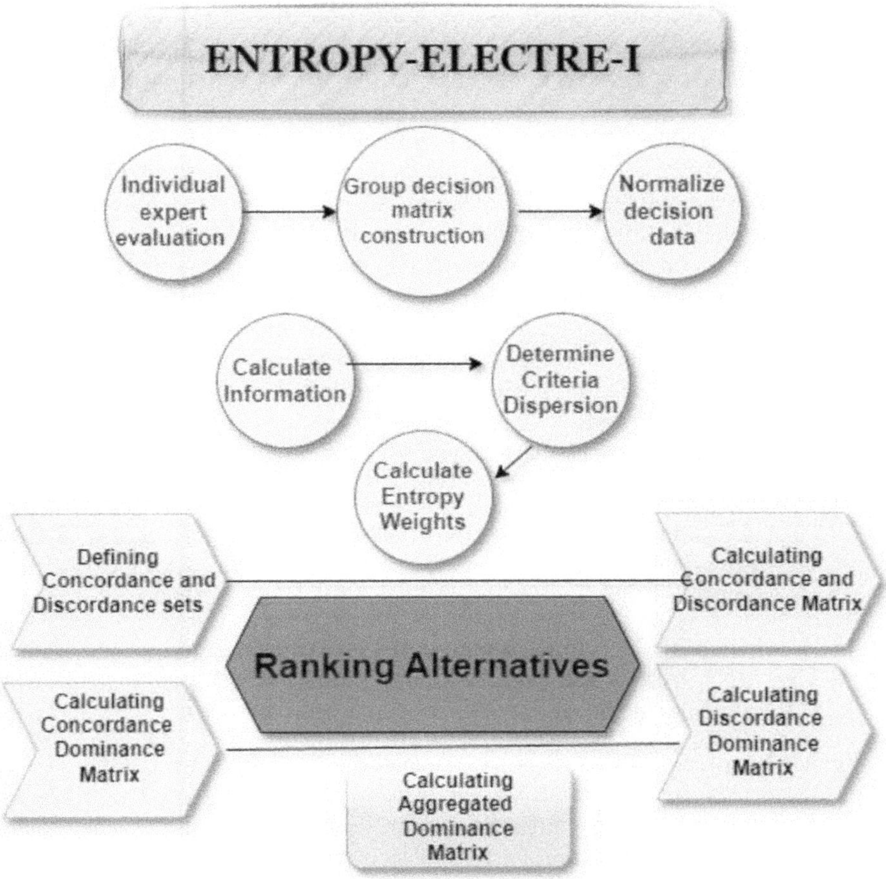


**FIGURE 8.3**: Entropy-ELECTRE-I.

## 8.4 Case Study

In terms of getting ready for future viral outbreaks, it is crucial to make timely decisions regarding treatment strategies. An increasingly important strategy is the repurposing of antiviral drugs, which entails assessing current antiviral treatments for potential new applications against emerging viruses. Global regulatory bodies including the U.S. Food and Drug Administration (FDA) and the European Medicines Agency (EMA), along with national public health authorities, have increasingly advocated for transparent frameworks to assess and authorize repurposed antivirals, acknowledging the significance of this approach [41, 42]. For instance, medications such as Remdesivir and

Favipiravir received emergency use authorizations derived from earlier data, compassionate use instances, and accelerated clinical trials. To ensure trust in the decision-making process, these approvals were frequently accompanied by public briefings, expert reviews, and data releases. To provide evaluations that drew from both clinical insight and practical feasibility, a panel of decision-makers with expertise in the relevant fields including public health professionals, regulatory experts, pharmacologists, and virologists was engaged. The decision problem entails evaluating a collection of repurposed antiviral medications based on factors like therapeutic efficacy, resistance potential, production scalability, and regulatory readiness. After an initial evaluation, let $\mathfrak{Y} = \mathfrak{Y}_1, \mathfrak{Y}_2,...,\mathfrak{Y}_7$ be a set of seven selected alternatives and let $\beth = \beth_1, \beth_2, \beth_3, \beth_4$ be a set of four attributes. Suppose seven selected alternatives are evaluated by four experts $\{\mathcal{D}_1, \mathcal{D}_2, \mathcal{D}_3, \mathcal{D}_4\}$ with weighting vector $\omega' = (0.2, 0.5, 0.3, 0.1)^T$ with parameter $q = 1$ and $g = 9$ for choosing the best decision matrices. Let's assume that four experts evaluate seven selected alternatives

- **Remdesivir ($\mathfrak{Y}_1$):** Originally developed for Ebola, Remdesivir is a nucleotide analog that inhibits RNA-dependent RNA polymerase (RdRp) in various RNA viruses. It has been repurposed for SARS-CoV-2 and showed modest efficacy in hospitalized COVID-19 patients.

- **Favipiravir ($\mathfrak{Y}_2$):** Initially developed for influenza, Favipiravir inhibits viral RNA polymerase and has demonstrated activity against multiple RNA viruses, including arenaviruses and bunyaviruses.

- **Molnupiravir ($\mathfrak{Y}_3$):** It is a ribonucleoside analog that introduces copying errors during viral RNA replication, reducing infectivity. It has received emergency use authorization in several countries. It is easy to administer orally, active in early infection stages, and promising for rapid deployment during future outbreaks.

- **Nitazoxanide ($\mathfrak{Y}_4$):** Originally an antiparasitic, Nitazoxanide has broad antiviral activity, including against respiratory syncytial virus, influenza, and coronaviruses. It modulates host immune responses and inhibits viral protein expression.

- **Baricitinib ($\mathfrak{Y}_5$):** It is an anti-inflammatory drug that also has antiviral properties via AAK1 inhibition, potentially reducing viral entry into cells. It is used in combination therapies. It's a strong candidate for combination therapy in severe viral infections due to dual antiviral and immunomodulatory effects.

- **Sofosbuvir ($\mathfrak{Y}_6$):** It's a hepatitis C virus (HCV) antiviral that has demonstrated potential RdRp inhibition in other RNA viruses, including flaviviruses and coronaviruses. It has high safety profile, oral bioavailability, and potential for cross-viral repurposing.

**TABLE 8.1**: Attributes that characterize the alternatives in the context of emergency pharmaceutical response.

| Attribute | Description |
|---|---|
| Therapeutic Efficacy ($\beth_1$) | Refers to the drug's clinical effectiveness in inhibiting or reducing viral replication and improving patient outcomes in previous studies or off-label use. |
| Regulatory Readiness ($\beth_2$) | Denotes the ease with which the drug can gain emergency use authorization or fast-track approval due to prior approvals, clinical trial data, or known safety profiles. |
| Storage and Transportation Feasibility ($\beth_3$) | Considers temperature sensitivity, shelf life, and logistics, especially for use in low-resource settings. |
| Pharmacoeconomic Viability ($\beth_4$) | Assesses the cost-effectiveness of the drug, including manufacturing cost, accessibility, and long-term affordability. |

- **Interferon ($\mathfrak{Y}_7$):** Interferon-based therapies are immunomodulatory agents that activate the body's innate antiviral defense mechanisms. These include Type I interferons (e.g., IFN-$\alpha$, IFN-$\beta$) and Type III interferons (e.g., IFN-$\lambda$), which stimulate the production of antiviral proteins and enhance immune cell activity. They are known to inhibit viral replication and modulate host responses, particularly during early stages of infection.

### 8.4.1 Decision-making procedure based on the entropy-ELECTRE-I methods

**Step 1:** We employ Eq. 8.10 to standardize the attribute values in the decision matrix, as presented in Table 8.8.

**Step 2:** The entropy degree for each attribute is computed using Eq. 8.11, and the results are recorded in Table 8.9.

**Step 3:** The degree of divergence for each attribute is then obtained through Eq. 8.12, as presented in Table 8.9.

**Step 4:** Subsequently, the entropy weights are derived using Eq. 8.13, with the outcomes listed in Table 8.9.

**TABLE 8.2**: Expert panel in the context of antiviral drug repurposing for future pandemics.

| DM expert | Expertise description |
|---|---|
| Virologists and Infectious Disease Experts ($\mathcal{D}_1$) | Virus-host interactions, mutation behavior, antiviral resistance; evaluate drug effectiveness against specific viral families and emerging variants |
| Regulatory Affairs Specialists ($\mathcal{D}_2$) | FDA, European Medicines Agency (EMA), WHO guidelines and emergency use authorization (EUA) processes; judge regulatory readiness and feasibility of fast-track approval |
| Epidemiologists ($\mathcal{D}_3$) | Outbreak dynamics, public health strategy, resource allocation; weigh scalability, global availability, and public impact |
| Pharmaceutical Manufacturing Experts ($\mathcal{D}_4$) | Assess production scalability, cost, and logistical feasibility; assess therapeutic efficacy, side effect profiles, and resistance potential. |

So, weight vector is $\xi = (0.1454, 0.1130, 0.6405, 0.1011)^T$ .

**Step 5:** Compute the aggregated weighted matrix that is shown in Table 8.10.

**Step 6:** The concordance matrix can be constructed as follows:
Concordance matrix ($C$):

$$C = \begin{bmatrix} 0 & 0.6018 & 0.7578 & 0.2602 & 0.1561 & 0.2422 & 0.1561 \\ 0.3982 & 0 & 0.7398 & 0.5024 & 0.3982 & 0.2422 & 0.3982 \\ 0.2422 & 0.2602 & 0 & 0.5024 & 0.1561 & 0.2422 & 0.1561 \\ 0.7398 & 0.4976 & 0.4976 & 0 & 0.1561 & 0.3982 & 0.4976 \\ 0.8439 & 0.6018 & 0.8439 & 0.8439 & 0 & 0.8439 & 0.7398 \\ 1.0000 & 0.7578 & 0.7578 & 0.6018 & 0.1561 & 0 & 0.4976 \\ 0.8439 & 0.6018 & 0.8439 & 0.5024 & 0.2602 & 0.5024 & 0 \end{bmatrix}$$

**Step 7:** The discordance matrix can be constructed as follows:
Disconcordance matrix ($D$):

$$D = \begin{bmatrix} 0 & 1.0000 & 0.4285 & 0.3341 & 0.6177 & 0.4063 & 0.5057 \\ 0.5844 & 0 & 0.1349 & 0.4666 & 0.8888 & 0.7441 & 0.6861 \\ 0.4472 & 0.7609 & 0 & 0.7130 & 0.9468 & 0.7504 & 0.8180 \\ 0.2351 & 0.9880 & 0.5069 & 0 & 0.6752 & 0.6345 & 0.7397 \\ 0.1922 & 0.4988 & 0.1924 & 0.4955 & 0 & 0.3564 & 0.1545 \\ 0 & 1.0000 & 0.4285 & 0.1946 & 0.6177 & 0 & 0.5057 \\ 0.3248 & 0.5094 & 0.2550 & 0.5892 & 0.6198 & 0.5106 & 0 \end{bmatrix}$$

**TABLE 8.3**: 2T-LSVNNs decision matrix by $\mathcal{D}_1$

| Alternative | $\daleth_1$ | $\daleth_2$ | $\daleth_3$ | $\daleth_4$ |
|---|---|---|---|---|
| $\mathfrak{Y}_1$ | $((\int_3,0),(\int_4,0),(\int_1,0))$ | $((\int_8,0),(\int_4,0),(\int_7,0))$ | $((\int_2,0),(\int_6,0),(\int_8,0))$ | $((\int_8,0),(\int_4,0),(\int_7,0))$ |
| $\mathfrak{Y}_2$ | $((\int_2,0),(\int_6,0),(\int_8,0))$ | $((\int_8,0),(\int_4,0),(\int_7,0))$ | $((\int_6,0),(\int_2,0),(\int_3,0))$ | $((\int_3,0),(\int_4,0),(\int_1,0))$ |
| $\mathfrak{Y}_3$ | $((\int_8,0),(\int_4,0),(\int_7,0))$ | $((\int_7,0),(\int_5,0),(\int_6,0))$ | $((\int_3,0),(\int_4,0),(\int_1,0))$ | $((\int_6,0),(\int_2,0),(\int_3,0))$ |
| $\mathfrak{Y}_4$ | $((\int_6,0),(\int_2,0),(\int_3,0))$ | $((\int_3,0),(\int_4,0),(\int_1,0))$ | $((\int_2,0),(\int_6,0),(\int_8,0))$ | $((\int_8,0),(\int_4,0),(\int_7,0))$ |
| $\mathfrak{Y}_5$ | $((\int_7,0),(\int_5,0),(\int_6,0))$ | $((\int_6,0),(\int_2,0),(\int_3,0))$ | $((\int_7,0),(\int_5,0),(\int_6,0))$ | $((\int_8,0),(\int_4,0),(\int_7,0))$ |
| $\mathfrak{Y}_6$ | $((\int_3,0),(\int_4,0),(\int_1,0))$ | $((\int_2,0),(\int_6,0),(\int_8,0))$ | $((\int_6,0),(\int_2,0),(\int_3,0))$ | $((\int_4,0),(\int_1,0),(\int_2,0))$ |
| $\mathfrak{Y}_7$ | $((\int_6,0),(\int_2,0),(\int_3,0))$ | $((\int_8,0),(\int_4,0),(\int_7,0))$ | $((\int_4,0),(\int_1,0),(\int_2,0))$ | $((\int_3,0),(\int_4,0),(\int_1,0))$ |

**TABLE 8.4**: 2T-LSVNNs decision matrix by $\mathcal{D}_2$

| Alternative | $\daleth_1$ | $\daleth_2$ | $\daleth_3$ | $\daleth_4$ |
|---|---|---|---|---|
| $\mathfrak{Y}_1$ | $((\int_3,0),(\int_4,0),(\int_1,0))$ | $((\int_2,0),(\int_6,0),(\int_8,0))$ | $((\int_6,0),(\int_2,0),(\int_3,0))$ | $((\int_4,0),(\int_1,0),(\int_2,0))$ |
| $\mathfrak{Y}_2$ | $((\int_8,0),(\int_4,0),(\int_7,0))$ | $((\int_4,0),(\int_1,0),(\int_2,0))$ | $((\int_3,0),(\int_4,0),(\int_1,0))$ | $((\int_6,0),(\int_2,0),(\int_3,0))$ |
| $\mathfrak{Y}_3$ | $((\int_4,0),(\int_1,0),(\int_2,0))$ | $((\int_3,0),(\int_4,0),(\int_1,0))$ | $((\int_2,0),(\int_6,0),(\int_8,0))$ | $((\int_4,0),(\int_1,0),(\int_2,0))$ |
| $\mathfrak{Y}_4$ | $((\int_3,0),(\int_4,0),(\int_1,0))$ | $((\int_2,0),(\int_6,0),(\int_8,0))$ | $((\int_6,0),(\int_2,0),(\int_3,0))$ | $((\int_7,0),(\int_5,0),(\int_6,0))$ |
| $\mathfrak{Y}_5$ | $((\int_7,0),(\int_5,0),(\int_6,0))$ | $((\int_8,0),(\int_4,0),(\int_7,0))$ | $((\int_7,0),(\int_5,0),(\int_6,0))$ | $((\int_3,0),(\int_4,0),(\int_1,0))$ |
| $\mathfrak{Y}_6$ | $((\int_3,0),(\int_4,0),(\int_1,0))$ | $((\int_8,0),(\int_4,0),(\int_7,0))$ | $((\int_2,0),(\int_6,0),(\int_8,0))$ | $((\int_8,0),(\int_4,0),(\int_7,0))$ |
| $\mathfrak{Y}_7$ | $((\int_2,0),(\int_6,0),(\int_8,0))$ | $((\int_8,0),(\int_4,0),(\int_7,0))$ | $((\int_6,0),(\int_2,0),(\int_3,0))$ | $((\int_3,0),(\int_4,0),(\int_1,0))$ |

**TABLE 8.5**: 2T-LSVNNs decision matrix by $\mathcal{D}_3$

| Alternative | $\daleth_1$ | $\daleth_2$ | $\daleth_3$ | $\daleth_4$ |
|---|---|---|---|---|
| $\mathfrak{Y}_1$ | $((\int_2,0),(\int_6,0),(\int_8,0))$ | $((\int_4,0),(\int_6,0),(\int_3,0))$ | $((\int_2,0),(\int_6,0),(\int_8,0))$ | $((\int_4,0),(\int_6,0),(\int_3,0))$ |
| $\mathfrak{Y}_2$ | $((\int_2,0),(\int_6,0),(\int_8,0))$ | $((\int_4,0),(\int_6,0),(\int_3,0))$ | $((\int_8,0),(\int_4,0),(\int_7,0))$ | $((\int_8,0),(\int_4,0),(\int_7,0))$ |
| $\mathfrak{Y}_3$ | $((\int_4,0),(\int_6,0),(\int_3,0))$ | $((\int_8,0),(\int_4,0),(\int_7,0))$ | $((\int_4,0),(\int_6,0),(\int_3,0))$ | $((\int_2,0),(\int_6,0),(\int_8,0))$ |
| $\mathfrak{Y}_4$ | $((\int_3,0),(\int_4,0),(\int_1,0))$ | $((\int_4,0),(\int_6,0),(\int_3,0))$ | $((\int_7,0),(\int_5,0),(\int_6,0))$ | $((\int_2,0),(\int_6,0),(\int_8,0))$ |
| $\mathfrak{Y}_5$ | $((\int_4,0),(\int_1,0),(\int_2,0))$ | $((\int_7,0),(\int_5,0),(\int_6,0))$ | $((\int_2,0),(\int_6,0),(\int_8,0))$ | $((\int_3,0),(\int_4,0),(\int_1,0))$ |
| $\mathfrak{Y}_6$ | $((\int_2,0),(\int_6,0),(\int_8,0))$ | $((\int_7,0),(\int_5,0),(\int_6,0))$ | $((\int_8,0),(\int_4,0),(\int_7,0))$ | $((\int_4,0),(\int_6,0),(\int_3,0))$ |
| $\mathfrak{Y}_7$ | $((\int_4,0),(\int_6,0),(\int_3,0))$ | $((\int_2,0),(\int_6,0),(\int_8,0))$ | $((\int_8,0),(\int_4,0),(\int_7,0))$ | $((\int_3,0),(\int_4,0),(\int_1,0))$ |

**TABLE 8.6**: 2T-LSVNNs decision matrix by $\mathcal{D}_4$

| Alternative | $\daleth_1$ | $\daleth_2$ | $\daleth_3$ | $\daleth_4$ |
|---|---|---|---|---|
| $\mathfrak{Y}_1$ | $((\int_4,0),(\int_1,0),(\int_2,0))$ | $((\int_3,0),(\int_4,0),(\int_1,0))$ | $((\int_6,0),(\int_2,0),(\int_3,0))$ | $((\int_8,0),(\int_4,0),(\int_7,0))$ |
| $\mathfrak{Y}_2$ | $((\int_7,0),(\int_5,0),(\int_6,0))$ | $((\int_6,0),(\int_2,0),(\int_3,0))$ | $((\int_8,0),(\int_4,0),(\int_7,0))$ | $((\int_4,0),(\int_6,0),(\int_3,0))$ |
| $\mathfrak{Y}_3$ | $((\int_6,0),(\int_2,0),(\int_3,0))$ | $((\int_7,0),(\int_5,0),(\int_6,0))$ | $((\int_4,0),(\int_6,0),(\int_3,0))$ | $((\int_7,0),(\int_5,0),(\int_6,0))$ |
| $\mathfrak{Y}_4$ | $((\int_2,0),(\int_6,0),(\int_8,0))$ | $((\int_7,0),(\int_5,0),(\int_6,0))$ | $((\int_8,0),(\int_4,0),(\int_7,0))$ | $((\int_4,0),(\int_6,0),(\int_3,0))$ |
| $\mathfrak{Y}_5$ | $((\int_4,0),(\int_6,0),(\int_3,0))$ | $((\int_2,0),(\int_6,0),(\int_8,0))$ | $((\int_8,0),(\int_4,0),(\int_7,0))$ | $((\int_3,0),(\int_4,0),(\int_1,0))$ |
| $\mathfrak{Y}_6$ | $((\int_4,0),(\int_1,0),(\int_2,0))$ | $((\int_3,0),(\int_4,0),(\int_1,0))$ | $((\int_6,0),(\int_2,0),(\int_3,0))$ | $((\int_8,0),(\int_4,0),(\int_7,0))$ |
| $\mathfrak{Y}_7$ | $((\int_7,0),(\int_5,0),(\int_6,0))$ | $((\int_6,0),(\int_2,0),(\int_3,0))$ | $((\int_8,0),(\int_4,0),(\int_7,0))$ | $((\int_4,0),(\int_6,0),(\int_3,0))$ |

**TABLE 8.7**: Aggregated matrix obtained by WA operator

| Alternative | $\daleth_1$ | $\daleth_2$ |
|---|---|---|
| $\mathfrak{Y}_1$ | $((\int_3, 0.0747), (\int_4, -0.3737), (\int_2, -0.3945))$ | $((\int_7, -0.2763), (\int_5, -0.3807), (\int_7, 0.1860))$ |
| $\mathfrak{Y}_2$ | $((\int_6, -0.2469), (\int_2, -0.0567), (\int_3, -0.4001))$ | $((\int_4, -0.0937), (\int_3, 0.3438), (\int_1, 0.2311))$ |
| $\mathfrak{Y}_3$ | $((\int_7, -0.4825), (\int_3, -0.0374), (\int_4, -0.1336))$ | $((\int_3, 0.0747), (\int_4, -0.3737), (\int_2, -0.3945))$ |
| $\mathfrak{Y}_4$ | $((\int_4, 0.4044), (\int_5, -0.4585), (\int_5, -0.2957))$ | $((\int_5, -0.1058), (\int_3, -0.0196), (\int_4, 0.3893))$ |
| $\mathfrak{Y}_5$ | $((\int_7, -0.4495), (\int_3, 0.2110), (\int_2, 0.1779))$ | $((\int_3, 0.2145), (\int_5, 0.3128), (\int_4, -0.4765))$ |
| $\mathfrak{Y}_6$ | $((\int_6, 0.4738), (\int_3, 0.0243), (\int_4, 0.3824))$ | $((\int_7, -0.3379), (\int_5, -0.1303), (\int_6, 0.3785))$ |
| $\mathfrak{Y}_7$ | $((\int_6, 0.0470), (\int_4, -0.3307), (\int_6, -0.3404))$ | $((\int_7, 0.0816), (\int_2, -0.0234), (\int_3, 0.4784))$ |

| Alternative | $\daleth_3$ | $\daleth_4$ |
|---|---|---|
| $\mathfrak{Y}_1$ | $((\int_5, -0.1175), (\int_5, 0.1017), (\int_5, -0.3412))$ | $((\int_6, -0.2469), (\int_2, -0.0567), (\int_3, -0.4001))$ |
| $\mathfrak{Y}_2$ | $((\int_7, -0.4208), (\int_4, -0.0562), (\int_2, 0.4635))$ | $((\int_4, -0.2790), (\int_5, 0.2168), (\int_4, -0.2236))$ |
| $\mathfrak{Y}_3$ | $((\int_8, -0.4956), (\int_4, -0.4247), (\int_6, -0.4242))$ | $((\int_7, 0.2553), (\int_4, 0.2769), (\int_6, -0.4895))$ |
| $\mathfrak{Y}_4$ | $((\int_7, 0.3937), (\int_4, -0.1134), (\int_7, -0.4719))$ | $((\int_6, 0.0909), (\int_2, 0.0828), (\int_3, -0.1702))$ |
| $\mathfrak{Y}_5$ | $((\int_7, -0.3370), (\int_3, -0.2840), (\int_3, -0.2179))$ | $((\int_5, -0.2965), (\int_2, -0.1459), (\int_3, 0.1569))$ |
| $\mathfrak{Y}_6$ | $((\int_7, -0.3907), (\int_5, -0.1501), (\int_6, 0.0438))$ | $((\int_5, -0.0264), (\int_4, -0.3116), (\int_1, 0.1847))$ |
| $\mathfrak{Y}_7$ | $((\int_7, 0.2050), (\int_3, 0.1569), (\int_4, 0.1208))$ | $((\int_3, 0.3425), (\int_4, -0.1589), (\int_1, -0.1040))$ |

**TABLE 8.8**: Combined matrix.

| | $\gimel_1$ | $\gimel_2$ | $\gimel_3$ | $\gimel_4$ |
|---|---|---|---|---|
| $\mathfrak{Y}_1$ | 0.0953 | 0.1470 | 0.0704 | 0.1065 |
| $\mathfrak{Y}_2$ | 0.1898 | 0.1873 | 0.3152 | 0.1332 |
| $\mathfrak{Y}_3$ | 0.1695 | 0.0788 | 0.1140 | 0.1510 |
| $\mathfrak{Y}_4$ | 0.1189 | 0.1356 | 0.0540 | 0.0874 |
| $\mathfrak{Y}_5$ | 0.2001 | 0.1201 | 0.1696 | 0.1810 |
| $\mathfrak{Y}_6$ | 0.0953 | 0.1470 | 0.0704 | 0.1696 |
| $\mathfrak{Y}_7$ | 0.1312 | 0.1841 | 0.2063 | 0.1713 |

**TABLE 8.9**: Weights calculated by entropy method.

| | **Degree of entropy** | **Degree of differences** | **Entropy weights** |
|---|---|---|---|
| $\gimel_1$ | 0.9794 | 0.0206 | 0.1454 |
| $\gimel_2$ | 0.9840 | 0.0160 | 0.1130 |
| $\gimel_3$ | 0.9093 | 0.0907 | 0.6405 |
| $\gimel_4$ | 0.9857 | 0.0143 | 0.1011 |

**Step 8:** The concordance dominance matrix is defined as:
Concordance dominance matrix ($L$):

$$L = \begin{pmatrix} 0 & 1 & 1 & 0 & 0 & 0 & 0 \\ 0 & 0 & 1 & 0 & 0 & 0 & 0 \\ 0 & 0 & 0 & 0 & 0 & 0 & 0 \\ 1 & 0 & 0 & 0 & 0 & 0 & 0 \\ 1 & 1 & 1 & 1 & 0 & 1 & 1 \\ 1 & 1 & 1 & 1 & 0 & 0 & 0 \\ 1 & 1 & 1 & 0 & 0 & 0 & 0 \end{pmatrix}$$

**Step 9:** The discordance dominance matrix is defined as:
Disconcordance dominance matrix ($P$):

$$P = \begin{pmatrix} 0 & 0 & 1 & 1 & 0 & 1 & 1 \\ 0 & 0 & 1 & 1 & 0 & 0 & 0 \\ 1 & 0 & 0 & 0 & 0 & 0 & 0 \\ 1 & 0 & 1 & 0 & 0 & 0 & 0 \\ 1 & 1 & 1 & 1 & 0 & 1 & 1 \\ 1 & 0 & 1 & 1 & 0 & 0 & 1 \\ 1 & 1 & 1 & 0 & 0 & 1 & 0 \end{pmatrix}$$

**Step 10:** The definition of the aggregated dominance matrix is:

**TABLE 8.10**: Weighted aggregated matrix

| Alternative | $\gimel_1$ | $\gimel_2$ |
|---|---|---|
| $\mathfrak{Y}_1$ | $((\int_1,-0.2554),\ (\int_1,-0.1218),\ (\int_0,0.3888))$ | $((\int_2,-0.3717),\ (\int_1,0.1186),\ (\int_2,-0.2598))$ |
| $\mathfrak{Y}_2$ | $((\int_1,0.3932),\ (\int_0,0.4706),\ (\int_1,-0.3704))$ | $((\int_1,-0.0540),\ (\int_1,-0.1902),\ (\int_0,0.2981))$ |
| $\mathfrak{Y}_3$ | $((\int_2,-0.4217),\ (\int_1,-0.2826),\ (\int_1,-0.0637))$ | $((\int_1,-0.2554),\ (\int_1,-0.1218),\ (\int_0,0.3888))$ |
| $\mathfrak{Y}_4$ | $((\int_1,0.0666),\ (\int_1,0.0998),\ (\int_1,0.1392))$ | $((\int_2,-0.3284),\ (\int_1,0.0179),\ (\int_1,0.4991))$ |
| $\mathfrak{Y}_5$ | $((\int_2,0.2372),\ (\int_1,0.0967),\ (\int_1,-0.2562))$ | $((\int_1,0.0979),\ (\int_2,-0.1855),\ (\int_1,0.2034))$ |
| $\mathfrak{Y}_6$ | $((\int_2,0.2110),\ (\int_1,0.0329),\ (\int_1,0.4967))$ | $((\int_2,0.2754),\ (\int_2,-0.3368,\ (\int_2,0.1785))$ |
| $\mathfrak{Y}_7$ | $((\int_2,0.0653),\ (\int_1,0.2532),\ (\int_2,-0.0670))$ | $((\int_2,0.4186),\ (\int_1,-0.3249),\ (\int_1,0.1880))$ |

| Alternative | $\gimel_3$ | $\gimel_4$ |
|---|---|---|
| $\mathfrak{Y}_1$ | $((\int_1,0.2707),\ (\int_1,0.3277),\ (\int_1,0.2124))$ | $((\int_1,0.4972),\ (\int_1,-0.4943),\ (\int_1,-0.3234))$ |
| $\mathfrak{Y}_2$ | $((\int_2,-0.2878),\ (\int_1,0.0264),\ (\int_1,-0.3589))$ | $((\int_1,-0.0316),\ (\int_1,0.3577),\ (\int_1,-0.0172))$ |
| $\mathfrak{Y}_3$ | $((\int_2,-0.0470),\ (\int_1,-0.0695),\ (\int_1,0.4511))$ | $((\int_2,-0.1118),\ (\int_1,0.1131),\ (\int_1,0.4341))$ |
| $\mathfrak{Y}_4$ | $((\int_2,-0.0758),\ (\int_1,0.0115),\ (\int_2,-0.3011))$ | $((\int_1,-0.0495),\ (\int_0,0.3250),\ (\int_0,0.4416))$ |
| $\mathfrak{Y}_5$ | $((\int_1,0.0398),\ (\int_0,0.4238),\ (\int_0,0.4341))$ | $((\int_1,-0.2660),\ (\int_0,0.2893),\ (\int_0,0.4926))$ |
| $\mathfrak{Y}_6$ | $((\int_1,0.0314),\ (\int_1,-0.2432),\ (\int_1,-0.0569))$ | $((\int_1,-0.2239),\ (\int_1,-0.4244),\ (\int_0,0.1849))$ |
| $\mathfrak{Y}_7$ | $((\int_1,0.1243),\ (\int_0,0.4926),\ (\int_1,-0.3570))$ | $((\int_1,-0.4784),\ (\int_1,-0.4006),\ (\int_0,0.1398))$ |

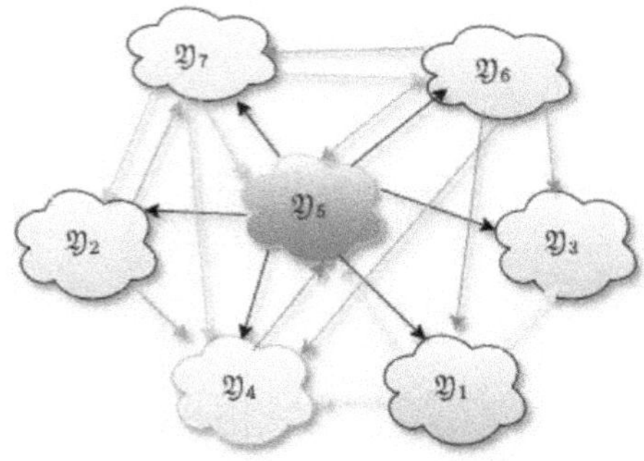


**FIGURE 8.4**: Ranking of alternatives.

Aggregated dominance matrix ($P$):

$$Z = \begin{pmatrix} 0 & 0 & 1 & 0 & 0 & 0 & 0 \\ 0 & 0 & 1 & 0 & 0 & 0 & 0 \\ 0 & 0 & 0 & 0 & 0 & 0 & 0 \\ 1 & 0 & 0 & 0 & 0 & 0 & 0 \\ 1 & 1 & 1 & 1 & 0 & 1 & 1 \\ 1 & 0 & 1 & 1 & 0 & 0 & 0 \\ 1 & 1 & 1 & 0 & 0 & 0 & 0 \end{pmatrix}$$

**Step 11:** We arranged the score functions of alternatives in descending order as:

$$\mathfrak{Y}_5 > \mathfrak{Y}_6 > \mathfrak{Y}_7 > \mathfrak{Y}_1 > \mathfrak{Y}_2 > \mathfrak{Y}_4 > \mathfrak{Y}_3.$$

The highest score functions of alternative $\mathfrak{Y}_5$ represent the best choice. The ranking of these alternatives are presented in Figure 8.4.

#### 8.4.1.1 Advantages of proposed methodology

The proposed 2TL-SVN MAGDM approach, integrated with entropy weighting and the ELECTRE I outranking method, offers a powerful, comprehensive framework for evaluating repurposed antiviral drugs under uncertainty. The key advantages of this methodology are outlined below:

1. By employing the 2TL-SVNS, the methodology effectively captures the inherent vagueness, hesitation, and conflicting opinions present in expert evaluations. This allows for more accurate modeling of real-world judgment in uncertain, high-stakes environments such as pandemic response.

2. The use of the entropy weighting method eliminates subjective bias in assigning importance to criteria. It provides an objective, data-driven

mechanism based on the inherent variability of expert evaluations across each attribute.

3. The ELECTRE-I outranking technique allows for pairwise comparison of alternatives, taking both concordance and discordance factors into account. This ensures a robust and transparent ranking of drug candidates, especially when criteria conflict or when compensatory trade-offs are inappropriate.

4. The proposed methodology is flexible and can be tailored to other public health decision-making contexts, such as vaccine selection, treatment protocol design, or medical resource allocation.

5. By structuring the decision process in a formal, repeatable, and explainable manner, the framework supports transparency and accountability in antiviral drug selection essential for public trust and global collaboration.

---

## 8.5 Comparative Analysis with Other Existing Techniques

The robustness and effectiveness of the proposed decision-making framework were evaluated by comparing it to the ELECTRE-I approach using three well-known MAGDM techniques: VIKOR [43], TOPSIS [44], and WASPAS [31]. The alternative set $\mathfrak{Y} = \mathfrak{Y}_1$, $\mathfrak{Y}_2$,...,$\mathfrak{Y}_7$ was evaluated using each technique. The sequence generated by ELECTRE-I shows that $\mathfrak{Y}_5$ is the best option, although it is also the best option according to WASPAS and VIKOR. However, $\mathfrak{Y}_3$ was offered by TOPSIS as the best substitute as shown in Table 8.11 and Figure 8.5. When evaluating and prioritizing options in complex and uncertain situations, such as antiviral drug repurposing, the choice of an appropriate MAGDM technique is essential. High levels of ambiguity, contradictory standards, and the coexistence of both quantitative and qualitative data are all part of antiviral drug repurposing. In this context, we evaluate the efficacy of four well-known MAGDM techniques ELECTRE I, WASPAS, TOPSIS, and VIKOR in assisting with strategic decision-making in antiviral drug repurposing. The multifaceted uncertainties and qualitative subtleties inherent for antiviral drug repurposing for future pandemics are not well represented by TOPSIS despite its continued usefulness for organized, quantitative problems with well-defined data and deterministic criteria. When expanded within the 2TL-SVNS framework, ELECTRE-I becomes a more suitable and potent instrument for this domain due to its flexibility and outranking philosophy. It is in a unique position to direct sound, practical, and forward-looking decisions in the construction of quantum communication infrastructures because of its

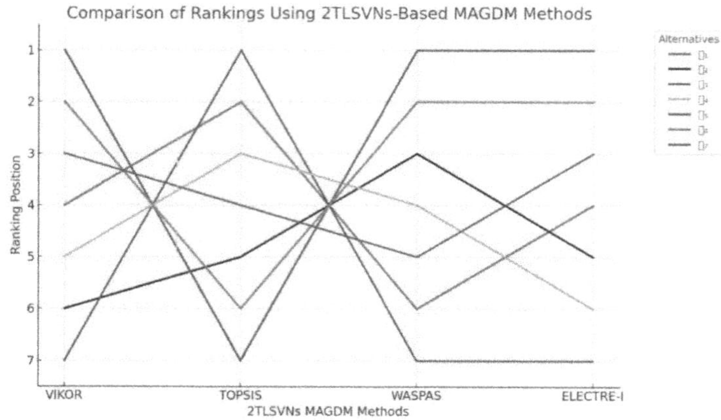


**FIGURE 8.5**: Comparison analysis with other existing techniques.

capacity to handle hesitation and incomparability, impose threshold-based decision logic, and enable partial preferences. where the parameters $p$ and $v$ are specified by the decision maker. The dominated matrix is as Eq. 12.4.

$$D_{m\times m} = \left(d'_{\zeta\kappa}\right)_{m\times m}; \quad d'_{\zeta\kappa} = \frac{\sum\limits_{\hbar=1}^{n} d_{\hbar}(\zeta,\kappa)\cdot \mathfrak{w}}{\sum\limits_{\hbar=1}^{n} \mathfrak{w}}, \quad 0 \le d'_{\zeta\kappa} \le 1 \tag{8.14}$$

Then, the credibility matrix is formed as Eq. 12.5.

$$S_{m\times m} = (s_{\zeta\kappa}); \quad s_{\zeta\kappa}\begin{cases} g_{\zeta\kappa} & \text{if } d'_{\zeta\kappa} \le g_{\zeta\kappa} \\ g_{\zeta\kappa}\cdot \prod\limits_{\hbar\in J_{\zeta\kappa}} \frac{1-d'_{\zeta\kappa}}{1-g_{\zeta\kappa}}, & \text{otherwise} \end{cases} \tag{8.15}$$

where $J_{\zeta\kappa}$ represents those attributes which are $d'_{\zeta\kappa} > g_{\zeta\kappa}$. The aggregated dominance matrix is determined as Eq. 12.6.

$$P_{m\times m} = (p_{\zeta\kappa}); \quad p_{\zeta\kappa}\begin{cases} 1 & \text{if } s_{\zeta\kappa} \ge \lambda - S(\lambda), \\ 0 & \text{otherwise} \end{cases} \tag{8.16}$$

where the values of $\lambda$ and $S(\lambda)$ are specified by Eq. 12.7.

$$\begin{cases} \lambda = \max(s_{\zeta\kappa}), \\ S(\lambda) = \alpha + \beta\lambda, \quad \zeta, k \in \{1,\ldots,m\} \end{cases} \tag{8.17}$$

The final ranking is done with respect to the aggregate dominance matrix.

**TABLE 8.11**: Comparison analysis with other methods

| Methods | Score values | Ranking |
|---|---|---|
| 2T-LSVNs-VIKOR [43] | $\mathfrak{Y}_1 = 0.5784$, $\mathfrak{Y}_2 = 0.6318$, $\mathfrak{Y}_3 = 1.0000$, $\mathfrak{Y}_4 = 0.6301$, $\mathfrak{Y}_5 = 0$, $\mathfrak{Y}_6 = 0.3355$, $\mathfrak{Y}_7 = 0.5321$ | $\mathfrak{Y}_5 > \mathfrak{Y}_6 > \mathfrak{Y}_7 > \mathfrak{Y}_1 > \mathfrak{Y}_4 > \mathfrak{Y}_2 > \mathfrak{Y}_3$ |
| 2T-LSVNs-TOPSIS [45] | $\mathfrak{Y}_1 = 0.6481$, $\mathfrak{Y}_2 = 0.5253$, $\mathfrak{Y}_3 = 0.6925$, $\mathfrak{Y}_4 = 0.5696$, $\mathfrak{Y}_5 = 0.4541$, $\mathfrak{Y}_6 = 0.4823$, $\mathfrak{Y}_7 = 0.5488$ | $\mathfrak{Y}_3 > \mathfrak{Y}_1 > \mathfrak{Y}_4 > \mathfrak{Y}_7 > \mathfrak{Y}_2 > \mathfrak{Y}_6 > \mathfrak{Y}_5$ |
| 2T-LSVNs-WASPAS [31] | $\mathfrak{Y}_1 = 0.9998$, $\mathfrak{Y}_2 = 1.0044$, $\mathfrak{Y}_3 = 0.9981$, $\mathfrak{Y}_4 = 1.0014$, $\mathfrak{Y}_5 = 1.0073$, $\mathfrak{Y}_6 = 1.0045$, $\mathfrak{Y}_7 = 1.0000$ | $\mathfrak{Y}_5 > \mathfrak{Y}_6 > \mathfrak{Y}_2 > \mathfrak{Y}_4 > \mathfrak{Y}_7 > \mathfrak{Y}_1 > \mathfrak{Y}_3$ |

## 8.6 Conclusion

The capacity to make prompt, well-informed, and open therapeutic decisions becomes vital in the face of erratic and recurrent worldwide pandemics. By utilizing already-existing pharmacological molecules, drug repurposing especially of antiviral agents offers a practical and effective response to new viral threats while reducing the time and expense involved in developing new drugs. However, because there are several, frequently incompatible criteria and a great deal of ambiguity, choosing appropriate repurposed antivirals is a difficult task. This chapter presents a thorough MAGDM framework that combines the ELECTRE-I outranking method for robust alternative ranking, an entropy-based weighting mechanism for objectively determining criterion importance, and the 2T-LSVNS for modeling uncertain expert judgments. As illustrated by a case study based on practical concerns, the suggested technique successfully manages unclear and inconsistent expert input, encourages transparency in high-stakes decision processes, and produces logical prioritizing of repurposed antiviral options. The evaluation methodology effectively integrates clinical, regulatory, and operational viewpoints by enlisting specialists from a variety of disciplines. Additionally, sensitivity and comparison assessments demonstrate the approach's resilience and dependability while pointing out both its advantages and possible shortcomings. Looking ahead, a number of study directions become apparent. The model may be more flexible in the face of changing epidemic conditions if dynamic, real-time data are included, such as trends in viral mutations, interim clinical trial results, and reports of antiviral resistance. Candidate selection may also be improved by incorporating machine learning methods for forecasting drug-virus interactions. The framework's applicability and worldwide relevance can also be increased by extending stakeholder engagement to include pharmaceutical producers, international health agencies, and frontline medical staff. The suggested MAGDM framework provides a reproducible approach for managing difficult group decisions in the face of uncertainty and improves the caliber of decision-making in public health preparedness. This strategy makes a significant contribution to pandemic response tactics and the larger objective of global health resilience by combining methodological rigor with practical feasibility.

## Bibliography

[1] Michael S Kinch, Eric Patridge, Mark Plummer, and Denton Hoyer. An analysis of FDA-approved drugs for infectious disease: antibacterial agents. *Drug Discovery Today*, 19(9):1283–1287, 2014.

[2] Thakur Uttam Singh, Subhashree Parida, Madhu Cholenahalli Lingaraju, Manickam Kesavan, Dinesh Kumar, and Raj Kumar Singh. Drug repurposing approach to fight Covid-19. *Pharmacological Reports*, 72(6):1479–1508, 2020.

[3] Totan Garai and Harish Garg. Multi-criteria decision making of Covid-19 vaccines (in india) based on ranking interpreter technique under single valued bipolar neutrosophic environment. *Expert Systems with Applications*, 208:118160, 2022.

[4] Esra Çakır, Mehmet Ali Taş, and Ziya Ulukan. Neutrosophic fuzzy weighted saving heuristic for Covid-19 vaccination. In *2021 systems and information engineering design symposium (SIEDS)*, pages 1–4. IEEE, 2021.

[5] Sumera Naz, Muhammad Akram, Arsham Borumand Saeid, and Aniqa Saadat. Models for MAGDM with dual hesitant q-rung orthopair fuzzy 2-tuple linguistic MSM operators and their application to Covid-19 pandemic. *Expert Systems*, 39(8):e13005, 2022.

[6] Sumera Naz, Muhammad Muneeb ul Hassan, Areej Fatima, Diaz Jorge Martinez, Elisa Ospino Mendoza, and Shariq Aziz Butt. A decision-making mechanism for multi-attribute group decision-making using 2-tuple linguistic t-spherical fuzzy maximizing deviation method. *Granular Computing*, 8(6):1659–1687, 2023.

[7] Lotfi A Zadeh. Fuzzy logic and approximate reasoning: In memory of grigore moisil. *Synthese*, 30:407–428, 1975.

[8] Krassimir Atanassov. Review and new results on intuitionistic fuzzy sets. *Preprint IM-MFAIS-1-88, Sofia*, 5(1), 1988.

[9] Miguel Delgado, Francisco Herrera, Enrique Herrera-Viedma, and Luis Martinez. Combining numerical and linguistic information in group decision making. *Information Sciences*, 107(1-4):177–194, 1998.

[10] Yejun Xu, José M Merigó, and Huimin Wang. Linguistic power aggregation operators and their application to multiple attribute group decision making. *Applied Mathematical Modelling*, 36(11):5427–5444, 2012.

[11] Qun Wu, Peng Wu, Yuanyuan Zhou, Ligang Zhou, Huayou Chen, and Xiyuan Ma. Some 2-tuple linguistic generalized power aggregation operators and their applications to multiple attribute group decision making. *Journal of Intelligent & Fuzzy Systems*, 29(1):423–436, 2015.

[12] Francisco Herrera and Luis Martínez. A 2-tuple fuzzy linguistic representation model for computing with words. *IEEE Transactions on Fuzzy Systems*, 8(6):746–752, 2002.

[13] Yucheng Dong, Cong-Cong Li, and Francisco Herrera. Connecting the linguistic hierarchy and the numerical scale for the 2-tuple linguistic model and its use to deal with hesitant unbalanced linguistic information. *Information Sciences*, 367:259–278, 2016.

[14] Francisco Herrera and Luis Martinez. An approach for combining linguistic and numerical information based on the 2-tuple fuzzy linguistic representation model in decision-making. *International Journal of Uncertainty, Fuzziness and Knowledge-Based Systems*, 8(05):539–562, 2000.

[15] JQ Wang and JJ Li. The multi-criteria group decision making method based on multi-granularity intuitionistic two semantics. *Science & Technology Information*, 33(1):8–9, 2009.

[16] Sumera Naz, Muhammad Akram, Mohammed M Ali Al-Shamiri, Mohammed M Khalaf, and Gohar Yousaf. A new MAGDM method with 2-tuple linguistic bipolar fuzzy heronian mean operators. *Mathematical Biosciences and Engineering*, 19(4):3843–3878, 2022.

[17] Zichun Chen, Penghui Liu, and Zheng Pei. An approach to multiple attribute group decision making based on linguistic intuitionistic fuzzy numbers. *International Journal of Computational Intelligence Systems*, 8(4):747–760, 2015.

[18] Li Zou, Xin Wen, and Yingxin Wang. Linguistic truth-valued intuitionistic fuzzy reasoning with applications in human factors engineering. *Information Sciences*, 327:201–216, 2016.

[19] James Finch. The single valued extension property on a banach space. *Pacific Journal of Mathematics*, 58(1):61–69, 1975.

[20] Pietro Aiena and Osmin Monsalve. Operators which do not have the single valued extension property. *Journal of Mathematical Analysis and Applications*, 250(2):435–448, 2000.

[21] Said Broumi and Florentin Smarandache. Several similarity measures of neutrosophic sets. *Infinite Study*, 410(1), 2013.

[22] Florentin Smarandache. A unifying field in logics: Neutrosophic logic. In *Philosophy*, pages 1–141. American Research Press, 1999.

[23] Haibin Wang, Florentin Smarandache, Yanqing Zhang, and Rajshekhar Sunderraman. *Single Valued Neutrosophic Sets.* Infinite study, 2010.

[24] Fei Wang and Xiaodong Zhao. Similarity and Pythagorean reliability measures of multivalued neutrosophic cubic set and its application to multiple-criteria decision-making. *International Journal of Intelligent Systems*, 37(1):105–134, 2022.

[25] Sumera Naz and Muhammad Aslam Malik. Single-valued neutrosophic line graphs. *TWMS Journal of Applied and Engineering Mathematics*, 8(2):483–494, 2018.

[26] Samina Ashraf, Sumera Naz, Hossein Rashmanlou, and M Aslam Malik. Regularity of graphs in single valued neutrosophic environment. *Journal of Intelligent & Fuzzy Systems*, 33(1):529–542, 2017.

[27] Rıdvan Şahin and Ahmet Küçük. Subsethood measure for single valued neutrosophic sets. *Journal of Intelligent & Fuzzy Systems*, 29(2):525–530, 2015.

[28] Irfan Deli and Said Broumi. Neutrosophic soft matrices and NSM-decision making. *Journal of Intelligent & Fuzzy Systems*, 28(5):2233–2241, 2015.

[29] Haibin Wang, Florentin Smarandache, Rajshekhar Sunderraman, and Yan-Qing Zhang. *Interval Neutrosophic Sets and Logic: Theory and Applications in Computing: Theory and Applications in Computing*, Volume 5. Infinite Study, 2005.

[30] Jun Ye. Multiple attribute group decision making based on interval neutrosophic uncertain linguistic variables. *International Journal of Machine Learning and Cybernetics*, 8:837–848, 2017.

[31] Sumera Naz, Syeda Saba Fatima, Shariq Aziz Butt, Mehwish Majeed, and Areej Fatima. A novel MAGDM approach for software quality assessment: A focus on microsoft devops transformation using 2-tuple linguistic single-valued neutrosophic set. In *Neutrosophic Paradigms: Advancements in Decision Making and Statistical Analysis: Neutrosophic Principles for Handling Uncertainty*, pages 403–432. Springer, 2025.

[32] Sumera Naz, Shariq Aziz Butt, Muhammad Muneeb ul Hassan, José Escorcia-Gutierrez, Areej Fatima, and Farhat ul Ain. A hybrid novel swara-ELECTRE-I method using probabilistic uncertain linguistic information for feature selection in image recognition. *Neurocomputing*, 612:128615, 2025.

[33] Xiaolu Long, Lizhi Liu, Can Xiao, Pengfei Cheng, and Chengxun Fu. Restoration methods selection for wood components of Chinese ancient architectures based on todim with single-valued neutrosophic sets. *Mathematical Problems in Engineering*, 2020(1):5049360, 2020.

[34] Shahzaib Ashraf, Saleem Abdullah, Florentin Smarandache, and Noor ul Amin. Logarithmic hybrid aggregation operators based on single valued neutrosophic sets and their applications in decision support systems. *Symmetry*, 11(3):364, 2019.

[35] Dongsheng Xu, Yanran Hong, and Kaili Xiang. A method of determining multi-attribute weights based on single-valued neutrosophic numbers and its application in todim. *Symmetry*, 11(4):506, 2019.

[36] Sumera Naz, Muhammad Akram, Afia Sattar, and Mohammed M Ali Al-Shamiri. 2-tuple linguistic q-rung orthopair fuzzy codas approach and its application in arc welding robot selection. *AIMS Math*, 7(9):17529–17569, 2022.

[37] Shengjun Wu, Jie Wang, Guiwu Wei, and Yu Wei. Research on construction engineering project risk assessment with some 2-tuple linguistic neutrosophic hamy mean operators. *Sustainability*, 10(5):1536, 2018.

[38] Jie Wang, Hui Gao, and Guiwu Wei. Some 2-tuple linguistic neutrosophic number muirhead mean operators and their applications to multiple attribute decision making. *Journal of Experimental & Theoretical Artificial Intelligence*, 31(3):409–439, 2019.

[39] Rudolf Julius Emanuel Clausius. *Ueber die wärmeleitung gasförmiger körper*. Verlag nicht ermittelbar, 1862.

[40] Rudolf Clausius. *The Mechanical Theory of Heat*. Macmillan, 1879.

[41] Büşra Meniz and E Mehmet Özkan. Vaccine selection for Covid-19 by AHP and novel VIKOR hybrid approach with interval type-2 fuzzy sets. *Engineering Applications of Artificial Intelligence*, 119:105812, 2023.

[42] Gourangajit Borah and Palash Dutta. Aggregation operators of quadripartitioned single-valued neutrosophic z-numbers with applications to diverse Covid-19 scenarios. *Engineering Applications of Artificial Intelligence*, 119:105748, 2023.

[43] Jingzhu Wei and Xiangyi Lin. The multiple attribute decision-making VIKOR method and its application. In *2008 4th international conference on wireless communications, networking and mobile computing*, pages 1–4. IEEE, 2008.

[44] Muhammad Akram, Sumera Naz, and Florentin Smarandache. Generalization of maximizing deviation and topsis method for MADM in simplified neutrosophic hesitant fuzzy environment. *Symmetry*, 11(8):1058, 2019.

[45] Sumera Naz, Aqsa Tasawar, Areej Fatima, Shariq Aziz Butt, and Zhoe Comas Gonzalez. An efficient 2-tuple linguistic cubic q-rung orthopair fuzzy cilos-topsis method: evaluating the hydrological geographical regions for watershed management in pakistan. *The Journal of Supercomputing*, 81(1):1–43, 2025.

# Chapter 9

## *Weighted Adjustable Level and Graded Utility Decision-Making Frameworks Based on Neutrosophic Soft Sets*

**Adem Yolcu and Taha Yasin Öztürk**

### 9.1 Introduction

Uncertainty is one of the most frequently encountered elements in today's decision-making problems. Especially in situations where multiple and conflicting criteria are evaluated simultaneously, it becomes difficult for decision-makers to reach rational and consistent results. Such complex problems are referred to as multi-criteria decision-making (MCDM) and have been solved using various methods for many years. Methods such as the analytic hierarchy process (AHP) [1], TOPSIS [2], and VIKOR [3] are widely used due to their systematic approaches. However, these classical models cannot always produce satisfactory results because real-world data is often incomplete, uncertain, or contradictory. These shortcomings have led researchers to more flexible structures that can directly model uncertainty. Fuzzy set theory, developed by Zadeh [4], has made it possible to define uncertain information using membership degrees. Subsequently, Atanassov [5] defined fuzzy sets and introduced the concept of "degree of hesitation" to measure uncertainty. Molodtsov [6] presented an approach based on parameters that does not require membership functions with soft set theory. Neutrosophic sets, proposed by Smarandache [7], provided a more detailed representation than previous theories by treating the components of truth, uncertainty, and falsity independently. As a synthesis of these theoretical developments, Maji and colleagues [8] developed the concept of neutrosophic soft set (NSS), and this structure has begun to yield highly effective results in multi-criteria decision problems. Over time, the NSS approach has been expanded with various extensions to become more comprehensive. Single-valued [9], interval-valued [10], bipolar [11],

DOI: 10.1201/9781003683339-9

Fermatean [12], and q-rung orthopair fuzzy sets [13] have made significant progress in defining different aspects of uncertainty. Additionally, fuzzy soft sets and Bonferroni-based aggregation operators [14] have increased preference flexibility in multi-criteria evaluations. The combination of interval-valued fuzzy soft sets with energy-based algorithms [15] reduces the effect of uncertainty, while intuitive fuzzy rough soft sets [16] extend the decision domain to a broader representation capability. With the development of the neutrosophic approach, richer and more systematic modeling forms have emerged. For example, complex single-valued neutrosophic soft sets (csvNSSs) [17] define uncertainty in greater detail, while matrix-based NSS models [18] enable the comprehensive application of classical decision rules such as Laplace or Savage. Energy-based decision models [19] propose new evaluation criteria by addressing both uncertainty and weighting elements together. Similarly, generalized $(m, n, o)$ T-spherical fuzzy sets developed with Hamacher joining operators [20] provide flexibility in group decision processes, while artificial intelligence-supported interval-valued approaches [21] integrate dynamic weighting systems into the MCDM structure. However, two fundamental gaps still stand out in the literature. First, most existing NSS and fuzzy soft-based methods focus either solely on binary elimination (pass/fail) or solely on continuous scoring (rating) approaches, failing to offer a structure that integrates both mechanisms. Second, reproducibility is limited because most models do not include open source code or accessible data sets.

This chapter proposes a unified decision-making framework based on a neutrosophic soft set to address these shortcomings. The proposed structure consists of two complementary algorithms:

- Weighted adjustable level-set screening (W-ALSS): It is a transparent and easily interpretable pre-screening mechanism that tests whether alternatives meet the specified criteria thresholds.
- Weighted graded utility fusion (W-GUF): It is a continuous scoring-based method that measures the extent to which alternatives meet the evaluation parameters using numerical scores.

By combining these two approaches, a comprehensive decision-making process has been achieved that can both apply strict constraints and reveal subtle distinctions. The application to renewable energy projects demonstrates the practical validity and explainability of the method. Furthermore, all steps have been explicitly coded in the Python environment, providing a transparent and reproducible analysis environment. The main contributions of the chapter can be summarized as follows:

1. A new decision-making approach based on NSS theory, combining binary and continuous evaluation, has been presented.
2. The reusability of the method has been increased thanks to step-by-step defined algorithms.

3. The Python-based application makes the method easily adaptable to different scenarios.

4. The example application in the field of renewable energy demonstrates the potential for integrating the method into practical decision-support systems.

In this context, a comparative overview of the proposed W-ALSS and W-GUF methods versus classical MCDM techniques is presented in Table 9.1.

This chapter consists of eight main sections. The first section outlines the general framework of the topic, discussing the fundamental approaches in the literature and the existing gaps. The second section presents the basic concepts and necessary definitions of neutrosophic soft set theory. The third and fourth sections present, in turn, the two proposed decision-making methods, W-ALSS and W-GUF in detail. The fifth section explains the application steps performed in the Python environment, while the sixth section presents a case study on renewable energy project selection. The seventh section includes sensitivity and comparison analyses, and the eighth section presents general evaluations, comparative results of the methods, and suggestions for future research. With this structure, the chapter aims to provide a comprehensive contribution to neutrosophic soft set-based decision-making models at both theoretical and practical levels.

---

## 9.2 Preliminaries

**Definition 9.1:** *[22]: Let $U$ be an initial universe set and $E$ be a set of parameters. Let $P(U)$ denote the set of all neutrosophic sets of $U$. Then, a neutrosophic soft set $(H, E)$ over $U$ is a set defined by a set valued function $F$ representing a mapping $F : E \to P(U)$ where $F$ is called approximate function of the neutrosophic soft set $(H, E)$. In other words, the neutrosophic soft set is a parameterized family of some elements of the set $P(U)$ and therefore it can be written as a set of ordered pairs,*

$$(H, E) = \{(e, < x, T_{F(e)}(x), I_{F(e)}(x), F_{F(e)}(x) >) : e \in E\}$$

*where , $T_{F(e)}(x), I_{F(e)}(x), F_{F(e)}(x) \in [0, 1]$, respectively, called the truth-membership, indeterminacy-membership, and falsity-membership function of $F(e)$. Since supremum of each $T$, $I$, $F$ is so, the inequality $0 \leq T_{F(e)}(x) + I_{F(e)}(x) + F_{F(e)}(x) \leq 3$ is obvious.*

**Definition 9.2:** *[23]: Let $(H, E)$ and $(G, E)$ be two neutrosophic soft sets over the universe set $U$. $(H, E)$ is said to be neutrosophic soft subset of $(G, E)$ if $T_{F(e)}(x) \leq T_{G(e)}(x)$, $I_{F(e)}(x) \leq I_{G(e)}(x)$ , $F_{F(e)}(x) \geq F_{G(e)}(x)$, $\forall e \in E$, $\forall x \in U$. It is denoted by $(H, E) \subseteq (G, E)$.*

TABLE 9.1: Comparison of the proposed methods (W-ALSS, W-GUF) with classical MCDM techniques

| **Criteria** | **W-ALSS** | **W-GUF** | **TOPSIS** [2] | **VIKOR** [3] | **MAUT** [1] |
|---|---|---|---|---|---|
| **Main Purpose** | Threshold-based screening using adjustable decision levels | Continuous scoring based on graded utilities | Ranks alternatives by distance from ideal and anti-ideal solutions | Provides compromise solutions balancing group utility and regret | Aggregates utilities based on preferences and probabilities |
| **Decision Type** | Binary pass/fail decisions | Fine-grained continuous ranking | Continuous ranking | Continuous ranking with compromise solution | Continuous ranking based on additive utilities |
| **Input Data Type** | NSS with T, I, F values | NSS with weighted graded utilities | Crisp or fuzzy numerical values | Crisp or fuzzy numerical values | Crisp probabilities and deterministic utilities |
| **Uncertainty Handling** | Explicitly models uncertainty via NSS thresholds | Handles uncertainty via graded neutrosophic utilities | Limited: requires defuzzification if fuzzy inputs exist | Moderate: integrates fuzzy numbers into optimization | Weak: uncertainty modeling not directly supported |
| **Weight Integration** | Weighted thresholds applied to screening levels | Weighted aggregation of graded utilities | Weights applied to distance calculations | Weights used in compromise programming | Weights multiplied directly with utilities |
| **Complexity** | Low to moderate (depends on number of thresholds) | Moderate (graded scoring increases calculations) | Low, $O(n \cdot m)$ | Moderate, $O(n \cdot m)$ with optimization steps | High if utility functions are non-linear |
| **Interpretability** | High: direct "accept/reject" decisions | High: intuitive continuous scoring | High: distance-based ranking is easy to interpret | Moderate: compromise interpretation can be subjective | Low to moderate: depends on elicitation of utilities |
| **Reproducibility** | Full Python implementation; threshold tuning | Full Python implementation with sensitivity analysis | Widely reproducible libraries available | Moderate reproducibility | Limited due to utility elicitation |
| **Best Application Scenarios** | Regulatory compliance, quality control, initial screening | Strategic planning, prioritization, resource allocation | Supplier selection, ranking universities, benchmarking | Conflict resolution, compromise-based decisions | Investment analysis, risk-based decision-making |
| **Limitations** | Less effective when nuanced ranking is required | More computation than W-ALSS | Assumes independence, scaling sensitive | Requires careful choice of $v$ | Utility elicitation subjective/error-prone |

*$(H, E)$ is said to be neutrosophic soft equal to $(G, E)$ if $(H, E)$ is neutrosophic soft subset of $(G, E)$ and $(G, E)$ is neutrosophic soft subset of $(H, E)$. It is denoted by $(H, E) = (G, E)$.*

**Definition 9.3:** *[24]: Let $(H_1, E)$ and $(H_2, E)$ be two neutrosophic soft sets over the universe set $U$. Then their union is denoted by $(H_1, E) \cup (H_2, E) = (H_3, E)$ and is defined by:*

$$(H_3, E) = \{(e, < x, T_{H_3(e)}(x), I_{H_3(e)}(x), F_{H_3(e)}(x) >: x \in U) : e \in E\}.$$

*where*

$$\begin{aligned} T_{H_3(e)}(x) &= \max\{T_{H_1(e)}(x), T_{H_2(e)}(x)\} \\ I_{H_3(e)}(x) &= \max\{I_{H_1(e)}(x), I_{H_2(e)}(x)\} \\ F_{H_3(e)}(x) &= \min\{F_{H_1(e)}(x), F_{H_2(e)}(x)\}. \end{aligned}$$

**Definition 9.4:** *[24]: Let $(H_1, E)$ and $(H_2, E)$ be two neutrosophic soft sets over the universe set $U$. Then their intersection is denoted by $(H_1, E) \cap (H_2, E) = (H_3, E)$ and is defined by:*

$$(H_3, E) = \{(e, < x, T_{H_3(e)}(x), I_{H_3(e)}(x), F_{H_3(e)}(x) >: x \in U) : e \in E\}.$$

*where*

$$\begin{aligned} T_{H_3(e)}(x) &= \min\{T_{H_1(e)}(x), T_{H_2(e)}(x)\} \\ I_{H_3(e)}(x) &= \min\{I_{H_1(e)}(x), I_{H_2(e)}(x)\} \\ F_{H_3(e)}(x) &= \max\{F_{H_1(e)}(x), F_{H_2(e)}(x)\}. \end{aligned}$$

**Definition 9.5:** *[25]: Let $(H, A)$ be a neutrosophic soft set over $U$, where $A \subseteq E$ and $E$ is a parameter set. For $s, t, h \in [0, 1]$, the $(s, t, h)-$level neutrosophic soft set of $(H, A)$ is a crisp soft set $L((H, A); s, t, h) = (H_{(s,t,h)}, A)$ defined by*

$$\begin{aligned} H_{(s,t,h)}(e) &= L(H(e); s, t, h) \\ &= \{x \in U : T_{H(e)}(x) \geq s,\ I_{H(e)}(x) \geq t \text{ and } F_{H(e)}(x) \leq h\} \end{aligned}$$

*for all $e \in A$.*

**Definition 9.6:** *[25]: Let $(H, A)$ be a neutrosophic soft set over $U$, where $A \subseteq E$ and $E$ is a parameter set. Let $\lambda : A \to [0, 1] \times [0, 1] \times [0, 1]$ be a neutrosophic set in $A$ which is called a threshold neutrosophic set. The level soft set of $(H, A)$ with respect to $\lambda$ is a crisp soft set $L((H, A); \lambda) = (F_\lambda, A)$ defined by $F_\lambda(e) = L(H(e); \lambda(e)) = \{x \in U : T_{H(e)}(x) \geq T_\lambda(e),\ I_{H(e)}(x) \geq I_\lambda(e) \text{ and } F_{H(e)}(x) \leq F_\lambda(e)\}$ for all $e \in A$.*

**Definition 9.7:** *[25]: Let $(H, A)$ be a neutrosophic soft set over $U$, where $A \subseteq E$ and $E$ is a parameter set. Based on the neutrosophic soft set $(H, A)$, we can define a neutrosophic set $mid_{(H,A)} : A \to [0,1] \times [0,1] \times [0,1]$ by*

$$
\begin{aligned}
T_{mid_{(H,A)}}(e) &= \frac{1}{|U|} \sum_{x \in U} T_{H(e)}(x),\ I_{mid_{(H,A)}}(e) = \frac{1}{|U|} \sum_{x \in U} I_{H(e)}(x) \text{ and} \\
F_{mid_{(H,A)}}(e) &= \frac{1}{|U|} \sum_{x \in U} F_{H(e)}(x)
\end{aligned}
$$

*for all $e \in A$. The mid-threshold of the neutrosophic soft set $(H, A)$ is known as the neutrosophic set $mid_{(H,A)}$. Furthermore, the mid-level soft set of $(H, A)$ is identified as $L((H, A); mid_{(H,A)})$, which is the level soft set of $(H, A)$ with respect to the mid-threshold neutrosophic set $mid_{(H,A)}$. The mid-level decision rule that follows refers to the use of the mid-threshold and the mid-level soft set in decision-making based on neutrosophic soft sets.*

**Definition 9.8:** *[25]: Let $(H, A)$ be a neutrosophic soft set over $U$, where $A \subseteq E$ and $E$ is a parameter set. Based on the neutrosophic soft set $(H, A)$, we can define a neutrosophic set $topbottom_{(H,A)} : A \to [0,1] \times [0,1] \times [0,1]$ by*

$$
\begin{aligned}
T_{topbottom_{(H,A)}}(e) &= \max_{x \in U} T_{H(e)}(x), I_{topbottom_{(H,A)}}(e) = \max_{x \in U} I_{H(e)}(x) \text{ and} \\
F_{topbottom_{(H,A)}}(e) &= \min_{x \in U} F_{H(e)}(x) \text{ for all } e \in A.
\end{aligned}
$$

*We can define a neutrosophic set $toptop_{(H,A)} : A \to [0,1] \times [0,1] \times [0,1]$ by*

$$
\begin{aligned}
T_{toptop_{(H,A)}}(e) &= \max_{x \in U} T_{H(e)}(x), I_{toptop_{(H,A)}}(e) = \max_{x \in U} I_{H(e)}(x) \text{ and} \\
F_{toptop_{(H,A)}}(e) &= \max_{x \in U} F_{H(e)}(x) \text{ for all } e \in A.
\end{aligned}
$$

*We can define a neutrosophic set $bottombottom_{(H,A)} : A \to [0,1] \times [0,1] \times [0,1]$ by*

$$
\begin{aligned}
T_{bottombottom_{(H,A)}}(e) &= \min_{x \in U} T_{H(e)}(x), I_{bottombottom_{(H,A)}}(e) = \min_{x \in U} I_{H(e)}(x) \text{ and} \\
F_{bottombottom_{(H,A)}}(e) &= \min_{x \in U} F_{H(e)}(x) \text{ for all } e \in A.
\end{aligned}
$$

*We can define a neutrosophic set $bottomtop_{(H,A)} : A \to [0,1] \times [0,1] \times [0,1]$ by*

$$
\begin{aligned}
T_{bottomtop_{(H,A)}}(e) &= \min_{x \in U} T_{H(e)}(x), I_{bottomtop_{(H,A)}}(e) = \min_{x \in U} I_{H(e)}(x) \text{ and} \\
F_{bottomtop_{(H,A)}}(e) &= \max_{x \in U} F_{H(e)}(x) \text{ for all } e \in A.
\end{aligned}
$$

*The neutrosophic set $topbottom_{(H,A)}$ is called the top-bottom-threshold of the neutrosophic soft set $(H, A)$; the neutrosophic set $toptop_{(H,A)}$ is called the top-top-threshold of the neutrosophic soft set $(H, A)$; the neutrosophic set $bottombottom_{(H,A)}$ is called the bottom-bottom-threshold of the neutrosophic soft set $(H, A)$. Also, the level soft set of $(H, A)$ w.r.t.*

*the top-bottom-threshold of the neutrosophic set* $topbottom_{(H,A)}$, *namely* $L((H,A);topbottom_{(H,A)})$ *is called top-bottom-level soft set of* $(H,A)$ *and simply denoted as* $L((H,A);topbottom)$. *The level soft set of* $(H,A)$ *w.r.t. the top-top-threshold of the neutrosophic set* $toptop_{(H,A)}$, *namely* $L((H,A);toptop_{(H,A)})$ *is called top-top-level soft set of* $(H,A)$ *and simply denoted as* $L((H,A);toptop)$. *The level soft set of* $(H,A)$ *w.r.t. the bottom-bottom-threshold of the neutrosophic set* $bottombottom_{(H,A)}$, *namely* $L((H,A);bottombottom_{(H,A)})$ *is called bottom-bottom-level soft set of* $(H,A)$ *and simply denoted as* $L((H,A);bottombottom)$.*The level soft set of* $(H,A)$ *w.r.t. the bottom-top-threshold of the neutrosophic set* $bottomtop_{(H,A)}$, *namely* $L((H,A);bottomtop_{(H,A)})$ *is called bottom-top-level soft set of* $(H,A)$ *and simply denoted as* $L((H,A);bottomtop)$.

*The use of the top-bottom threshold and the top-bottom soft set in neutrosophic soft-sets based decision-making is what the top-bottom-level decision rule entails in the following. Using the top-top-threshold and taking into account the top-top-level soft set in neutrosophic soft sets-based decision-making is what the top-top-level decision rule entails. When making decisions based on neutrosophic soft sets, the bottom-bottom-level decision rule will need applying the bottom-bottom-threshold and taking the bottom-bottom-level soft set into account.*

**Theorem 9.1:** *[25]: Let* $(H,A)$ *be a neutrosophic soft set over* $U$, *where* $A \subseteq E$ *and* $E$ *is a parameter set.* $L((H,A);s_1,t_1,h_1)$ *and* $L((H,A);s_2,t_2,h_2)$ *are* $(s_1,t_1,h_1)$ *level neutrosophic soft set and* $(s_2,t_2,h_2)$ *level neutrosophic soft set of* $(H,A)$, *respectively, where* $s_1,t_1,h_1,s_2,t_2,h_2 \in [0,1]$. *If* $s_2 \geq s_1, t_2 \geq t_1$ *and* $h_2 \leq h_1$, *then we have* $L((H,A);s_2,t_2,h_2) \subseteq L((H,A);s_1,t_1,h_1)$.

**Theorem 9.2:** *[25]: Let* $(H,A)$ *and* $(G,B)$ *be neutrosophic soft sets over* $U$, *where* $A,B \subseteq E$ *and* $E$ *is a parameter set.* $L((H,A);s,t,h)$ *and* $L((G,B);s,t,h)$ *are* $(s,t,h)$ *level neutrosophic soft sets of* $(H,A)$ *and* $(G,B)$, *respectively, where* $s,t,h \in [0,1]$. *If* $(H,A) \subseteq (G,B)$, *then we have* $L((H,A);s,t,h) \subseteq L((G,B);s,t,h)$.

**Theorem 9.3:** *Let* $(H,A)$ *be a neutrosophic soft set over the universe* $U$, *where* $A \subseteq E$ *and* $E$ *is the set of parameters. Let* $L((H,A);\lambda)$ *denote the level soft set of* $(H,A)$ *based on the threshold vector* $\lambda = (s,t,h)$, *where* $s,t,h \in [0,1]$.

*For each parameter* $e_j \in A$, *define four different threshold configurations as follows:*

- *Top-Top Threshold* ($\lambda_{tt}$): $s_j = \max T(e_j)$, $t_j = \max I(e_j)$, $h_j = \max F(e_j)$.
- *Top-Bottom Threshold* ($\lambda_{tb}$): $s_j = \max T(e_j)$, $t_j = \max I(e_j)$, $h_j = \min F(e_j)$.
- *Bottom-Bottom Threshold* ($\lambda_{bb}$): $s_j = \min T(e_j)$, $t_j = \min I(e_j)$, $h_j = \min F(e_j)$.
- *Mid-Level Threshold* ($\lambda_{mid}$): $s_j = \mathrm{avg}(T(e_j))$, $t_j = \mathrm{avg}(I(e_j))$, $h_j = \mathrm{avg}(F(e_j))$.

**TABLE 9.2**: Threshold configurations and inclusion relationships for $I \geq t$ case.

| **Threshold type** | **T** | **I** | **F** | **Strictness level** |
|---|---|---|---|---|
| Top-Top ($\lambda_{tt}$) | max($T$) | max($I$) | max($F$) | Most relaxed |
| Top-Bottom ($\lambda_{tb}$) | max($T$) | max($I$) | min($F$) | Moderately strict |
| Mid-Level ($\lambda_{mid}$) | avg($T$) | avg($I$) | avg($F$) | Medium strictness |
| Bottom-Bottom ($\lambda_{bb}$) | min($T$) | min($I$) | min($F$) | Most strict |

*Assume the acceptance rule for any alternative $x_i \in U$ under parameter $e_j$ is given by:*

$$T_{ij} \geq s_j, \quad I_{ij} \geq t_j, \quad F_{ij} \leq h_j,$$

*where higher $T$ and $I$ values are preferred, and lower $F$ values are desirable.*

*Then, the following inclusion relationships hold among the level soft sets:*

$$L((H,A);\lambda_{tb}) \subseteq L((H,A);\lambda_{mid}),$$
$$L((H,A);\lambda_{tb}) \subseteq L((H,A);\lambda_{tt}),$$
$$L((H,A);\lambda_{tb}) \subseteq L((H,A);\lambda_{bb}).$$

**Remark 9.1:** *The top-top threshold configuration ($\lambda_{tt}$) uses the maximum permissible values for $T$, $I$, and $F$, which make it the most relaxed threshold and result in selecting the largest possible subset of alternatives. The top-bottom configuration ($\lambda_{tb}$) is stricter than top-top because it still requires high $T$ and $I$ values but enforces a low $F$ threshold. The mid-level thresholds ($\lambda_{mid}$) use average values and therefore produce an inclusion set size between $\lambda_{tb}$ and $\lambda_{bb}$. Finally, the bottom-bottom configuration ($\lambda_{bb}$) is the most restrictive, selecting only those alternatives where $T$ and $I$ are sufficiently high and $F$ is minimal.*

*Table 9.2 presents the threshold configurations and the corresponding inclusion relationships for the $I \geq t$ scenario.*

**Definition 9.9:** *Let $(H, A)$ be a neutrosophic soft set over the universe $U$, where $A = \{e_1, e_2, \ldots, e_n\}$ is the set of parameters and*

$$H(e_j) = \Big\{\langle x,\, T_F(e_j)(x),\, I_F(e_j)(x),\, F_F(e_j)(x)\rangle : x \in U\Big\}.$$

*For each parameter $e_j$, let $(s_j, t_j, h_j)$ be the threshold vector, where:*

- *$s_j$ is the minimum acceptable truth-membership degree,*
- *$t_j$ is the minimum required indeterminacy-membership degree,*
- *$h_j$ is the maximum allowable falsity-membership degree.*

*Given the acceptance rule:*

$$T_{ij} \geq s_j, \qquad I_{ij} \geq t_j, \qquad F_{ij} \leq h_j,$$

*where higher $T$ and $I$ values are preferred and lower $F$ values are desirable.*

*The graded component utilities of an alternative $a_i$ with respect to parameter $e_j$ are defined as:*

$$u_{ij}^T = \begin{cases} \dfrac{T_{ij} - s_j}{1 - s_j}, & T_{ij} \geq s_j, \\ 0, & T_{ij} < s_j, \end{cases} \qquad u_{ij}^I = \begin{cases} \dfrac{I_{ij} - t_j}{1 - t_j}, & I_{ij} \geq t_j, \\ 0, & I_{ij} < t_j, \end{cases}$$

$$u_{ij}^F = \begin{cases} \dfrac{h_j - F_{ij}}{h_j}, & F_{ij} \leq h_j, \\ 0, & F_{ij} > h_j. \end{cases}$$

*The per-criterion graded utility is then obtained using a component-weighted geometric mean:*

$$U_{ij} = \left(u_{ij}^T\right)^{\alpha_T} \left(u_{ij}^I\right)^{\alpha_I} \left(u_{ij}^F\right)^{\alpha_F}, \qquad \alpha_T, \alpha_I, \alpha_F \in [0,1], \quad \alpha_T + \alpha_I + \alpha_F = 1.$$

*Finally, the overall graded utility score of $a_i$ is computed by aggregating across all parameters using the criterion weights $w_j$:*

$$S(a_i) = \sum_{j=1}^{n} w_j \, U_{ij}, \qquad w_j \geq 0, \quad \sum_{j=1}^{n} w_j = 1.$$

**Remark 9.2:** *Definition 9.9 generalizes the classical $(s,t,h)$-level soft set by introducing continuous utility degrees instead of binary selection $(0/1)$. If $U_{ij} = 1$, the alternative fully satisfies parameter $e_j$; if $U_{ij} = 0$, it completely fails. The formulation integrates concepts from multi-attribute utility theory (MAUT) [26] and normalized benefit scoring [2], adapted to neutrosophic environments.*

---

## 9.3 Weighted Decision-Making Framework Based on Neutrosophic Soft Sets

### 9.3.1 Weighted adjustable level soft set (W-ALSS)

In decision-making environments involving uncertainty, transforming incomplete, ambiguous, or conflicting data into meaningful and interpretable results requires powerful analytical tools. Developed to address this need, the W-ALSS approach extends the classical neutrosophic soft set structure, offering a flexible evaluation mechanism supported by both adjustable threshold values and criterion weights. The W-ALSS method is designed especially for situations where decision options must be tested for meeting certain conditions before proceeding to the final ranking stage. Thus, the method acts as a reliable filter in the initial stage of complex and multidimensional decision problems, advancing only those alternatives that meet the specified standards to subsequent evaluation steps. This structure both increases the transparency of the decision-making process and enables subsequent analyses to be conducted in a more focused and meaningful manner. In this sense, W-ALSS performs as a transparent filtering mechanism that preserves the interpretability of neutrosophic information while simplifying the early evaluation stage. Let $(H, A)$ be a neutrosophic soft set over the universe $U$, where $A = \{e_1, e_2, \ldots, e_n\}$ is the set of parameters and:

$$F(e_j) = \{\langle \Theta, T_{F(e_j)}(\Theta), I_{F(e_j)}(\Theta), F_{F(e_j)}(\Theta)\rangle : \Theta \in U\}.$$

For each parameter $e_j$, a threshold vector $(s_j, t_j, h_j)$ is defined, where:

- $s_j$ = minimum acceptable truth-membership,
- $t_j$ = minimum required indeterminacy-membership,
- $h_j$ = maximum allowable falsity-membership.

For each alternative $a_i$ and parameter $e_j$, the binary decision value $L_{ij}$ is computed as:

$$L_{ij} = \begin{cases} 1, & \text{if } T_{ij} \geq s_j,\ I_{ij} \geq t_j,\ F_{ij} \leq h_j \\ 0, & \text{otherwise} \end{cases}.$$

The weighted score of an alternative $a_i$ is calculated as:

$$S(a_i) = \sum_{j=1}^{n} w_j \cdot L_{ij},$$

where $w_j$ is the weight of parameter $e_j$ and $\sum_{j=1}^{n} w_j = 1$. Alternatives are ranked in descending order according to $S(a_i)$.

**Algorithm Steps**

1. Input: Universe of alternatives $U = \{a_1, a_2, \ldots, a_m\}$, parameter set $A = \{e_1, e_2, \ldots, e_n\}$, neutrosophic decision matrix $M = [(T_{ij}, I_{ij}, F_{ij})]$, parameter weights $w_j$, adjustable thresholds $(s_j, t_j, h_j)$.

2. Construct the Decision Matrix: Build the neutrosophic decision matrix $M$ containing the values $(T_{ij}, I_{ij}, F_{ij})$.

3. Apply Thresholds: For each $a_i$ and $e_j$, determine $L_{ij}$ based on $(s_j, t_j, h_j)$.

4. Calculate Weighted Scores: Compute $S(a_i) = \sum_{j=1}^{n} w_j \cdot L_{ij}$ for all alternatives.

5. Rank all alternatives in descending order of $S(a_i)$.

6. Output: Weighted scores $S(a_i)$ and ranking of alternatives.

**Remark 9.3:** *The W-ALSS algorithm:*

- *Integrates adjustable thresholds and parameter weights.*
- *Provides a flexible structure for multi-criteria decision-making.*
- *Produces an interpretable binary decision matrix.*
- *However, it does not distinguish between alternatives that just meet thresholds and those that exceed them significantly.*

**Example 9.1:** *Input: Consider* $m = 4$ *alternatives* $A_1, \ldots, A_4$ *and* $n = 3$ *criteria* $e_1, e_2, e_3$. *The neutrosophic decision matrix is* $M = [(T_{ij}, I_{ij}, F_{ij})]$. *Let the adjustable thresholds be*

$$(s_j, t_j, h_j) = (0.6, 0.5, 0.3) \quad \textit{for each } j \in \{1, 2, 3\},$$

*and the criterion weights be*

$$w = (w_1, w_2, w_3) = (0.40, 0.35, 0.25), \qquad \sum_{j=1}^{3} w_j = 1.$$

*Step 1: Neutrosophic decision matrix M*

| ***Alt.*** | $e_1$ | $e_2$ | $e_3$ |
|---|---|---|---|
| $A_1$ | (0.72, 0.55, 0.22) | (0.68, 0.50, 0.28) | (0.58, 0.52, 0.18) |
| $A_2$ | (0.65, 0.48, 0.25) | (0.60, 0.55, 0.40) | (0.75, 0.51, 0.20) |
| $A_3$ | (0.58, 0.60, 0.20) | (0.70, 0.45, 0.22) | (0.64, 0.50, 0.34) |
| $A_4$ | (0.62, 0.52, 0.35) | (0.66, 0.60, 0.26) | (0.62, 0.49, 0.22) |

*Step 2: Level decision matrix L (pass/fail)*
*For W-ALSS, an entry passes if* $T_{ij} \geq s_j$, $I_{ij} \geq t_j$, *and* $F_{ij} \leq h_j$. *Thus,*

$$L_{ij} = \begin{cases} 1, & T_{ij} \geq 0.6,\ I_{ij} \geq 0.5,\ F_{ij} \leq 0.3, \\ 0, & \textit{otherwise}. \end{cases}$$

*Evaluating each cell:*

| ***Alt.*** | $e_1$ | $e_2$ | $e_3$ |
|---|---|---|---|
| $A_1$ | 1 | 1 | 0 |
| $A_2$ | 0 | 0 | 1 |
| $A_3$ | 0 | 0 | 0 |
| $A_4$ | 0 | 1 | 0 |

*Step 3: Weighted scores and ranking*

$$S(A_i) = \sum_{j=1}^{3} w_j\, L_{ij}.$$

*Hence*

$$\begin{aligned} S(A_1) &= 0.40 \cdot 1 + 0.35 \cdot 1 + 0.25 \cdot 0 = 0.75, \\ S(A_2) &= 0.40 \cdot 0 + 0.35 \cdot 0 + 0.25 \cdot 1 = 0.25, \\ S(A_3) &= 0.00, \\ S(A_4) &= 0.40 \cdot 0 + 0.35 \cdot 1 + 0.25 \cdot 0 = 0.35. \end{aligned}$$

*Step 4: Final ranking*

$$A_1 \succ A_4 \succ A_2 \succ A_3.$$

### 9.3.2 Weighted graded utility fusion (W-GUF)

The W-GUF method extends level-based decision rules by replacing binary pass/fail outcomes with *graded utilities* in $[0,1]$. Each alternative $a_i$ is evaluated under parameters $e_j$ with neutrosophic tuples $(T_{ij}, I_{ij}, F_{ij}) \in [0,1]^3$. For each parameter $e_j$, a threshold vector $(s_j, t_j, h_j)$ and a parameter weight $w_j$ are given. Component-importance weights $\alpha_T, \alpha_I, \alpha_F \geq 0$ with $\alpha_T + \alpha_I + \alpha_F = 1$ control the relative impact of truth, indeterminacy, and falsity.

Let $(H, A)$ be the neutrosophic soft representation over universe $U$, with parameters $A = \{e_1, \dots, e_n\}$ and alternatives $U = \{a_1, \dots, a_m\}$. For thresholds $(s_j, t_j, h_j)$:

For each $i, j$,

$$u_{ij}^T = \begin{cases} \dfrac{T_{ij} - s_j}{1 - s_j}, & T_{ij} \geq s_j, \\ 0, & T_{ij} < s_j, \end{cases} \qquad u_{ij}^I = \begin{cases} \dfrac{I_{ij} - t_j}{1 - t_j}, & I_{ij} \geq t_j,\ t_j > 0, \\ 0, & I_{ij} < t_j \text{ or } t_j = 0, \end{cases}$$

$$u_{ij}^F = \begin{cases} \dfrac{h_j - F_{ij}}{h_j}, & F_{ij} \leq h_j,\ h_j > 0, \\ 0, & F_{ij} > h_j \text{ or } h_j = 0. \end{cases}$$

*Per-parameter graded utility (weighted geometric mean)*

$$U_{ij} = \left(u_{ij}^T\right)^{\alpha_T} \left(u_{ij}^I\right)^{\alpha_I} \left(u_{ij}^F\right)^{\alpha_F}.$$

*Alternative score (parameter-weighted aggregation)*

$$S(a_i) = \sum_{j=1}^{n} w_j\, U_{ij}, \qquad \sum_{j=1}^{n} w_j = 1, \quad w_j \geq 0.$$

Alternatives are ranked in descending order of $S(a_i)$.

**Algorithm Steps:**

- Input:
  - Neutrosophic matrix $M = [(T_{ij}, I_{ij}, F_{ij})]$ for $i = 1..m$, $j = 1..n$.
  - Thresholds $(s_j, t_j, h_j)$ for each parameter $e_j$.
  - Parameter weights $w_j$ with $\sum_j w_j = 1$.
  - Component weights $(\alpha_T, \alpha_I, \alpha_F)$ with $\alpha_T + \alpha_I + \alpha_F = 1$.
- Compute Component Utilities: For all $i, j$, compute $u_{ij}^T, u_{ij}^I, u_{ij}^F$ using the piecewise definitions above.
- Fuse Components Per Parameter: For all $i, j$, compute $U_{ij} = (u_{ij}^T)^{\alpha_T}(u_{ij}^I)^{\alpha_I}(u_{ij}^F)^{\alpha_F}$.
- Aggregate across parameters: For all $i$, compute $S(a_i) = \sum_{j=1}^{n} w_j U_{ij}$.
- Ranking: Sort alternatives by $S(a_i)$ in descending order. Report scores and ranking.

**Remark 9.4:**

- If any threshold has $t_j = 0$ or $h_j = 0$, the corresponding utility terms default to 0 unless $I_{ij} = 0$ or $F_{ij} = 0$ respectively; this enforces strict compliance.
- Replacing the geometric mean with min yields a more conservative fusion; replacing it with a weighted arithmetic mean yields a more compensatory behavior.
- Sensitivity analysis can be performed by varying $(s_j, t_j, h_j)$, $w_j$, or $(\alpha_T, \alpha_I, \alpha_F)$.

**Example 9.2:** *Step 1: Data and thresholds alternatives: $A_1, \ldots, A_4$; Criteria: $e_1, e_2, e_3$.*
*Neutrosophic matrix $M = [(T_{ij}, I_{ij}, F_{ij})]$:*

| ***Alt.*** | $e_1$ | $e_2$ | $e_3$ |
|---|---|---|---|
| $A_1$ | (1.00, 0.25, 0.15) | (0.80, 0.00, 0.15) | (0.60, 0.25, 0.00) |
| $A_2$ | (0.80, 0.50, 0.30) | (1.00, 0.00, 0.00) | (0.80, 0.25, 0.15) |
| $A_3$ | (0.60, 0.00, 0.00) | (0.80, 0.25, 0.00) | (1.00, 0.25, 0.15) |
| $A_4$ | (1.00, 0.50, 0.30) | (0.80, 0.00, 0.30) | (1.00, 0.00, 0.00) |

*Thresholds : $T_{ij} \geq s_j$, $I_{ij} \geq t_j$, $F_{ij} \leq h_j$.*
*Original (Example 9.2) $s, h$: $(s_1, s_2, s_3) = (0.6, 0.6, 0.6)$, $(h_1, h_2, h_3) = (0.3, 0.3, 0.3)$.*
*$\lambda_{mid}$ for $t$ (column-wise mean of $I$):*

$$(t_1, t_2, t_3) = \left(\frac{0.25+0.50+0.00+0.50}{4}, \frac{0.00+0.00+0.25+0.00}{4}, \frac{0.25+0.25+0.25+0.00}{4}\right)$$
$$= (0.3125,\ 0.0625,\ 0.1875).$$

*Parameter weights:* $w = (0.40,\ 0.35,\ 0.25)$ *with* $\sum_j w_j = 1$.
*Component weights (geometric mean):* $(\alpha_T, \alpha_I, \alpha_F) = \left(\frac{1}{3}, \frac{1}{3}, \frac{1}{3}\right)$.

*Step 2: Utilities.*

$$u_{ij}^T = \begin{cases} \dfrac{T_{ij} - s_j}{1 - s_j}, & T_{ij} \geq s_j, \\ 0, & \textit{otherwise}, \end{cases} \qquad u_{ij}^I = \begin{cases} \dfrac{I_{ij} - t_j}{1 - t_j}, & I_{ij} \geq t_j, \\ 0, & \textit{otherwise}, \end{cases}$$

$$u_{ij}^F = \begin{cases} \dfrac{h_j - F_{ij}}{h_j}, & F_{ij} \leq h_j, \\ 0, & \textit{otherwise}. \end{cases}$$

*Per-criterion fusion:* $U_{ij} = (u_{ij}^T u_{ij}^I u_{ij}^F)^{1/3}$. *Score:* $S(a_i) = \sum_{j=1}^{3} w_j U_{ij}$.

*Step 3: Per-criterion utilities. Using* $s = (0.6, 0.6, 0.6)$, $t = (0.3125, 0.0625, 0.1875)$, $h = (0.3, 0.3, 0.3)$, *the nonzero* $U_{ij}$ *entries are:*

$U_{23} = (u^T, u^I, u^F) = (0.5,\ 0.0816327,\ 0.5) \Rightarrow (0.5 \cdot 0.0816327 \cdot 0.5)^{1/3} = 0.26792$,
$U_{32} = (0.5,\ 0.2133333,\ 1.0) \Rightarrow (0.5 \cdot 0.2133333 \cdot 1.0)^{1/3} = 0.46416$,
$U_{33} = (1.0,\ 0.1230769,\ 0.5) \Rightarrow (1.0 \cdot 0.1230769 \cdot 0.5)^{1/3} = 0.33755$.
*All other* $U_{ij} = 0$.

*Step 4:* $U_{ij}$*-matrix and scores.*

$$U = \begin{pmatrix} 0 & 0 & 0 \\ 0 & 0 & 0.26792 \\ 0 & 0.46416 & 0.33755 \\ 0 & 0 & 0 \end{pmatrix},$$

$$\begin{aligned} S(A_1) &= 0.40 \cdot 0 + 0.35 \cdot 0 + 0.25 \cdot 0 = 0 \\ S(A_2) &= 0.40 \cdot 0 + 0.35 \cdot 0 + 0.25 \cdot 0.26792 = 0.06698 \\ S(A_3) &= 0.40 \cdot 0 + 0.35 \cdot 0.46416 + 0.25 \cdot 0.33755 = 0.24684 \\ S(A_4) &= 0.40 \cdot 0 + 0.35 \cdot 0 + 0.25 \cdot 0 = 0 \end{aligned}$$

*Step 5: Ranking.*

$$\boxed{A_3 \succ A_2 \succ A_1 \equiv A_4.}$$

---

## 9.4 Practical Implementation in Python

This section shows how the proposed methods—Algorithm 1 (W-ALSS) and Algorithm 2 (W-GUF)—can be executed in a small Python script. The aim is to clarity rather than build a full software package: all inputs are placed directly in the code, and the procedures follow the same sequence introduced in the theory. With this setup, one can try different scenarios and run quick sensitivity checks without any external files or special libraries.

### 9.4.1 Data definition

To demonstrate the applicability of the method, an example scenario was created by evaluating four alternatives ($A_1$–$A_4$) under three different criteria ($e_1, e_2, e_3$). Each alternative was represented by neutrosophic triples consisting of accuracy ($T$), uncertainty ($I$), and falsity ($F$) components. Furthermore, the weight vector $w_j$ for the criteria and the threshold values ($s_j, t_j, h_j$) are defined within the model's own structure, thus presenting the method in a completely independent and internally integrated form.

```
# Define neutrosophic decision matrix
M = [
[(0.72,0.55,0.22), (0.68,0.50,0.28), (0.58,0.52,0.18)],  # A1
[(0.65,0.48,0.25), (0.60,0.55,0.40), (0.75,0.51,0.20)],  # A2
[(0.58,0.60,0.20), (0.70,0.45,0.22), (0.64,0.50,0.34)],  # A3
[(0.62,0.52,0.35), (0.66,0.60,0.26), (0.62,0.49,0.22)],  # A4
]
# Parameter weights and thresholds
w = [0.40, 0.35, 0.25]
s = [0.60, 0.60, 0.60]
t = [0.50, 0.50, 0.50]
h = [0.30, 0.30, 0.30]
```

### 9.4.2 Implementation of Algorithm 1 (W-ALSS)

At this stage, Algorithm 1 applies a simple yet effective binary decision rule to evaluate each alternative. For each alternative–criterion pair, the defined neutrosophic triplet $(T, I, F)$ is compared with the relevant threshold values $(s, t, h)$. If the degree of truth exceeds its own limit while the uncertainty and falsity values remain below the specified thresholds, the condition for this criterion is considered satisfied ($L_{ij} = 1$); otherwise, it is considered unsatisfied ($L_{ij} = 0$). Thus, the decision matrix, initially defined in a fuzzy manner, is converted into a binary form indicating whether each condition is met or not.

The short Python function below automatically performs this evaluation process. The function iterates over all alternatives and criteria, updating the cumulative scores at each step, and finally returns a list containing the weighted elimination results. This structure ensures that the method is both repeatable and practical to use.

```
def w_alss(M, w, s, t, h):
m, n = len(M), len(M[0])
scores = [0.0] * m
for i in range(m):
for j in range(n):
T, I, F = M[i][j]
```

```
passed = (T >= s[j]) and (I >= t[j]) and (F <= h[j])
scores[i] += w[j] * int(passed)
return scores

scores_alss = w_alss(M, w, s, t, h)
print("W-ALSS Scores:", scores_alss)
```

Each line of code directly reflects the logic of the mathematical definition. Within this structure, the passed variable stores the result of the threshold check performed for each criterion, while the `scores_alss` list contains the weighted results for all alternatives. The output values obtained range from 0 to 1, showing both how many criteria were successfully met and the effect of each criterion's weight on the overall score. This simple and systematic structure increases the transparency of the algorithm, while also making it easily adaptable when necessary. Thus, the scores obtained in the W-ALSS phase clearly reveal the first-level suitability analysis of the alternatives before proceeding to the subsequent W-GUF evaluation.

### 9.4.3 Implementation of Algorithm 2 (W-GUF)

Algorithm 2 follows a similar logical structure to W-ALSS but eliminates the rigid pass/fail rule, instead offering a flexible evaluation approach based on graded benefit values. Each criterion is defined by its own neutrosophic components $(T, I, F)$, which contribute to the overall result based on how close these values are to the relevant threshold levels $(s, t, h)$. This allows alternatives to be quantitatively evaluated based on the extent to which they meet the conditions, rather than simply being classified as "suitable" or "unsuitable." Such a structure provides a more continuous, realistic, and discriminative performance measurement by also taking partial suitability into account.

```
def guf_components(T, I, F, s, t, h, eps=1e-12):
uT = max(0.0, (T - s) / max(1 - s, eps)) if T >= s else 0.0
uI = max(0.0, (I - t) / max(1 - t, eps)) if I >= t else 0.0
uF = max(0.0, (h - F) / max(h, eps))        if F <= h else 0.0
return uT, uI, uF

def w_guf(M, w, s, t, h, alpha=(1/3, 1/3, 1/3)):
m, n = len(M), len(M[0])
scores = [0.0] * m
for i in range(m):
for j in range(n):
T, I, F = M[i][j]
uT, uI, uF = guf_components(T, I, F, s[j], t[j], h[j])
Uij = (uT**alpha[0]) * (uI**alpha[1]) * (uF**alpha[2])
scores[i] += w[j] * Uij
return scores
```

```
scores_guf = w_guf(M, w, s, t, h)
print("W-GUF Scores:", scores_guf)
```

The code provided above calculates the extent to which each criterion contributes to the overall performance for each alternative. Each line of the code is directly related to the mathematical formulation explained earlier; thus, the connection between the theoretical structure and the application is clearly visible. The results obtained are in the form of a list of continuous values: high scores indicate that the alternative conforms more strongly to the decision threshold, while low scores indicate that this conformity is weak. This step-by-step structured algorithm can be easily adapted to different data sets without changing the theoretical framework and can be flexibly adjusted for additional sensitivity analyses.

### 9.4.4 Advantages of the Python implementation

The developed Python application operates entirely within its own defined structure without requiring external libraries. Since all data, parameters, and threshold values are defined directly within the code, the system has an independent and self-sufficient design. Thanks to this structure, both algorithms can be run repeatedly with different thresholds, weights, or membership values, allowing users to quickly test various scenarios. The simple and modular structure of the code is highly suitable for sensitivity analyses, reproducibility studies, and ranking comparisons.

Python's flexibility enhances the usability of the proposed framework and enables the two algorithms to work interactively with each other. The first algorithm, W-ALSS, explicitly subjects alternatives to a binary evaluation, while the second algorithm, W-GUF, refines this process using a graded scoring approach to make a clearer distinction between feasible alternatives. The application's modular design allows the system to be easily updated, expanded, or integrated into more comprehensive decision support systems. This provides researchers with a powerful tool to analyze real-world multi-criteria decision problems in a flexible, repeatable, and practical manner.

---

## 9.5 Case Study: Renewable Energy Project Selection

In this section, we demonstrate the application of both proposed algorithms—Algorithm 1 (W-ALSS) and Algorithm 2 (W-GUF)—to a renewable energy project selection problem. The analysis evaluates five potential projects under multiple criteria using neutrosophic decision-making. All computations are performed step by step and do not require any external data files.

**TABLE 9.3**: Neutrosophic decision matrix $M$ for five renewable energy projects.

| Alt. | $E_1$ | $E_2$ | $E_3$ | $E_4$ | $E_5$ |
|---|---|---|---|---|---|
| A1 | (0.78,0.22,0.20) | (0.75,0.20,0.25) | (0.68,0.22,0.32) | (0.68,0.24,0.30) | (0.70,0.24,0.26) |
| A2 | (0.90,0.28,0.10) | (0.55,0.25,0.45) | (0.50,0.30,0.50) | (0.66,0.26,0.28) | (0.58,0.28,0.36) |
| A3 | (0.70,0.20,0.30) | (0.82,0.18,0.18) | (0.72,0.20,0.28) | (0.78,0.20,0.20) | (0.80,0.20,0.18) |
| A4 | (0.74,0.25,0.22) | (0.70,0.22,0.30) | (0.60,0.26,0.40) | (0.82,0.22,0.16) | (0.62,0.26,0.34) |
| A5 | (0.65,0.18,0.35) | (0.68,0.24,0.32) | (0.76,0.22,0.24) | (0.60,0.24,0.38) | (0.68,0.24,0.30) |

Alternatives ($A$): A1: Onshore Wind Farm, A2: Offshore Wind Farm, A3: Utility-Scale Solar PV, A4: Run-of-River Hydro, A5: Biomass CHP

Criteria ($E$):

- $E_1$: Annual Net Energy Yield (benefit, higher $T$ preferred, $F$ = low production risk)
- $E_2$: Levelized Cost of Energy (cost, lower $F$ preferred, $T$ = low LCOE)
- $E_3$: Grid Integration Complexity (cost, lower complexity = higher $T$)
- $E_4$: Environmental & Social Impact (benefit, lower impact = higher $T$)
- $E_5$: Permitting & Land/Sea Access (benefit, readiness = higher $T$)

Weights:

$$w = (0.25,\ 0.25,\ 0.20,\ 0.15,\ 0.15)$$

All weights sum to 1.0.

Neutrosophic decision matrix is given by Table 9.3.
Mid-level thresholds $(s_j, t_j, h_j)$ are per-criterion averages of $T$, $I$, and $F$ over all alternatives:

$$(s_1, t_1, h_1) = (0.754,\ 0.226,\ 0.234), \quad (s_2, t_2, h_2) = (0.700,\ 0.218,\ 0.300),$$
$$(s_3, t_3, h_3) = (0.652,\ 0.240,\ 0.348), \quad (s_4, t_4, h_4) = (0.708,\ 0.232,\ 0.264),$$
$$(s_5, t_5, h_5) = (0.676,\ 0.244,\ 0.288).$$

### 9.5.1 Algorithm 1 (W-ALSS)

A cell passes if and only if

$$T_{ij} \geq s_j, \quad I_{ij} \geq t_j, \quad F_{ij} \leq h_j,$$

and $L_{ij} \in \{0, 1\}$ denotes pass/fail. The weighted score is $S(a_i) = \sum_{j=1}^{5} w_j L_{ij}$.

- $A_1, E_1$: $(0.78, 0.22, 0.20)$ vs $(0.754, 0.226, 0.234)$: $0.78 \geq 0.754$, $0.22 \not\geq 0.226$, $0.20 \leq 0.234 \Rightarrow L_{A_1,E_1} = 0$.

- $A_2, E_1$: $(0.90, 0.28, 0.10)$ vs $(0.754, 0.226, 0.234)$: All three conditions hold $\Rightarrow L_{A_2,E_1} = 1$.
- $A_3, E_2$: $(0.82, 0.18, 0.18)$ vs $(0.700, 0.218, 0.300)$: $0.18 \ngeq 0.218 \Rightarrow L_{A_3,E_2} = 0$.

*Pass matrix L (mid-level)*

| | $E_1$ | $E_2$ | $E_3$ | $E_4$ | $E_5$ |
|---|---|---|---|---|---|
| $A_1$ | 0 | 0 | 0 | 0 | 0 |
| $A_2$ | 1 | 0 | 0 | 0 | 0 |
| $A_3$ | 0 | 0 | 0 | 0 | 0 |
| $A_4$ | 0 | 1 | 0 | 0 | 0 |
| $A_5$ | 0 | 0 | 0 | 0 | 0 |

Weighted scores

$$S(A_2) = 0.25 \cdot 1 = 0.25, \qquad S(A_4) = 0.25 \cdot 1 = 0.25,$$
$$S(A_1) = 0.00, \quad S(A_3) = 0.00, \quad S(A_5) = 0.00.$$

*Ranking (W-ALSS, mid-level)*

$$A_2 \succ A_4 \succ (A_1 = A_3 = A_5).$$

### 9.5.2 Algorithm 2 (W-GUF)

$$u_{ij}^T = \max\Big\{0, \frac{T_{ij} - s_j}{1 - s_j}\Big\}, \quad u_{ij}^I = \max\Big\{0, \frac{I_{ij} - t_j}{1 - t_j}\Big\}, \quad u_{ij}^F = \max\Big\{0, \frac{h_j - F_{ij}}{h_j}\Big\}.$$

Per-criterion fusion with component weights $(\alpha_T, \alpha_I, \alpha_F) = (0.5, 0.2, 0.3)$:

$$U_{ij} = (u_{ij}^T)^{0.5}\,(u_{ij}^I)^{0.2}\,(u_{ij}^F)^{0.3}, \qquad S(a_i) = \sum_{j=1}^{5} w_j\, U_{ij}.$$

*Illustrative computations*

- $A_3, E_2$ : $(0.82, 0.18, 0.18)$ vs $(0.700, 0.218, 0.300)$
  $u^T = \frac{0.82-0.70}{0.30} = 0.40$, $u^I = \max\{0, \frac{0.18-0.218}{0.782}\} = 0$, $u^F = \frac{0.300-0.18}{0.300} = 0.40$.
  $U_{A_3,E_2} = 0.40^{0.5} \cdot 0^{0.2} \cdot 0.40^{0.3} = 0$.
- $A_1, E_3$ : $(0.68, 0.22, 0.32)$ vs $(0.652, 0.240, 0.348)$
  $u^T = \frac{0.68-0.652}{0.348} \approx 0.0805$, $u^I = \max\{0, \frac{0.22-0.240}{0.760}\} = 0$, $u^F = \frac{0.348-0.32}{0.348} \approx 0.0805$.
  $U_{A_1,E_3} = 0.0805^{0.5} \cdot 0^{0.2} \cdot 0.0805^{0.3} = 0$.

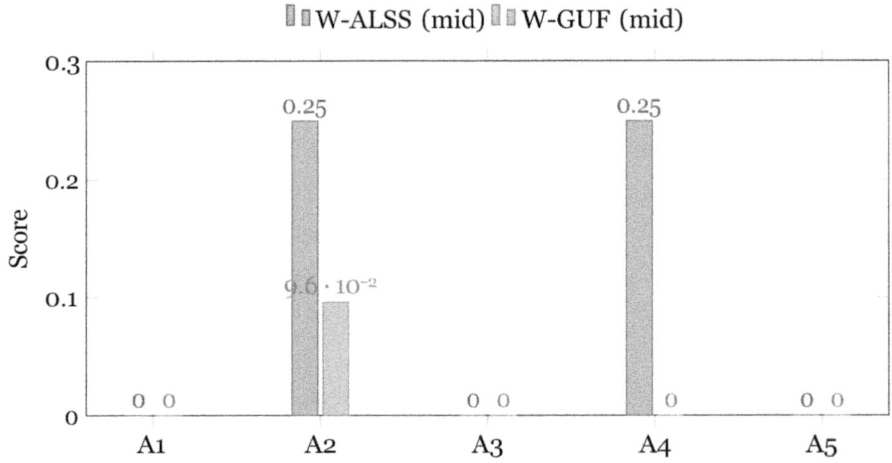


**FIGURE 9.1**: Comparison of W-ALSS and W-GUF scores under mid-level thresholds.

*Per-criterion utilities (rounded to 3 d.p.)*

| | $U_{\cdot,E_1}$ | $U_{\cdot,E_2}$ | $U_{\cdot,E_3}$ | $U_{\cdot,E_4}$ | $U_{\cdot,E_5}$ |
|---|---|---|---|---|---|
| $A_1$ | 0.000 | 0.000 | 0.000 | 0.000 | 0.000 |
| $A_2$ | 0.383 | 0.000 | 0.000 | 0.000 | 0.000 |
| $A_3$ | 0.000 | 0.000 | 0.000 | 0.000 | 0.000 |
| $A_4$ | 0.000 | 0.000 | 0.000 | 0.000 | 0.000 |
| $A_5$ | 0.000 | 0.000 | 0.000 | 0.000 | 0.000 |

*Weighted scores (mid-level)*

$$S(A_2) = 0.25(0.383) = 0.096 \quad \text{(rounded)},$$
$$S(A_1) = 0.000, \quad S(A_3) = 0.000, \quad S(A_4) = 0.000, \quad S(A_5) = 0.000.$$

*Ranking (W-GUF, mid-level)*

$$A_2 \succ (A_1 = A_3 = A_4 = A_5).$$

Scores for both methods are illustrated in Figure 9.1.

---

## 9.6 Comparison of Algorithm 1 (W-ALSS) and Algorithm 2 (W-GUF)

This section compares Algorithm 1 (W-ALSS) and Algorithm 2 (W-GUF) based on the corrected mid-level computations. The analysis focuses on score magnitudes, ranking stability, and methodological differences.

**TABLE 9.4**: Scores and ranks for both methods under mid-level thresholds.

| **Alternative** | **W-ALSS (mid)** | | **W-GUF (mid)** | |
|---|---|---|---|---|
| | Score | Rank | Score | Rank |
| A1 | 0.000 | 3 | 0.000 | 2 |
| A2 | 0.250 | 1 | 0.096 | 1 |
| A3 | 0.000 | 3 | 0.000 | 2 |
| A4 | 0.250 | 1 | 0.000 | 2 |
| A5 | 0.000 | 3 | 0.000 | 2 |

- Both methods consistently identify $A_2$ as the best-performing project.
- W-ALSS also identifies $A_4$ as a strong candidate with the same score as $A_2$, but W-GUF ranks $A_4$ lower. The scores and ranks for both methods under mid-level thresholds are summarized in Table 9.4.
- Alternatives $A_1$, $A_3$, and $A_5$ receive near-zero scores in both methods.
- W-ALSS produces binary-oriented scores based on pass/fail decisions, resulting in higher absolute values (e.g., $A_2$ and $A_4$ scoring 0.25).
- W-GUF produces lower but more discriminative scores due to graded utility aggregation, distinguishing $A_2$ from all other projects.
- W-GUF effectively separates top-performing alternatives from weaker ones, while W-ALSS can create ties when multiple projects pass the same criteria.

- Both methods now clearly highlight $A_2$ as the top-performing project.
- W-ALSS is more suitable for strict regulatory checks, where binary compliance decisions are needed.
- W-GUF is better for strategic prioritization since it differentiates alternatives more precisely. Table 9.5 provides an objective methodological comparison between W-ALSS and W-GUF, highlighting differences in scoring approach, sensitivity to thresholds, parameterization, interpretability, and score ranges.

---

## 9.7 Sensitivity Analysis

This section examines the robustness of Algorithm 1 (W-ALSS) and Algorithm 2 (W-GUF) by analyzing how rankings change under variations in criterion weights and threshold levels. The analysis focuses on three scenarios:

**TABLE 9.5**: Objective methodological comparison between W-ALSS and W-GUF.

| Aspect | Algorithm 1: W-ALSS | Algorithm 2: W-GUF |
|---|---|---|
| **Scoring approach** | Binary decision: $L_{ij} = 1$ if $T \geq s,\ I \geq t,\ F \leq h$, else 0. Final score is weighted sum of passes. | Graded utilities per criterion; combines $u^T, u^I, u^F$ using a weighted geometric mean, producing continuous $U_{ij} \in [0, 1]$. |
| **Sensitivity to thresholds** | High: small changes in $(s, t, h)$ can flip pass/fail outcomes and rankings. | Moderate: utilities vary smoothly with $(s, t, h)$, resulting in more stable rankings. |
| **Parameterization** | Criterion weights $w_j$. | Criterion weights $w_j$ and component weights $(\alpha_T, \alpha_I, \alpha_F)$. |
| **Interpretability** | Simple and intuitive pass/fail framework. | Provides richer insight into partial satisfaction of criteria. |
| **Score range (dataset)** | [0.00, 0.25] for this dataset. | [0.000, 0.096] for this dataset. |
| **Ranking granularity** | Can yield ties if multiple alternatives achieve the same number of passes. | Lower chance of ties due to continuous scoring. |
| **Computational cost** | $\mathcal{O}(m \cdot n)$. | $\mathcal{O}(m \cdot n)$, same as W-ALSS. |

**TABLE 9.6**: Baseline results under mid-level thresholds.

| Alternative | W-ALSS | | W-GUF | |
|---|---|---|---|---|
| | Score | Rank | Score | Rank |
| A1 | 0.000 | 3 | 0.000 | 2 |
| A2 | 0.250 | 1 | 0.096 | 1 |
| A3 | 0.000 | 3 | 0.000 | 2 |
| A4 | 0.250 | 1 | 0.000 | 2 |
| A5 | 0.000 | 3 | 0.000 | 2 |

1. Baseline configuration (mid-level thresholds, original weights).
2. Adjusted weights scenario ($\pm 10\%$ variation).
3. Strict thresholds scenario (tptp policy).

### 9.7.1 Baseline (mid-level thresholds)

Under the baseline configuration, results are shown in Table 9.6:

### 9.7.2 Adjusted weights ($\pm 10\%$)

To test weight sensitivity, we increase $E_1$ and $E_2$ weights by $+10\%$ each and proportionally decrease the others to maintain $\sum w_j = 1$:

TABLE 9.7: Results under adjusted weights.

| Alternative | W-ALSS (adjusted $w'$) | | W-GUF (adjusted $w'$) | |
|---|---|---|---|---|
| | Score | Rank | Score | Rank |
| A1 | 0.000 | 3 | 0.000 | 2 |
| A2 | 0.275 | 1 | 0.105 | 1 |
| A3 | 0.000 | 3 | 0.000 | 2 |
| A4 | 0.275 | 1 | 0.000 | 2 |
| A5 | 0.000 | 3 | 0.000 | 2 |

$$w' = (0.275,\ 0.275,\ 0.185,\ 0.135,\ 0.130).$$

- W-ALSS: Since only $E_1$ and $E_2$ generate non-zero passes, $A_2$ and $A_4$ benefit equally; their ranking remains unchanged.
- W-GUF: Because $A_2$ performs well on $E_1$, its score increases slightly, reinforcing its dominance.

Table 9.7 presents the results of W-ALSS and W-GUF for all alternatives under adjusted weights.

Table 9.8 presents the scores and ranks of all alternatives under strict thresholds (tptp) for both W-ALSS and W-GUF.

### 9.7.3 Strict thresholds (tptp policy)

Strict thresholds set higher selectivity levels per criterion:

$$s_j = \max_i T_{ij}, \quad t_j = \max_i I_{ij}, \quad h_j = \min_i F_{ij}.$$

*Observed effects.*

- W-ALSS: Under strict thresholds, only $A_2$ passes $E_1$; all other alternatives fail completely.
- W-GUF: Scores decrease overall, but $A_2$ remains the highest-ranked alternative.

### 9.7.4 Ranking stability analysis

Ranking stability under different scenarios is shown in Table 9.9.

### 9.7.5 Insights from sensitivity analysis

- Under moderate weight adjustments, both methods exhibit stable rankings; $A_2$ consistently dominates.

**TABLE 9.8**: Results under strict thresholds (tptp).

| Alternative | W-ALSS (tptp) | | W-GUF (tptp) | |
|---|---|---|---|---|
| | Score | Rank | Score | Rank |
| A1 | 0.000 | 2 | 0.041 | 2 |
| A2 | 0.250 | 1 | 0.051 | 1 |
| A3 | 0.000 | 3 | 0.000 | 3 |
| A4 | 0.000 | 4 | 0.000 | 3 |
| A5 | 0.000 | 5 | 0.000 | 3 |

**TABLE 9.9**: Ranking stability under different scenarios.

| Alternative | Baseline rank | Adjusted weights | Strict thresholds |
|---|---|---|---|
| A1 | 3 (W-ALSS) / 2 (W-GUF) | Stable | Drops slightly in W-GUF |
| A2 | 1 (both) | Stable | Remains best in both methods |
| A3 | 3 (W-ALSS) / 2 (W-GUF) | Stable | Loses rank under strict thresholds |
| A4 | 1 (W-ALSS) / 2 (W-GUF) | Stable | Loses rank under strict thresholds |
| A5 | 3 (both) | Stable | Remains low-ranked |

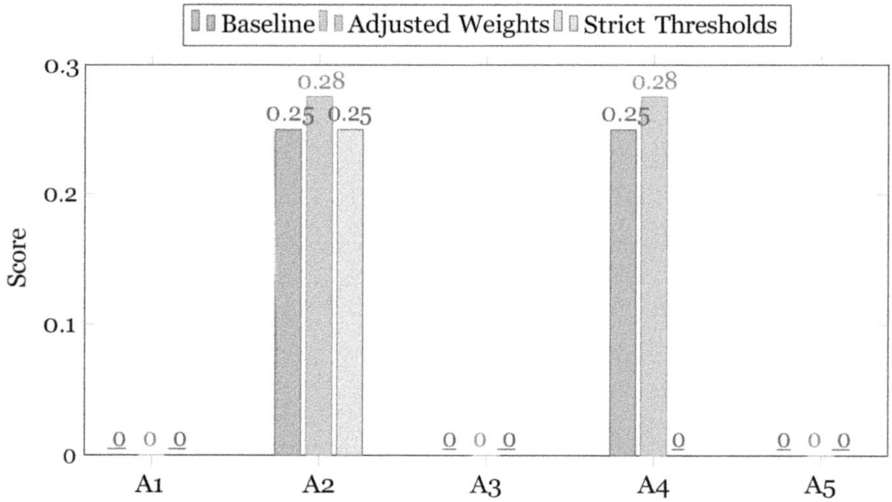


**FIGURE 9.2**: W-ALSS score sensitivity across scenarios.

- Under strict thresholds, W-ALSS scores collapse for all but $A_2$, while W-GUF maintains smoother, more discriminative scores.
- W-GUF demonstrates higher robustness to threshold variations, whereas W-ALSS is highly sensitive due to its binary pass/fail structure.

## 9.8 Conclusion

This chapter combines two complementary methods based on neutrosophic soft set theory as an alternative to classical approaches used in decision-making processes under uncertainty. The developed W-ALSS and W-GUF algorithms provide decision-makers with a flexible and interpretable evaluation structure. The W-ALSS method acts as a filter in the initial stage of the decision process, eliminating alternatives that do not meet a certain threshold, while the W-GUF method analyzes the performance of the remaining options in a graded manner after this stage. Thus, the gap between the binary decision mechanism and continuous evaluation has been closed, creating a multi-layered decision support framework that can be applied in different sectors. The pilot study conducted on renewable energy projects has demonstrated the practical validity of the proposed approach. The results show that the offshore wind project achieved the highest suitability value in both the W-ALSS and W-GUF methods. This finding proves that the methods produce consistent results when used together and increase decision reliability. Figure 9.2 illustrates the score sensitivity of W-ALSS across the different scenarios. Furthermore, the open and reproducible code structure developed in the Python environment demonstrates that the study is not only theoretical but also a practical tool. In the future, this framework can be expanded, particularly by supporting it with group decision-making, dynamic weighting, or artificial intelligence-based optimization modules. Thus, neutrosophic-based decision systems can be effectively used not only in energy planning but also in areas with high uncertainty, such as health technologies, environmental sustainability, or policy analysis.

## Bibliography


[1] T. L. Saaty. *The Analytic Hierarchy Process.* McGraw-Hill, 1980.

[2] C.-L. Hwang and K. Yoon. *Multiple Attribute Decision Making: Methods and Applications.* Springer, 1981.

[3] S. Opricovic and G.-H. Tzeng. Compromise solution by mcdm methods: Vikor. *European Journal of Operational Research*, 156(2):445–455, 2004.

[4] L. A. Zadeh. Fuzzy sets. *Information and Control*, 8(3):338–353, 1965.

[5] K. T. Atanassov. Intuitionistic fuzzy sets. *Fuzzy Sets and Systems*, 20(1):87–96, 1986.

[6] D. Molodtsov. Soft set theory—first results. *Computers & Mathematics with Applications*, 37(4–5):19–31, 1999.

[7] F. Smarandache. Neutrosophic set: A generalization of the intuitionistic fuzzy set. *International Journal of Pure and Applied Mathematics*, 24(3):287–297, 2005.

[8] P. K. Maji, R. Biswas, and A. R. Roy. Soft set theory. *Computers & Mathematics with Applications*, 45(4-5):555–562, 2009.

[9] H. Wang, F. Smarandache, Y. Q. Zhang, and R. Sunderraman. Single valued neutrosophic sets. *Multispace and Multistructure*, 4:410–413, 2010.

[10] J. Ye. Multiple attribute decision-making method using the interval-valued neutrosophic linguistic normalized weighted bonferroni mean. *Journal of Intelligent & Fuzzy Systems*, 27(2):901–907, 2014.

[11] Q. Luo and P. Liu. Bipolar neutrosophic sets and their application in multi-attribute decision making. *International Journal of Machine Learning and Cybernetics*, 9(1):99–111, 2018.

[12] T. K. Senapati and M. Sarkar. Fermatean fuzzy sets and its application in decision making. *Journal of Intelligent & Fuzzy Systems*, 38(3):2775–2786, 2020.

[13] R. R. Yager. Generalized orthopair fuzzy sets. *IEEE Transactions on Fuzzy Systems*, 25(5):1222–1230, 2017.

[14] V. Jayakumar and R. Ramakrishnan. Insights from $L_q^*$ q-rung orthopair multi-fuzzy soft sets in mcdm contexts. *Soft Computing*, 29:345–362, 2025.

[15] L. Djurović, M. Laković, and N. Stojanović. Decision-making algorithm based on the energy of interval-valued fuzzy soft sets. *Fuzzy Sets and Systems*, 421:1–19, 2024.

[16] Q. T. Bui. Intuitionistic fuzzy rough soft sets: A decision-making framework. *Information Sciences*, 600:100–115, 2025.

[17] K. A. Khan. Complex single-valued neutrosophic soft sets in mcdm. *Neutrosophic Sets and Systems*, 60:75–89, 2024.

[18] P. Jayaprakash and S. Sundar. Neutrosophic soft set matrix and multi-criteria decision-making. *Multidisciplinary Science Journal*, 6:e2024ss0110, 2024.

[19] Z. Baser. Energy-based neutrosophic soft set decision applications. *Neutrosophic Sets and Systems*, 68:22–38, 2025.

[20] Y. Akhtar, M. Ali, and M.-S. Yang. Generalized orbicular t-spherical fuzzy sets with hamacher operators for group decision-making. *IEEE Transactions on Fuzzy Systems*, 32(5):1–14, 2024.

[21] M. Taghizadeh. Ai-integrated interval-valued fuzzy mcdm for dynamic weight adaptation. *Journal of Computational Intelligence*, 58(2):114–128, 2025.

[22] I. Deli and S. Broumi. Neutrosophic soft matrices and their applications. *International Journal of Circuits, Systems, and Signal Processing*, 8:218–226, 2014.

[23] P. K. Maji. Neutrosophic soft sets. *Annals of Fuzzy Mathematics and Informatics*, 5(1):157–168, 2013.

[24] T. Ozturk, C. Gunduz Aras, and S. Bayramov. A new approach to operations on neutrosophic soft sets and to neutrosophic soft topological spaces. *Communications in Mathematics and Applications*, 10(3):481–493, 2019.

[25] A. Yolcu and T. Y. Öztürk. An adjustable method for decision making problems with neutrosophic soft sets. *International Journal of System Assurance Engineering and Management*, 2025.

[26] R. L. Keeney and H. Raiffa. *Decisions with Multiple Objectives: Preferences and Value Trade-offs*. Cambridge University Press, 1993.

# Chapter 10

## Neutrosophic ISOCOV in Smart and Sustainable Material Selection

**Nivetha Martin, Said Broumi, and Florentin Smarandache**

### 10.1 Introduction

The consumers of this era demand multi-utility and eco-friendly products. The metamorphosis in the purchasing behavior of the consumers has escalated the demand for both smart and sustainable materials. The green market is blooming with the increasing buyers of eco-friendly products across the globe. In addition, the current advancements in the technology have also fostered the exploration of smart materials. The industrial sectors are redefining their production process to manufacture products with multi-attributes. One of the primary challenges faced by the manufacturing units lies in material selection which is constrained by multiple and conflicting criteria. The decision-making on material selection is also hindered by uncertainty, indeterminacy, and ambiguity. To handle such complications, the researchers develop neutrosophic decision-making models.

Smarandache [1] conceptualized neutrosophic theory and contributed to the development of neutrosophic sets. These sets are basically characterized as a three valued logic with truth, indeterminate, and falsity membership values. In general, the multi-criteria decision-making (MCDM) method involves sequential steps to choose the finest alternative subjected to predefined criteria. The MCDM methods are classified into three major classes based on (i) aggregation of performance rating using utility functions, (ii) determination of preference relation between the alternatives, and (iii) positive and negative ideal solutions using maximum and minimum values. Abdelli et al. [2] presented the method of ISOCOV (ideal solution with constraint on values) as the enhancement of the third class of MCDM methods by considering value constraints of the criteria. Martin and Edalatpanah [3] discussed ISOCOV using fuzzy representations and demonstrated the efficacy of the method in nanomaterial selection. This method has been proven competent and effective

DOI: 10.1201/9781003683339-10

in practice in material selection as the constraint values are represented using linguistic variables with fuzzy numbers. This has motivated the authors to extend fuzzy ISOCOV to neutrosophic framework to explore the advantages of neutrosophic representations in smart and sustainable material selection.

The other contents are structured. The review of works is discussed in Section 10.2. The fundamentals of neutrosophic sets are presented in Section 10.3. The steps followed in neutrosophic ISOCOV (N-ISOCOV) are sketched in Section 10.4. The application of N-ISOCOV in smart and sustainable material selection is demonstrated in Section 10.5 using a case study. The outcomes of the model are discoursed in Section 10.6. The final section outlines the industrial applications of N-ISOCOV and summarizes the work with further discussions and directions.

## 10.2 Review of Related Works

This section summarizes the related works to this study of material selection. The utility of MCDM methods in the domains of sustainable and smart material selection are discussed in general and in specific with respect to neutrosophic.

### 10.2.1 General MCDM methods in sustainable material selection

Academicians have employed different MCDM methods for sustainable material selection across diverse domains. To mention a few recent and noteworthy works, Haruna et al. [4] applied the method of analytical hierarchy process (AHP) in sustainable construction. Magdy et al. [5] employed life cycle assessment in selecting sustainable materials. Riaz et al. [6] used linear diophantine fuzzy soft rough sets in selecting sustainable material of equipment. Rajak et al. [7] utilized the method of VIKOR (VlseKriterijumska Optimizacija I Kompromisno Resenje) in selecting materials for sustainable construction. Hatefi et al. [8] employed additive ratio assessment method integrated with evidence theory. Bapat et al. [9] discussed the Internet of Things (IoT) enabled sustainable material selection for energy optimization in a railway project. Alkan et al. [10] discoursed the utility of Bayesian best worst method (BWM) in sustainable construction material selection. Chen et al. [11] initiated an integrated MCDM framework. Dhanalakshmi et al. [12] applied MOORA method in sustainable building material. Wakeel et al. [13] constructed a hybrid decision model integrating goal programming and BWM for choosing sustainable material in the manufacturing of complex automotive products.

Aires and Ferreira [14] developed a new multi-criteria approach in sustainable material selection. Nicolalde et al. [15] applied MCDM methods in sustainable structural beam material selection. Italia et al. [16] employed taxonomy method. Haq et al. [17] used interval-valued neutrosophic-based MAIRCA (multi-attribute ideal-real comparative analysis). Maity et al. [18] utilized in the selection of construction materials of aerodynamic wing. Sonia et al. [19] framed a fuzzy-based group decision-making model to make optimal material selection of sustainable composites. Tegegne et al. [20] used sustainable building materials in construction requires a multi-criteria approach that balances performance, social benefits, and environmental impact from the early design stage. Rana et al. [21] contributed to the development of hybrid fuzzy-Python MCDM to select storage material. Sudakova et al. [22] proposed sustainable material integrates hybrid engineering in advanced decision methods. Bulut et al. [23] applied different MCDM methods in sustainable thermoplastic material selection. Soni et al. [24] employed q-rung orthopair fuzzy SWARA-MABAC (stepwise weight assessment ratio analysis-multi-attributive border approximation area comparison) approach in the selection of sustainable construction materials. Kusnere et al. [25] used multi-criteria decision analysis in sustainable material selection for biomethanation. Saha et al. [26] discoursed intuitionistic fuzzy Dombi-Archimedean weighted aggregation operators in sustainable material selection. Amellal et al. [27] applied geographic information system (GIS)-based decision-support systems in the selection of sustainable material extraction sites. El Hallaoui [28] applied MCDM in material selection to the rooftop PV system design. Takano and Aiki [29] applied MCDM in sustainable material selection. Mazaheri et al. [30] contributed to sustainable project scheduling. Gligorić et al. [31] framed a novel approach to make decisions on sustainable materials. Dasgupt et al. [32] applied entropy-TOPSIS (technique for order of preference by similarity to ideal solution) in the material selection of sustainable polymer composite. El Mounla et al. [33] used enhanced decision-making method in construction material selection.

### 10.2.2 Neutrosophic MCDM in sustainable material selection

Researchers have discussed MCDM methods with neutrosophic representations in sustainable material selection. To mention a few, Haq et al. [34] compared the efficacy of single-valued neutrosophic-MEREC-MARCOS (method based on removal effects of criteria-measurement of alternatives and ranking according to compromise solution) approach over crisp and ambiguous data. Saeed et al. [35] developed sustainable material selection model using quantified neutrosophic set subjected to textile manufacturing. Simic et al. [36] modeled neutrosophic-based LOPCOW-ARAS (logarithmic percentage change-driven objective weighting- additive ratio assessment) for prioritizing sustainable strategies.

Furthermore, Pontarollo et al. [37] discussed material selection of nuclear fusion reactors. Seethalakshmi et al. [38] formulated deep learning-based smart material selection approach considering cost and sustainable strategies. Asim et al. [39] explored smart material selection for thermoplastic composites. The researchers have described the attributes to be considered in decision-ranking phenomenon. For better understanding, the above discussed works are presented in Table 10.1

### 10.2.3 Research gaps

From the aforementioned review of works, the following shortcomings are identified.

- **No exploration of neutrosophic ISOCOV**
  - The ISOCOV method has not been discussed using neutrosophic representations.
  - The comparative analysis of deterministic and neutrosophic ISOCOV is not unveiled to exhibit the efficiency of neutrosophic representations.
- **Devoid of applications of ISOCOV in material selection**
  - ISOCOV, in either crisp or fuzzy form, has not been used in material selection.
  - Factors such as sustainability and robustness have not been considered in the ISOCOV framework.
- **Lack of combined attributes of sustainable and smart materials**

  The existing MCDM models cater to either sustainable or smart material selection; however, the combination of both does not explore to the best of our knowledge.

### 10.2.4 Characteristics of the proposed model

These shortcomings have led to the development of a neutrosophic-based ISOCOV MCDM method for selecting both sustainable and smart materials. The attributes are relevant to both sustainability and smartness. To address these gaps, this study has proposed a neutrosophic-based ISOCOV model with the following characteristics:

- **Neutrosophic representation:** The proposed model includes truth, indeterminacy, and falsity to manage uncertainty in decision data.
- **Dual-dimensional criteria:** The model integrates sustainability and smartness attributes in material selection at the same time.

**TABLE 10.1**: Applications of MCDM in diverse contexts.

| **Sustainability-focused MCDM** | | |
|---|---|---|
| **Author and year** | **MCDM method** | **Domain of ranking alternatives** |
| Haruna et al., 2020 [4] | Analytical network process | Sustainable construction |
| Magdy et al., 2020 [5] | Life cycle assessment. | Sustainable material |
| Riaz et al., 2020 [6] | Linear diophantine fuzzy soft rough sets | Sustainable equipment |
| Rajak et al., 2020 [7] | VIKOR | Green material |
| Hatefi et al., 2021 [8] | Integrated Dempster-Shafer evidence theory ARAS | Sustainable material |
| Bapat et al., 2021 [9] | FCM-BIM (fuzzy cognitive map- building information modeling) | IoT in material choice and energy management |
| Alkan et al., 2021 [10] | Bayesian BWM-SAW (best worst method- simple additive weighting) hybrid model | Sustainable construction |
| Chen et al., 2021 [11] | General MCDM | Sustainable building material |
| Dhanalakshmi et al., 2021 [12] | MOORA (multi-objective optimization on the basis of ratio analysis) | Biomass material |
| Wakeel et al., 2021 [13] | Hybrid Goal Programming-BWM | Complex automotive products |
| Aires et al., 2022 [14] | New multi-criteria approach | Sustainable material |
| Nicolalde et al., 2022 [15] | General MCDM | Sustainable structural beam material |
| Italia et al., 2023 [16] | Taxonomy and systematization of design approach | Sustainable material |
| Haq et al., 2023 [17] | Interval-valued neutrosophic based MAIRCA | Sustainable material |
| Maity et al., 2023 [18] | General MCDM | Material for the construction of UAV aerodynamic wing |
| Soni et al., 2023 [19] | Fuzzy group decision-making approach | Sustainable composites of recycling |
| Tegegne et al., 2023 [20] | General MCDM | Sustainable building material |
| Rana et al., 2023 [21] | Hybrid fuzzy-Python MCDM | Sustainable latent heat storage material |
| Sudakova et al., 2024 [22] | General MCDM | Construction material |
| Bulut et al., 2024 [23] | Comparative MCDM | Sustainable thermoplastic material |

**TABLE 10.1**: Applications of MCDM in diverse contexts (*Continued*).

| **Sustainability-focused MCDM** | | |
|---|---|---|
| **Author and year** | **MCDM method** | **Domain of ranking alternatives** |
| Soni et al., 2024 [24] | q-rung orthopair fuzzy SWARA-MABAC approach | Construction material |
| Kusnere et al., 2024 [25] | General MCDM | Biomethanation material |
| Saha et al., 2024 [26] | Intuitionistic fuzzy Dombi-Archimedean weighted aggregation operators | Sustainable material |
| Amellal et al., 2024 [27] | GIS, AHP, & MACBETH (measuring attractiveness by a categorical based evaluation technique) | Sustainable material-extraction Sites |
| El Hallaoui., 2024 [28] | Combined MCDM | Rooftop material |
| Takano et al., 2025 [29] | General MCDM | Sustainable building material |
| Mazaheri et al., 2025 [30] | Combined MCDM | Sustainable project scheduling |
| Gligorić et al., 2025 [31] | Integrated MCDM | Sustainable construction material |
| Dasgupta et al., 2025 [32] | Entropy-TOPSIS | Sustainable polymer composite |
| El Mounla et alal., 2025 [33] | Integrating MCDM | Sustainable construction |
| **Neutrosophic-based MCDM** | | |
| **Author and year** | **MCDM method** | **Domain of ranking alternatives** |
| Haq et al ., 2022 [34] | Single-valued neutrosophic-MEREC-MARCOS framework | Sustainable material |
| Saeed et al., 2025 [35] | Quantified neutrosophic set (QtNS)-based MCDM algorithms | Sustainable material for anti-microbial bio-fabricated textile manufacturing. |
| Simic et al., 2023 [36] | Neutrosophic LOPCOW-ARAS model | Material handling technologies |
| **Smart MCDM** | | |
| **Author and Year** | **MCDM method** | **Areas of application** |
| Pontarollo et al., 2021 [37] | General MCDM | Smart material for nuclear fusion reactors. |
| Seethalakshmi et al., 2024 [38] | Deep learning | Smart material for concrete production |
| Asim et al., 2025 [39] | General MCDM | Smart material for ballistic thermoplastic composites |

- **Improved decision reliability:** It provides more dependable outcomes by tackling vagueness and inconsistency in expert judgments.
- **Comprehensive material evaluation:** The developed model ensures thorough selection by balancing environmental, economic, technological, and intelligent adaptability aspects.

---

## 10.3 Preliminaries

This section presents the basic definitions related to this study.

### 10.3.1 Neutrosophic set [40]

Let $X$ be a universal set. A neutrosophic set $A$ on $X$ is defined as $A = \{< T_A(x), I_A, (x), F_A(x),\ F_A(x > : x \in X)$, where $T_A(x), I_A, (x), F_A(x) : X \rightarrow {}^{-}]0,1[^{+}$ represents the degree of belongingness, degree of indeterministic, and degree of falsity, respectively, of the element $x \in X$ such that ${}^{-}0 \leq T_A(x), I_A, (x), F_A(x) \leq 3^{+}$.

### 10.3.2 Neutrosophic number [40]

A neutrosophic set $A$ defined on the universal set of real numbers $R$ is said to be neutrosophic number if it has the following properties.

(i) $A$ is normal if there exist $x_0 \in R$, such that $T_A(x_0) = ;\ I_A(x_0) = 0;\ F_A(x_0) = 0$.

(ii) $A$ is convex set for the truth membership function $T_A(x)$, *i.e.*, $T_A(\mu x_1 + (1-\mu)x_2) \geq \min(T_A(x_1),\ T_A(x_2))\ \forall\ x_1,\ x_2 \in R,\ \mu \in [0,1]$.

(iiii) $A$ is concave set for the indeterministic function and false function $I_A(x)$ and $F_A(x)$, i.e.,

$$I_A(\mu x_1 + (1-\mu)x_2) \geq \max(I_A(x_1),\ I_A(x_2))\ \forall\ x_1,\ x_2 \in R,\ \mu \in [0,1].$$
$$F_A(\mu x_1 + (1-\mu)x_2) \geq \max(F_A(x_1),\ F_A(x_2))\ \forall\ x_1,\ x_2 \in R,\ \mu \in [0,1].$$

### 10.3.3 Trapezoidal neutrosophic number [40]

Trapezoidal neutrosophic number $\tilde{N} =< (t_1, t_2, t_3, t_4),\ (i_1, i_2,\ i_3,\ i_4), (f_1, f_2, f_3, f_4) >$ defined on the set of real numbers $\mathbb{R}$ by its belongingness, indeterminacy, and falsity function [16]:

$$\mu_{\tilde{N}}(x) = \begin{cases} \frac{x-t_1}{t_2-t_1}; t_1 \leq x \leq t_2 \\ 1; t_2 \leq x \leq t_3 \\ \frac{t_4-x}{t_4-t_3}; t_3 \leq x \leq t_4 \\ 0; \quad otherwise \end{cases}$$

$$\delta_{\tilde{N}}(x) = \begin{cases} \frac{i_2-x}{i_2-i_1}; i_1 \leq x \leq i_2 \\ 1; i_2 \leq x \leq i_3 \\ \frac{x-i_3}{i_4-i_3}; i_3 \leq x \leq i_4 \\ 0; \quad otherwise \end{cases}$$

$$\upsilon_{\tilde{N}}(x) = \begin{cases} \frac{f_2-x}{f_2-f_1}; f_1 \leq x \leq f_2 \\ 1; f_2 \leq x \leq f_3 \\ \frac{x-f_3}{f_4-f_3}; f_3 \leq x \leq f_4 \\ 0; \quad otherwise \end{cases}$$

### 10.3.4 $(\alpha, \beta, \gamma)$-cuts of trapezoidal neutrosophic number [40]

Let $\tilde{N}$ be a single valued trapezoidal neutrosophic number. The crisp subset of real is defined by:

$$\begin{aligned} \tilde{N}_{\alpha,\beta,\gamma} = & \; x | \mu_{\tilde{N}}(x) \geq \alpha, \; \delta_{\tilde{N}}(x) \leq \beta, \; \nu_{\tilde{N}}(x) \leq \gamma\} \\ & = \{[t_1 + \alpha(t_2 - t_1), \; t_4 - \alpha(t_4 - t_3)], \; [i_2 + \beta(i_2 - i_1), \; i_3 + \beta(i_4 - i_3)], \\ & \quad [f_2 + \gamma(f_2 - f_1), f_3 + \gamma(f_4 - f_3)]\} \end{aligned}$$

It is called $(\alpha, \beta, \gamma)-$ cuts of $\tilde{N}$, where $0 \leq \alpha, \beta, \gamma \leq 1$ and $0 \leq \alpha + \beta + \gamma \leq 3$.

### 10.3.5 Neutrosophic score function [40]

The score function of the neutrosophic set $(T, I, F)$ is determined as

$$\frac{2 + T - I - F}{3}$$

## 10.4 Methodology of Neutrosophic ISOCOV

The steps followed in neutrosophic ISOCOV are outlined in this section. The methodology of fuzzy ISOCOV proposed by Martin and Edalatpanah is extended to neutrosophic and presented below with three-valued logic representations.

*Step 1: Problem formulation*

The decision-making problem is clearly formulated by specifying $m$ alternatives and $n$ evaluation criteria. An initial decision matrix is then constructed, where each element $s_{ij}$ represents the degree of satisfaction of criterion $j$ corresponding to alternative $i$.

*Step 2: Computation of criteria weights*

The criterion weights $W = [w_1,\ w_2,..w_n]$ are determined based on their significance and contribution to the decision-making problem.

*Step 3: Normalization of the matrix*

The values of the primary matrix are normalized using vector normalization form.

For benefit criteria,

$$n_{ij} = 1 - \frac{s_{ij}}{\sqrt{\sum_{i=1}^{m} s_{ij}}} \tag{10.1}$$

For cost criteria,

$$n_{ij} = \frac{s_{ij}}{\sqrt{\sum_{i=1}^{m} s_{ij}}} \tag{10.2}$$

*Step 4: Determining neutrosophic constraint satisfaction values*

At this stage, neutrosophic trapezoidal fuzzy numbers of the form $\langle (\mathrm{t}_1, \mathrm{t}_2, \mathrm{t}_3, \mathrm{t}_4),\ (\mathrm{i}_1, \mathrm{i}_2, \mathrm{i}_3, \mathrm{i}_4),\ (\mathrm{f}_1, \mathrm{f}_2, \mathrm{f}_3, \mathrm{f}_4)\ \rangle$ is instrumental in determining the neutrosophic constraint satisfaction values. The $\alpha,\ \beta, \gamma$-cut is given by

$$\begin{aligned} &= \{[\mathrm{t}_1 + \alpha(\mathrm{t}_2 - \mathrm{t}_1), \mathrm{t}_4 - \alpha(\mathrm{t}_4 - \mathrm{t}_3)], [\ \mathrm{i}_2 + \beta(\mathrm{i}_2 - \mathrm{i}_1), \mathrm{i}_3 + \beta(\mathrm{i}_4 - \mathrm{i}_3)], \\ &\quad [\ \mathrm{f}_2 + \gamma(\mathrm{f}_2 - \mathrm{f}_1), \mathrm{f}_3 + \gamma(\mathrm{f}_4 - \mathrm{f}_3)]\} \end{aligned} \tag{10.3}$$

In this case, the values of $\alpha,\ \beta, \gamma$ are taken as 1, and the respective representation thus obtained is $[t_2, t_3]$, $[2i_2 - i_1,\ i_4]$, $[2f_2, f_4]$.

For better understanding, the $\alpha,\ \beta, \gamma$-cut is given by $[\mathrm{T}^U, \mathrm{T}^L]$, $[\mathrm{I}^U, \mathrm{I}^L]$, $[F^U, F^L]$.

The truth value of the constraint satisfaction value $T_{CS}$ is

$$T_{CS} = \begin{cases} 1 & if\ s_{ij}\ \in [T^U, T^L] \\ \frac{100-e}{100}\ if\ \left|s_{ij} - T^U\right| \leq e \\ \frac{100-e}{100}\ if\ \left|s_{ij} - T^L\right| \leq e \\ 0 & if\ s_{ij} = 0 \end{cases} \tag{10.4}$$

The indeterminate value of the constraint satisfaction value $I_{CS}$ is

$$T_{CS} = \begin{cases} 1 & if\ s_{ij} \in [I^U, I^L] \\ \frac{100-e}{100}\ if\ \left|s_{ij} - I^U\right| \leq e \\ \frac{100-e}{100}\ if\ \left|s_{ij} - I^L\right| \leq e \\ 0 & if\ s_{ij} = 0 \end{cases} \tag{10.5}$$

The falsity value of the constraint satisfaction value $F_{CS}$ is

$$T_{CS} = \begin{cases} 1 & if\ s_{ij} \in [F^U, F^L] \\ \frac{100-e}{100}\ if\ \left|s_{ij} - F^U\right| \leq e \\ \frac{100-e}{100}\ if\ \left|s_{ij} - F^L\right| \leq e \\ 0 & if\ s_{ij} = 0 \end{cases} \tag{10.6}$$

The constraint values thus obtained in the form of a neutrosophic values are converted to a single value using score function. Here $e$ refers to the tolerance level and in general it is assumed to be a minimum value.

*Step 5: Weighted normalized matrix with deneutrosophied constraints values*

Using the values obtained from step 2 and step 3, the weighted matrix thus obtained is $[D_{ij}]$.

*Step 6: Finding ideal solutions*

The positive ideal solution ($PI$) and negative ideal solution ($NI$) are obtained. For benefit-based criteria, $PI$ = Max $\{d_{ij}\}$, $NI$ = Min $\{d_{ij}\}$, for $i = 1$ to $u$. Conversely, for non-benefit criteria, $PI$ = Min $\{d_{ij}\}$, $NI$ = Max $\{d_{ij}\}$.

*Step 7: Calculation of distance measures*

The differences between $dij$ values from each alternative's $PI$ and $NI$ are obtained using Euclidean distance measures.

$$E_i^+ = \sqrt{\sum_{i=1}^{v} (PI - d_{ij})^2}, E_i^- = \sqrt{\sum_{i=1}^{v} (NI - d_{ij})^2} \tag{10.7}$$

*Step 8: Determination of closeness measures*

The closeness measure $C_i$ for each of the alternative is computed as follows:

$$C_i = \frac{E_i^-}{E_i^- + E_i^+} \tag{10.8}$$

*Step 9: Ordering of alternatives*

Using the score values obtained in step 9, the alternatives are positioned. The alternative with maximum score value is given first priority.

## 10.5 Application of Neutrosophic ISOCOV in Material Selection

Let us consider the decision-making problem of ranking sustainable and smart materials. The attributes or the criteria considered for ranking the alternatives are presented as follows:

CR1: Cost-effectiveness

CR2: Durability

CR3: Energy efficiency

CR4: Environmental impact

CR5: Adaptability

CR6: Compatibility

The criteria presented above shall be categorized as benefit and cost. CR1 and CR4 are the cost criteria and the other criteria are considered as benefit criteria. Let us consider the materials as alternatives. The values in each cell of the decision-making matrix presents the percentage score of alternatives satisfying the criteria as in Table 10.2.

The above matrix is normalized using step 3 and the normalized matrix $(N_{ij})$ thus obtained is presented in Table 10.3.

The constraint satisfaction values are presented as linguistic variable and quantified using neutrosophic trapezoidal numbers (NTrNs). The constraint satisfaction linguistic variable for each of the criteria and its equivalent $\alpha$, $\beta, \gamma$-cut of NTrN are presented in Table 10.4.

**TABLE 10.2**: Primary decision-making matrix.

| **Alternative** | **CR1** | **CR2** | **CR3** | **CR4** | **CR5** | **CR6** |
|---|---|---|---|---|---|---|
| M1 | 85 | 70 | 75 | 80 | 65 | 70 |
| M2 | 75 | 80 | 70 | 85 | 70 | 75 |
| M3 | 90 | 60 | 80 | 70 | 60 | 65 |
| M4 | 80 | 85 | 65 | 75 | 75 | 80 |
| M5 | 70 | 75 | 85 | 80 | 80 | 70 |

TABLE 10.3: Normalized matrix.

| Alternative | CR1 | CR2 | CR3 | CR4 | CR5 | CR6 |
|---|---|---|---|---|---|---|
| M1 | .997364 | .002523 | .002643 | .997381 | .002626 | .002687 |
| M2 | .997674 | .002883 | .002467 | .997218 | .002828 | .002879 |
| M3 | .997209 | .002162 | .002819 | .997709 | .002424 | .002495 |
| M4 | .997519 | .003063 | .002291 | .997545 | .00303 | .003071 |
| M5 | .997829 | .002703 | .002996 | .997381 | .003232 | .002687 |

TABLE 10.4: $\alpha$, $\beta, \gamma$ of neutrosophic constraint satisfaction values.

| Criteria | Preferred linguistic constraint value | Neutrosophic trapezoidal number (NTrN) | $\alpha, \beta, \gamma$ of NTrN |
|---|---|---|---|
| CR1 | Moderate | $\left\langle \begin{matrix} (0.1,\ 0.2,\ 0.3,\ 0.4),\ (0.0,\ 0.2,\ 0.3,\ 0.5), \\ (0.0,\ 0.1,\ 0.4,\ 0.5) \end{matrix} \right\rangle$ | {[0.2,0.3], [0.2,0.7], [0.1,0.9]} |
| CR2 | High | $\left\langle \begin{matrix} (0.3,\ 0.4,\ 0.5,\ 0.6),\ (0.2, 0.4, 0.5, 0.7), \\ (0.2, 0.3, 0.6, 0.7) \end{matrix} \right\rangle$ | {[0.4,0.5], [0.6,0.9], [0.4,1.0]} |
| CR3 | Very high | $\left\langle \begin{matrix} (0.5,\ 0.6,\ 0.7,\ 0.8,)\ (0.4,\ 0.6,\ 0.7,\ 0.9), \\ (0.4,\ 0.5,\ 0.8,\ 0.9) \end{matrix} \right\rangle$ | {[0.6,0.7], [0.8,1.1], [0.6,1.2]} |
| CR4 | Low | $\left\langle \begin{matrix} (0.0,\ 0.1,\ 0.2,\ 0.3),\ (0.0,\ 0.1,\ 0.2,\ 0.3), \\ (0.0,\ 0.1,\ 0.2,\ 0.3) \end{matrix} \right\rangle$ | {[0.1,0.1], [0.2,0.4], [0.1,0.4]} |
| CR5 | Very high | $\left\langle \begin{matrix} (0.5,\ 0.6,\ 0.7,\ 0.8,)\ (0.4,\ 0.6,\ 0.7,\ 0.9), \\ (0.4,\ 0.5,\ 0.8,\ 0.9) \end{matrix} \right\rangle$ | {[0.6,0.7], [0.8,1.1], [0.6,1.2]} |
| CR6 | High | $\left\langle \begin{matrix} (0.3,\ 0.4,\ 0.5,\ 0.6),\ (0.2, 0.4, 0.5, 0.7), \\ (0.2, 0.3, 0.6, 0.7) \end{matrix} \right\rangle$ | {[0.4,0.5], [0.6,0.9], [0.4,1.0]} |

TABLE 10.5: Deneutrosophied constraint satisfaction values.

| Alt | CR1 | CR2 | CR3 | CR4 | CR5 | CR6 |
|---|---|---|---|---|---|---|
| M1 | .6167 | .6667 | .5833 | .5333 | .8333 | .1333 |
| M2 | .6500 | .7000 | .9000 | .5167 | .9000 | .1500 |
| M3 | .6000 | .3667 | .3667 | .6000 | .8000 | .9333 |
| M4 | .6333 | .6833 | .8667 | .5333 | .9333 | .1667 |
| M5 | .6667 | .7167 | .1167 | .5333 | .3000 | .1667 |

The final deneutrosophied constraint satisfaction values obtained using step 4 are presented in Table 10.5.

Using step 5, the weighted normalized matrix considering equal weightage with constraint satisfaction values thus obtained is presented in Table 10.6.

The positive and negative ideal solutions of each alternative computed using step 7 is presented in Table 10.7.

The score values of the alternatives thus obtained using step 8 is presented in Table 10.8.

**TABLE 10.6**: Weighted normalized matrix.

| **Alt** | CR1 | CR2 | CR3 | CR4 | CR5 | CR6 |
|---|---|---|---|---|---|---|
| M1 | .102717 | .000281 | .000257 | .088828 | .000365 | .00005982 |
| M2 | .108298 | .000337 | .000371 | .086049 | .000425 | .00007212 |
| M3 | .09992 | .000132 | .000173 | .09997 | .000324 | .00038887 |
| M4 | .105499 | .00035 | .000332 | .088842 | .000472 | .00008549 |
| M5 | .111097 | .000324 | .0000084 | .088828 | .000162 | .00007480 |

**TABLE 10.7**: Ideal solutions of the alternatives.

| **Alternative** | $\boldsymbol{E_i^+}$ | $\boldsymbol{E_i^-}$ |
|---|---|---|
| M1 | .003961 | .013946 |
| M2 | .008384 | .014208 |
| M3 | .013925 | .011183 |
| M4 | .006247 | .012466 |
| M5 | .01153 | .011144 |

**TABLE 10.8**: Score values of alternatives.

| **Alternative** | $C_i$ |
|---|---|
| M1 | .778802 |
| M2 | .628895 |
| M3 | .445396 |
| M4 | .666168 |
| M5 | .491488 |

## 10.6 Results and Discussion

The alternatives are ranked based on the score values in Table 10.9.

The method neutrosophic ISOCOV is more beneficial in comparison with fuzzy ISOCOV as it accommodates the constrain satisfaction values in the neutrosophic representations. The ranking results are sensitively analyzed under different considerations to test the consistency of the ranking results.

**TABLE 10.9**: Ranking of alternatives with equal criterion weightage.

| **M1** | **M2** | **M3** | **M4** | **M5** |
|---|---|---|---|---|
| 1 | 3 | 5 | 2 | 4 |

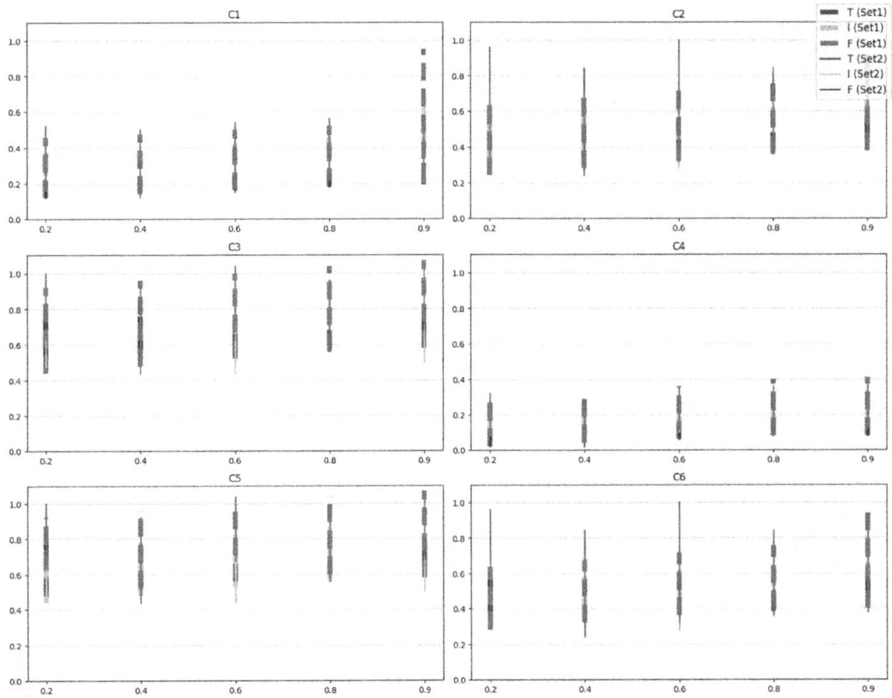


**FIGURE 10.1**: Graphical representation of $\alpha$, $\beta, \gamma$-cut of NTrN.

**Case 1**

The proposed neutrosophic ISOCOV method is discussed under different criterion weights.

The ranking results obtained for different criterion weights are presented in Table 10.10.

**Case 2a**

The ranking results shall also be analyzed using different values of $\alpha$, $\beta, \gamma$ of NTrN. Initially in the first set of $\alpha$, $\beta, \gamma$ values, all are assumed to be same. The rankings obtained are presented in Table 10.11.

**Case 2b**

In the second set of $\alpha$, $\beta, \gamma$, the values of these parameters are taken in combination. The rankings of alternatives are presented in Table 10.12.

The graphical representation of $\alpha$, $\beta, \gamma$-cut of NTrN with these two different sets of values of $\alpha$, $\beta, \gamma$ are presented in Figure 10.1 as sub-plots for each of the cases.

The score values considering the different values of $\alpha$, $\beta$, $\gamma$ are presented in Table 10.13.

**TABLE 10.10**: Ranking of alternatives using different criterion weights.

| | | | | | | |
|---|---|---|---|---|---|---|
| **Case (1a)** | | | | | | |
| **Criterion weights** | **CR1** | **CR2** | **CR3** | **CR4** | **CR5** | **CR6** |
| | .20 | .15 | .10 | .25 | .20 | .10 |
| **Score values** | **M1** | **M2** | **M3** | **M4** | **M5** | |
| | .59748 | .000048 | .37987 | .4181 | .43038 | |
| **Ranking results** | 1 | 5 | 4 | 3 | 2 | |
| **Case (1b)** | | | | | | |
| **Criterion weights** | **CR1** | **CR2** | **CR3** | **CR4** | **CR5** | **CR6** |
| | .30 | .10 | .05 | .15 | .25 | .15 |
| **Score values** | **M1** | **M2** | **M3** | **M4** | **M5** | |
| | .47095 | .000029407 | .42893 | .46587 | .47605 | |
| **Ranking results** | 3 | 5 | 4 | 2 | 1 | |
| **Case (1c)** | | | | | | |
| **Criterion weights** | **CR1** | **CR2** | **CR3** | **CR4** | **CR5** | **CR6** |
| | .10 | .20 | .15 | .20 | .25 | .10 |
| **Score values** | **M1** | **M2** | **M3** | **M4** | **M5** | |
| | .56267 | .00002411 | .2267009 | .29130 | .31534 | |
| **Ranking results** | 1 | 5 | 4 | 3 | 2 | |
| **Case (1d)** | | | | | | |
| **Criterion weights** | **CR1** | **CR2** | **CR3** | **CR4** | **CR5** | **CR6** |
| | .25 | .25 | .10 | .15 | .15 | .10 |
| **Score values** | **M1** | **M2** | **M3** | **M4** | **M5** | |
| | .55077 | .00002757 | .3709665 | .41512 | .429817 | |
| **Ranking results** | 1 | 5 | 4 | 3 | 2 | |
| **Case (1e)** | | | | | | |
| **Criterion weights** | **CR1** | **CR2** | **CR3** | **CR4** | **CR5** | **CR6** |
| | .15 | .10 | .20 | .20 | .15 | .20 |
| **Score values** | **M1** | **M2** | **M3** | **M4** | **M5** | |
| | .7371 | .00002228 | .2810634 | .33784 | .35817 | |
| **Ranking results** | 1 | 5 | 4 | 3 | 2 | |

**TABLE 10.11**: Ranking of alternatives using same $\alpha$, $\beta, \gamma$ values.

| Criteria | Case 1<br>$\alpha, \beta, \gamma = 0.2$ | Case 2<br>$\alpha, \beta, \gamma = 0.4$ | Case 3<br>$\alpha, \beta, \gamma = 0.6$ | Case 4<br>$\alpha, \beta, \gamma = 0.8$ | Case 5<br>$\alpha, \beta, \gamma = 0.9$ |
|---|---|---|---|---|---|
| **CR1 moderate** | {[0.12,0.38],<br>[0.24,0.34],<br>[0.12,0.46]} | {[0.14,0.36],<br>[0.28,0.38],<br>[0.14,0.48]} | {[0.16,0.34],<br>[0.32,0.42],<br>[0.16,0.50]} | {[0.18,0.32],<br>[0.36,0.46],<br>[0.18,0.52]} | {[0.28,0.31],<br>[0.38,0.69],<br>[0.19,0.95]} |
| **CR2 high** | {[0.32,0.58],<br>[0.28,0.54],<br>[0.24,0.64]} | {[0.34,0.56],<br>[0.36,0.58],<br>[0.28,0.68]} | {[0.36,0.54],<br>[0.44,0.62],<br>[0.32,0.72]} | {[0.38,0.52],<br>[0.52,0.66],<br>[0.36,0.76]} | {[0.46,0.55],<br>[0.56,0.88],<br>[0.38,0.97]} |
| **CR3 very high** | {[0.52,0.78],<br>[0.44,0.74],<br>[0.44,0.92]} | {[0.54,0.76],<br>[0.48,0.78],<br>[0.48,0.96]} | {[0.56,0.74],<br>[0.52,0.82],<br>[0.52,1.00]} | {[0.58,0.72],<br>[0.56,0.86],<br>[0.56,1.04]} | {[0.64,0.73],<br>[0.58,0.99],<br>[0.58,1.07]} |
| **CR4 low** | {[0.02,0.28],<br>[0.12,0.22],<br>[0.02,0.28]} | {[0.04,0.26],<br>[0.14,0.24],<br>[0.04,0.32]} | {[0.06,0.24],<br>[0.16,0.26],<br>[0.06,0.36]} | {[0.08,0.22],<br>[0.18,0.28],<br>[0.08,0.40]} | {[0.08,0.23],<br>[0.19,0.29],<br>[0.08,0.41]} |
| **CR5 very high** | {[0.52,0.78],<br>[0.44,0.74],<br>[0.48,0.92]} | {[0.54,0.76],<br>[0.48,0.78],<br>[0.52,0.96]} | {[0.56,0.74],<br>[0.52,0.82],<br>[0.56,1.00]} | {[0.58,0.72],<br>[0.56,0.86],<br>[0.60,1.04]} | {[0.64,0.73],<br>[0.58,0.99],<br>[0.58,1.07]} |
| **CR6 high** | {[0.32,0.58],<br>[0.28,0.54],<br>[0.28,0.64]} | {[0.34,0.56],<br>[0.36,0.58],<br>[0.32,0.68]} | {[0.36,0.54],<br>[0.44,0.62],<br>[0.36,0.72]} | {[0.38,0.52],<br>[0.52,0.66],<br>[0.40,0.76]} | {[0.46,0.55],<br>[0.56,0.88],<br>[0.40,0.94]} |

**TABLE 10.12**: Ranking of alternatives using different combination of $\alpha$, $\beta, \gamma$ values.

| Criteria | Case 1<br>**(0.2,0.4,0.6)** | Case 2<br>**(0.3,0.6,0.2)** | Case 3<br>**(0.5,0.2,0.8)** | Case 4<br>**(0.7,0.8,0.4)** | Case 5<br>**(0.9,0.5,0.7)** |
|---|---|---|---|---|---|
| **CR1 moder-ate** | {[0.12,0.38],<br>[0.28,0.38],<br>[0.16,0.52]} | {[0.13,0.37],<br>[0.32,0.42],<br>[0.12,0.50]} | {[0.15,0.35],<br>[0.24,0.32],<br>[0.18,0.54]} | {[0.18,0.32],<br>[0.36,0.46],<br>[0.22,0.56]} | {[0.28,0.31],<br>[0.30,0.50],<br>[0.20,0.52]} |
| **CR2 high** | {[0.32,0.58],<br>[0.32,0.58],<br>[0.32,0.96]} | {[0.33,0.57],<br>[0.44,0.66],<br>[0.24,0.84]} | {[0.36,0.54],<br>[0.28,0.50],<br>[0.32,1.0]} | {[0.38,0.52],<br>[0.52,0.66],<br>[0.36,0.84]} | {[0.46,0.55],<br>[0.50,0.80],<br>[0.38,0.91]} |
| **CR3 very high** | {[0.52,0.78],<br>[0.48,0.76],<br>[0.64,1.0]} | {[0.53,0.77],<br>[0.52,0.78],<br>[0.44,0.92]} | {[0.56,0.74],<br>[0.44,0.76],<br>[0.68,1.04]} | {[0.58,0.72],<br>[0.68,0.92],<br>[0.56,0.96]} | {[0.64,0.73],<br>[0.50,0.90],<br>[0.70,1.03]} |
| **CR4 low** | {[0.02,0.28],<br>[0.14,0.24],<br>[0.08,0.32]} | {[0.03,0.27],<br>[0.16,0.26],<br>[0.02,0.28]} | {[0.06,0.24],<br>[0.10,0.20],<br>[0.12,0.36]} | {[0.08,0.22],<br>[0.18,0.28],<br>[0.08,0.36]} | {[0.08,0.23],<br>[0.15,0.25],<br>[0.12,0.39]} |
| **CR5 very high** | {[0.52 0.78],<br>[0.48,0.78],<br>[0.64,1.0]} | {[0.53,0.77],<br>[0.52,0.78],<br>[0.44,0.92]} | {[0.56,0.74],<br>[0.44,0.76],<br>[0.68,1.04]} | {[0.58,0.72],<br>[0.68,0.92],<br>[0.56,0.96]} | {[0.64,0.73],<br>[0.50,0.90],<br>[0.70,1.03]} |
| **CR6 high** | {[0.32.0.58],<br>[0.32,0.58],<br>[0.32,0.96]} | {[0.33,0.57],<br>[0.44,0.66],<br>[0.24,0.84]} | {[0.36,0.54],<br>[0.28,0.50],<br>[0.32,1.0]} | {[0.38,0.52],<br>[0.52,0.66],<br>[0.36,0.84]} | {[0.46,0.55],<br>[0.50,0.80],<br>[0.38,0.91]} |

The graphical representation of the score values is presented in Figure 10.2.

From all the above ranking results, it is found that the alternative M1 occupies the first position and M2 occupies the last position. The consistency of the ranking results is validated across several cases.

**TABLE 10.13**: Score values of alternatives.

| $\alpha$, $\beta$, $\gamma$ cases | M1 | M2 | M3 | M4 | M5 |
|---|---|---|---|---|---|
| $\alpha = \beta = \gamma = 0.2$ | 0.042 | 0.038 | 0.039 | 0.041 | 0.040 |
| $\alpha = \beta = \gamma = 0.4$ | 0.084 | 0.075 | 0.076 | 0.080 | 0.082 |
| $\alpha = \beta = \gamma = 0.6$ | 0.125 | 0.113 | 0.115 | 0.120 | 0.122 |
| $\alpha = \beta = \gamma = 0.8$ | 0.167 | 0.151 | 0.153 | 0.160 | 0.164 |
| $\alpha = \beta = \gamma = 0.9$ | 0.188 | 0.170 | 0.172 | 0.180 | 0.184 |
| $\alpha$, $\beta$, $\gamma =$ (0.2,0.4,0.6) | 0.065 | 0.058 | 0.059 | 0.062 | 0.063 |
| $\alpha$, $\beta$, $\gamma =$ (0.3,0.6,0.2) | 0.072 | 0.064 | 0.065 | 0.069 | 0.070 |
| $\alpha$, $\beta$, $\gamma =$ (0.5,0.2,0.8) | 0.102 | 0.094 | 0.096 | 0.098 | 0.100 |
| $\alpha$, $\beta$, $\gamma =$ (0.7,0.8,0.4) | 0.140 | 0.126 | 0.128 | 0.134 | 0.136 |
| $\alpha$, $\beta$, $\gamma =$ (0.9,0.5,0.7) | 0.163 | 0.149 | 0.150 | 0.157 | 0.160 |

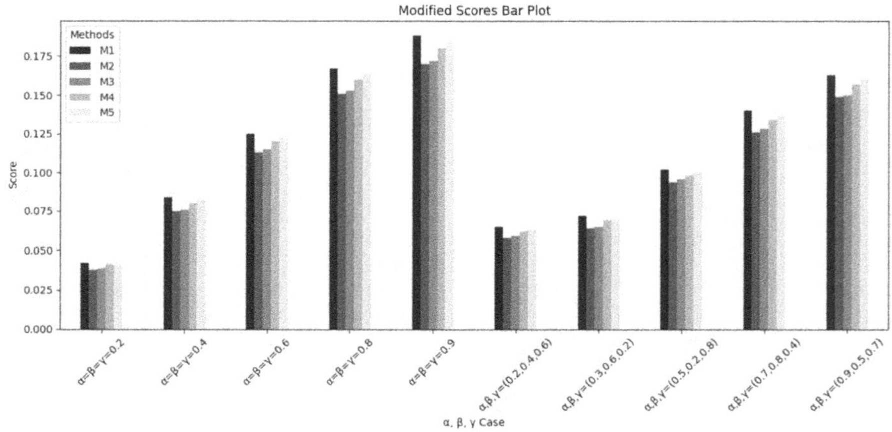


**FIGURE 10.2**: Graphical representation of the score values under different cases.

## 10.7 Conclusion

This chapter proposes a neutrosophic ISOCOV decision-making model as an extended version of existing fuzzy ISOCOV. The consideration of $\alpha$, $\beta$, $\gamma$-cut of neutrosophic trapezoidal number is the significance of this model. The constraint satisfaction values represented as neutrosophic forms present a vivid picture of each criteria. In addition to the criterion weights, the consideration of constraint satisfaction values contributes to the robustness and accuracy of the ranking results. The consistency of the ranking results is validated across different cases. The proposed neutrosophic model shall be applied to any intricate decision-making environment. This method shall also be used in combination with other methods and applied to handle industrial decision-making problems

## Abbreviations

| | |
|---|---|
| AHP | Analytic Hierarchy Process |
| ANP | Analytical Network Process |
| ARAS | Additive Ratio Assessment |
| BIM | Building Information Modeling |
| BWM | Best Worst Method |
| CRITIC | Criteria Importance Through Intercriteria Correlation |
| FCM | Fuzzy Cognitive Map |
| GP | Goal Programming |
| ISOCOV | Ideal Solution with Constraint on Values |
| LCA | Life Cycle Assessment |
| LOPCOW | Logarithmic Percentage Change-Driven Objective Weighting |
| MABAC | Multi-Attributive Border Approximation Area Comparison |
| MACBETH | Measuring Attractiveness by a Categorical Based Evaluation Technique |
| MAIRCA | Multi-Attribute Ideal-Real Comparative Analysis |
| MARCOS | Measurement of Alternatives and Ranking according to Compromise Solution |
| MCDA | Multi-Criteria Decision Analysis |
| MCDM | Multi-Criteria Decision Making |
| MEREC | Method based on Removal Effects of Criteria |
| MOORA | Multi-Objective Optimization on the Basis of Ratio Analysis |
| PIV | Proximity Indexed Value |
| SAW | Simple Additive Weighting |
| SWARA | Stepwise Weight Assessment Ratio Analysis |
| TOPSIS | Technique for Order of Preference by Similarity to Ideal Solution |
| VIKOR | VlseKriterijumska Optimizacija I Kompromisno Resenje (Multi-Criteria Optimization and Compromise Solution) |

## Bibliography

[1] Smarandache, F. (2010). Neutrosophic set–A generalization of the intuitionistic fuzzy set. *Journal of Defense Resources Management (JoDRM)*, *1*(1), 107–116.

[2] Abdelli, A., Mokdad, L., & Hammal, Y. (2020). Dealing with value constraints in decision making using MCDM methods. *Journal of Computational Science*, *44*, 101154. https://doi.org/10.1016/j.jocs.2020.101154.

[3] Martin, N., & Edalatpanah, S. A. (2023). Application of extended fuzzy ISOCOV methodology in nanomaterial selection based on performance measures. *Journal of Operational and Strategic* Analytics, *1*(2), 55–61. https: //doi.org/10.56578/josa010202.

[4] Haruna, A., Shafiq, N., Montasir, O. A., Haruna, S., & Mohammed, M. (2020, April). Design, material selection and manufacturing for sustainable construction: An analytical network process approach. In *IOP Conference Series: Earth and Environmental Science* (Vol. 476, No. 1, p. 012006). IOP Publishing.

[5] Magdy, M., Alamoudy, F. O., & Ismaeel, W. S. (2020). An approach for sustainable material selection using life cycle assessment. *Fayoum University Journal of Engineering*, *3*(2), 53–67.

[6] Riaz, M., Hashmi, M. R., Kalsoom, H., Pamucar, D., & Chu, Y. M. (2020). Linear Diophantine fuzzy soft rough sets for the selection of sustainable material handling equipment. *Symmetry*, *12*(8), 1215.

[7] Rajak, S., Vivek, P., & Jha, S. K. (2020). Application of VIKOR for the selection of material for the green and sustainable construction. In *Innovative Product Design and Intelligent Manufacturing Systems: Select Proceedings of ICIPDIMS 2019* (pp. 777–785). Springer Singapore.

[8] Hatefi, S. M., Asadi, H., Shams, G., Tamošaitienė, J., & Turskis, Z. (2021). Model for the sustainable material selection by applying integrated Dempster-Shafer evidence theory and additive ratio assessment (ARAS) method. *Sustainability*, *13*(18), 10438.

[9] Bapat, H., Sarkar, D., & Gujar, R. (2021). Application of integrated fuzzy FCM-BIM-IoT for sustainable material selection and energy management of metro rail station box project in western India. *Innovative Infrastructure Solutions*, *6*(2), 73.

[10] Alkan, R., Yucesan, M., & Gul, M. (2021, June). A multi-attribute decision-making to sustainable construction material selection: A

Bayesian BWM-SAW hybrid model. In *The International Workshop on Best-Worst Method* (pp. 67–78). Springer International Publishing.

[11] Chen, Z. S., Yang, L. L., Chin, K. S., Yang, Y., Pedrycz, W., Chang, J. P., & Skibniewski, M. J. (2021). Sustainable building material selection: An integrated multi-criteria large group decision making framework. *Applied Soft Computing, 113*, 107903.

[12] Dhanalakshmi, C. S., Mathew, M., & Madhu, P. (2021). Biomass material selection for sustainable environment by the application of multi-objective optimization on the basis of ratio analysis (MOORA). In *Materials, Design, and Manufacturing for Sustainable Environment: Select Proceedings of ICMDMSE 2020* (pp. 345–354). Springer Singapore.

[13] Wakeel, S., Bingol, S., Bashir, M. N., & Ahmad, S. (2021). Selection of sustainable material for the manufacturing of complex automotive products using a new hybrid goal programming model for best worst method–proximity indexed value method. *Proceedings of the Institution of Mechanical Engineers, Part L: Journal of Materials: Design and Applications, 235*(2), 385–399.

[14] Aires, R. F. D. F., & Ferreira, L. (2022). A new multi-criteria approach for sustainable material selection problem. *Sustainability, 14*(18), 11191.

[15] Nicolalde, J. F., Yaselga, J., & Martínez-Gómez, J. (2022). Selection of a sustainable structural beam material for rural housing in Latin América by multicriteria decision methods means. *Applied Sciences, 12*(3), 1393.

[16] Italia, M., Papile, F., Santi, R., & Del Curto, B. (2023). Sustainable material selection framework: Taxonomy and systematisation of design approaches to sustainable material selection. *Sustainability, 15*(24), 16689.

[17] Haq, R. S. U., Saeed, M., Mateen, N., Siddiqui, F., & Ahmed, S. (2023). An interval-valued neutrosophic based MAIRCA method for sustainable material selection. *Engineering Applications of Artificial Intelligence, 123*, 106177.

[18] Maity, R., Mishra, R., Pattnaik, P. K., & Pandey, A. (2023). Selection of sustainable material for the construction of UAV aerodynamic wing using MCDM technique. *Materials Today: Proceedings.*

[19] Soni, A., Chakraborty, S., Das, P. K., & Saha, A. K. (2023). Material selection of sustainable composites by recycling of waste plastics and agro-industrial waste for structural applications: A fuzzy group decision-making approach. *Journal of Building Engineering, 73*, 106787.

[20] Tegegne, D., Abera, M., & Alemayehu, E. (2023). Selection of sustainable building material using multicriteria decision-making model: A case of masonry work in Lideta Subcity, Addis Ababa. *Advances in Civil Engineering, 2023*(1), 9729169.

[21] Rana, V. S., Singh, R. K., Mathur, N., Bisht, Y. S., & Arya, M. K. (2023, July). A hybrid fuzzy-Python MCDM model used in Sheffler solar reflector for the selection of sustainable latent heat storage material. In *International Conference on Scientific and Technological Advances in Materials for Energy Storage and Conversions* (pp. 509–525). Springer Nature Singapore.

[22] Sudakova, A., Lalitha, Y. S., Reddy, K. R., Kataria, A., Prakash, S., Sharma, G., & Almusawi, M. (2024). Sustainable material selection in construction using multi criteria decision analysis (MCDA). In *E3S Web of Conferences* (Vol. 581, p. 01022). EDP Sciences.

[23] Bulut, M. S., Ordu, M., Der, O., & Basar, G. (2024). Sustainable thermoplastic material selection for hybrid vehicle battery packs in the automotive industry: A comparative multi-criteria decision-making approach. *Polymers*, *16*(19), 2768.

[24] Soni, A., Chakraborty, S., Das, P. K., & Saha, A. K. (2024). Selection of sustainable construction material from recycled waste plastics by q-rung orthopair fuzzy SWARA-MABAC approach. *Chemosphere*, *364*, 143166.

[25] Kusnere, Z., Lauka, D., & Spalvins, K. (2024). Multi-criteria decision analysis of wood waste ash and glass foam: Toward sustainable material selection for biomethanation. *Agronomy Research*, *22*(2), 863–875.

[26] Saha, A., & Sumitra, P. (2024). Intuitionistic fuzzy Dombi-Archimedean weighted aggregation operators and their applications in sustainable material selection. In *Deep Learning Applications in Operations Research* (pp. 162–173). Auerbach Publications.

[27] Amellal, I., Amellal, A., & Hattabi, S. (2024). Optimizing selection of sustainable material-extraction sites with GIS-based decision-support systems (AHP/MACBETH): In-depth case study from Settat, Morocco. *Geomatics and Environmental Engineering*, *18*(4), 27–54.

[28] El Hallaoui, Z. (2024). Design of a sustainable out-of-town office building in Morocco: Combining the multi-criteria decision making (MCDM) approach for material selection to the rooftop PV system design. *ESS Open Archive eprints*, *193*, 19383380.

[29] Takano, A., & Aiki, M. (2025). A multi-criteria approach to sustainable building material selection: A case study in a Japanese context. *Sustainability*, *17*(9), 4210.

[30] Mazaheri, S., Zahedi, F., Heidari, A., & Kebriyaii, O. (2025). Material ordering and supplier selection in sustainable project scheduling: A multi-criteria decision making and mathematical modelling approach. *Environment, Development and Sustainability*, *1*, 1–46.

[31] Gligorić, M., Gligorić, Z., Lutovac, S., & Štrbac-Savić, S. (2025). Sustainable construction material selection based on novel integrated criterion free-fall time–sphericity of alternative decision-making model. *Buildings, 15*(9), 1440.

[32] Dasgupta, A., Sen, B., Dutta, P., Rachchh, N., Patil, N., Mahapatro, A., & Karthikeyan, A. (2025). Entropy-TOPSIS-based material selection for sustainable polymer composite: An MCDM framework promoting circular economy. *Journal of Elastomers & Plastics, 57*(5), 703–729.

[33] El Mounla, K., Beladjine, D., & Beddiar, K. (2025). Integrating BIM with lean principles for enhanced decision-making: Optimizing insulation material selection in sustainable construction project. *Energy Informatics, 8*(1), 61.

[34] Haq, R. S. U., Saeed, M., Mateen, N., Siddiqui, F., Naqvi, M., Yi, J. B., & Ahmed, S. (2022). Sustainable material selection with crisp and ambiguous data using single-valued neutrosophic-MEREC-MARCOS framework. *Applied Soft Computing, 128*, 109546.

[35] Saeed, M., Khalid, N. A., & Smarandache, F. (2025). *Quantified Neutrosophic Set (QtNS)-Based MCDM Algorithms for Sustainable Material Selection for Anti-microbial Bio-fabricated Textile Manufacturing.* Infinite Study.

[36] Simic, V., Dabic-Miletic, S., Tirkolaee, E. B., Stević, Ž., Ala, A., & Amirteimoori, A. (2023). Neutrosophic LOPCOW-ARAS model for prioritizing Industry 4.0-based material handling technologies in smart and sustainable warehouse management systems. *Applied Soft Computing, 143*, 110400.

[37] Pontarollo, A., Gennaro, M. E., Hartwig, Z. S., Jepeal, S. J., Short, M. P., Dacus, B. R., & De Marchi, E. (2021, September). Smart material selection for nuclear fusion reactors. In *Offshore Mediterranean Conference and Exhibition* (pp. OMC-2021). OMC.

[38] Seethalakshmi, T., Murugan, M., & Maria Antony Sebastin Vimalan, P. (2024). Smart material selection strategies for sustainable and cost-effective high-performance concrete production using deep learning. *Automatika: časopis za automatiku, mjerenje, elektroniku, računarstvo i komunikacije, 65*(4), 1533–1544.

[39] Asim, A., Nasreen, A., Ahmed, S., Nawab, Y., Siddiqui, F., ul Haq, R. S., & Yi, J. (2025). Innovative 3D commingled weaving and smart material selection for ballistic thermoplastic composites. *Fibers and Polymers, 26*(6), 2595–2607.

[40] Zeina, M. B., Zeitouny, O., Masri, F., Kadoura, F., & Broumi, S. (2021). Operations on single-valued trapezoidal neutrosophic numbers using ($\alpha$, $\beta$, $\gamma$)-cuts "maple package". *International Journal of Neutrosophic Science (IJNS)*, *15*(2), 113–122.

# Chapter 11

## *Latin Square Designs under Indeterminacy: Data Analysis and Applications*

**Abdulrahman AlAita, Muhammad Aslam, and Yasser Al Zaim**

### 11.1 Introduction

The experimental design is fundamental in planning and analyzing experiments in a way that accurate and objective conclusions are drawn. The well-known Latin square design (LSD) is one of the wide spread experimental designs, in which numerous studies have been performed in that field of research. This design could be utilized when each treatment may occur only once, in each row and each column. As a result, the count of treatments, columns, and rows are identical. The LSDs were firstly introduced by Fisher [1]. A review and examples of LSDs and Graeco- LSDs are given in [2, 3]. Exact analysis of LSDs in the presence of a single missing value is discussed by Sirikasemsuk and Leerojanaprapa [4]. Further, the definition and the analysis of balanced incomplete LSD have been discussed in [5, 6].

The neutrosophic statistics (NS), as a measure of indeterminacy, was driven from the neutrosophic logic which is a normal straight generalization of fuzzy logic framework. Likewise, the NS is a straight expansion of classical statistics, but aims to analyze observations under indeterminacy. In effect, the outstanding work of Smarandache [7] was the formal first study to introduce the neutrosophic logic as a more efficient approach than fuzzy logic. Moreover, Smarandache [8] developed NS as an extension of classical statistics, in sense that NS is properly applied if the drawn sample includes uncertain observations. In [9, 10], neutrosophic numbers are introduced as a methodology for the analysis of rock engineering issues. Furthermore, Aslam [11] managed to investigate the existence of what he called a neutrosophic statistical quality

DOI: 10.1201/9781003683339-11 

control (NSQC). Again, Aslam [12] plotted an informative table that catches differences between classical, NS, and fuzzy statistics. The neutrosophic analysis of variance (NANOVA) procedure was given in Aslam [13], and it was followed by defining the neutrosophic ANCOVA with respect to completely randomized, randomized complete block, and split-plot designs in AlAita and Aslam [14]. On the other hand, after rejecting the underlying NANOVA null hypothesis, Aslam and Albassam [15] suggested and performed a series of post hoc comparisons. Other NS approaches raised in other directional researches, like Huang [16], where an innovative new distance measure was defined for single-valued neutrosophic sets. In addition, relying on the flexibility parameters and their reasonable properties, Chutia et al. [17] handled the problem of ordering single-valued neutrosophic numbers. Similarly and based on fuzzy neutrosophic sets, Nafei et al. [18] worked on optimizing score function in the framework of multi-attribute group decision-making. Under a single - valued bipolar neutrosophic environment, Garai and Garg [19] studied water resource management problems in sense of possibilistic multi-attribute decision-making. Concerning handling the problem of resilient supplier selection, the authors of [20] proposed a fuzzy hybrid neutrosophic-based decision-making procedure. There are more and more books and articles related to focusing on NS, and we may refer the interested reader to [21–28].

Today, LSDs are defined for both classic statistical thinking and fuzzy logic. However, those procedures fail to statistically test LSDs under uncertainty, and as a result, we aim in this chapter to fill that gap and introduce a solution that is consistent with imprecise, vague, and uncertain data. To be more precise, our motivation was dealing with complete and incomplete LSDs under NS, which in turn provides all beneficial additional information about the measure of indeterminacy. Obviously, all classical statistical approaches cannot provide such information, since they were designed differently. Our proposed designs can be implemented in a variety of real-life examples, in addition to those mentioned in Section 11.6 of this chapter. For example, the digestion trail for five cows of a particular breed such that each cow receives five different feeds (A, B, C, D, and E) in five successive periods; an evaluation of five major brands of room air cleaners in five homes over a period of five months, and so on. In these examples, we see that classical statistics is not able to treat or analyze this neutrosophic data, since a notable degree of indeterminacy in data does exist. In light of this, our proposed tests in this chapter are necessary for these researches. Objectives of this chapter are: (1) introducing the neutrosophic Latin square design (NLSD) and neutrosophic incomplete Latin square design (NILSD); (2) defining the methodology for NLSD and NILSD; and (3) comparing the performance of existing test statistics to the given test statistics based on Monte Carlo (MC) simulation and real data under NS.

We could not find any previous NS-based approaches that treat both complete and incomplete LSDs, and in our point of view, this is the first literature doing so. Two models would be introduced, namely, the NS-based complete and incomplete LSDs, denoted by NLSD and NILSD. The proposed tests can be applied in an uncertain environment. One may expect that the proposed NLSD and NILSD would perform better than classical LSD and ILSD approaches in terms of flexibility, efficiency, and informatively when testing the equality of means across various groups in uncertain framework. In fact, besides simulation studies reflect that fact, the real data analysis will also be employed to demonstrate that fact. Above all, the MC simulation study will compare both nominal level of significance and test power for the proposed tests with the previously existing classical tests. All codes are written in R language, and would be added online as a supplemental material. The flexibility of R language comes from the ability to build NLSD and NILSD, and to make a full comparison to LSD and ILSD.

Next, some basic and necessarily concepts related to NS are presented in short as preliminaries. In Section 11.2, the analysis of variance (ANOVA) test of NLSD is provided. Results in Section 11.2 are reproduced for NILSD in Section 11.3. In Section 11.4, statistical hypotheses are formulated, and accompanied decision rules are planned. Numerical examples and simulation study scenarios are presented in Sections 11.5 and 11.6 and are widely discussed in detail. In Section 11.7, there is a comparison analysis. The conclusion is presented in Section 11.8.

## 11.2 Preliminaries

Indeterminacy natural of a given study may lead to major false conclusions drawn from the classical statistical analysis, since it is built on the basis of full certain observations. And hence, the problem should be carefully treated so that it does not adversely affect the results, and in effect, it has attracted large group of researchers until, in recent years, they came up with a suitable solution using the neutrosophic statistics. Here are some basic concepts on neutrosophic statistics. Suppose that an neutrosophic random variable (NRV), namely, $X_N \in [X_L, X_U]$ follows an neutrosophic normal distribution (NND) with a neutrosophic grouped mean $\mu_N \in [\mu_L, \mu_U]$ and a neutrosophic variance $\sigma_N^2 \in [\sigma_L^2, \sigma_U^2]$, where $X_L$ and $X_U$ are, respectively, the smaller and larger values of indeterminacy interval. Assume that $X_N = X_L + X_U I_N$ is the neutrosophic formulation of NRV with indeterminate part $X_U I_N$ and determinate part $X_L$; $I_N \in [I_L, I_U]$, and $I_N \in [I_L, I_U]$ is the indeterminate interval. Let $n$ be a neutrosophic simple random sample drawn randomly from a population of $N$ individuals, that has indeterminate observations. Notably, one may express $\mu_N$ and $\sigma_N^2$ as

$$\mu_N \in \left[\frac{\sum_{i=1}^{N} X_{Li}}{N}, \frac{\sum_{i=1}^{N} X_{Ui}}{N}\right]; \mu_N \in [\mu_L, \mu_U]$$

$$\sigma_N^2 \in \left[\frac{\sum_{i=1}^{N} (X_{Li} - \mu_L)^2}{N}, \frac{\sum_{i=1}^{N} (X_{Ui} - \mu_U)^2}{N}\right]; \sigma_N^2 \in \left[\sigma_L^2, \sigma_U^2\right]$$

Of course, as $\mu_N$ and $\sigma_N^2$ are unknown parameters, they can be estimated on the basis of sample observations as numerical examples show later. That was said, the neutrosophic sample mean, denoted by $\bar{X}_N$, and the variance sample, denoted by $s_N^2$, are written as

$$\bar{X}_N \in \left[\frac{\sum_{i=1}^{n} X_{Li}}{n}, \frac{\sum_{i=1}^{n} X_{Ui}}{n}\right]; \bar{X}_N \in \left[\bar{X}_L, \bar{X}_U\right]$$

$$s_N^2 \in \left[\frac{\sum_{i=1}^{n} \left(X_{Li} - \bar{X}_L\right)^2}{n-1}, \frac{\sum_{i=1}^{n} \left(X_{Ui} - \bar{X}_U\right)^2}{n-1}\right]; s_N^2 \in \left[s_L^2, s_U^2\right]$$

### 11.2.1 NS-based post hoc multiple comparisons

This subsection is dedicated to recall post hoc tests presented by Aslam and Albassam [15] under NS.

#### 11.2.1.1 NLSD test (NLSDT)

To build the neutrosophic least significant difference (LSD) test, define

$$LSD_{(i_N - j_N)N} = \frac{\widehat{\mu}_{iN} - \widehat{\mu}_{jN}}{s_{N_{\left(\widehat{\mu}_{iN} - \widehat{\mu}_{jN}\right)}}}; LSD_{(i_N - j_N)N} \in \left[LSD_{(i_L - j_L)L}, LSD_{(i_U - j_U)U}\right] \tag{11.1}$$

where $\widehat{\mu}_{i_N}$ and $\widehat{\mu}_{j_N}$ are neutrosophic means of the $i$th and $j$th groups, and $s_{N_{\left(\widehat{\mu}_{iN} - \widehat{\mu}_{jN}\right)}}$ is the neutrosophic standard error (NSE) between $i$th and $j$th groups.

The proposed test, namely, NLSD can be expressed as follows:

$$LSD_{(i_N - j_N)N} = A_L + A_U I_N; I_N \in [I_L, I_U] \tag{11.2}$$

where $A_L$ and $A_U I_N$ represent determinate part and indeterminate one, respectively. It is obvious that for $I_N = 0$, the NLSD test reduces to ordinary LSD.

#### 11.2.1.2 Neutrosophic Bonferroni test (NBT)

Concerning building neutrosophic Bonferroni test (NBT), take the following quantities

$$BT_N = \min\left(\frac{j_N(j_N-1)}{2} \times LSD_{(i_N-j_N)N}, [1,1]\right); BT_N \in [BT_L, BT_U] \tag{11.3}$$

The proposed test, namely, NBT is given by:

$$LSD_{(i_N-j_N)N} = B_L + B_U I_N; I_N \in [I_L, I_U] \tag{11.4}$$

where $B_L$ and $B_U I_N$ represent determinate part and indeterminate one, respectively. It is obvious that for $I_N = 0$, the NBT reduces to familiar classical counterpart.

#### 11.2.1.3 Neutrosophic Scheffé test (NST)

First, take the quantity $ST_N$ as

$$ST_N = \frac{\left(LSD_{(i_N-j_N)N}\right)^2}{j_N - 1}; ST_N \in [ST_L, ST_U] \tag{11.5}$$

The proposed test, namely, NST is resulted as

$$LSD_{(i_N-j_N)N} = C_L + C_U I_N; I_N \in [I_L, I_U] \tag{11.6}$$

where $C_L$ and $C_U I_N$ represent determinate part and indeterminate one, respectively. Again, this test statistics reduces to the classical counterpart when $I_N = 0$.

### 11.2.2 Neutrosophic analysis of variance (NANOVA)

NANOVA is a new statistical technique that aims to analyze data in the NS framework, such that the overall aggregate variability within dataset is separated into two independent components: neutrosophic random factors and neutrosophic systematic factor. A neutrosophic systematic factor impacts, in statistical meaning, the studied dataset, whereas the neutrosophic random factor will not. Researchers would employ NANOVA methodology to check the effect of independent NRVs on another NRV, which is the dependent variable, and this is known as neutrosophic regression analysis (see [13, 29] for more information).

## 11.3 Neutrosophic Latin Square Design (NLSD)

The $\text{NLSD}_{b\times b}$ Latin square model is given by

$$y_{Nijk} = \mu_N + \alpha_{Ni} + \beta_{Nj} + \tau_{Nk} + \varepsilon_{Nijk};\begin{cases} i = 1, 2, \cdots, b \\ j = 1, 2, \cdots, b \\ k = 1, 2, \cdots, b \end{cases} \tag{11.7}$$

where $y_{Nijk}$ is the neutrosophic observation in the $i$th row and $j$th column for the $k$th treatment, $\mu_N$ is a neutrosophic grand mean, $\alpha_{Ni}$ is the neutrosophic effect of the $i$th row, $\beta_{Nj}$ is the neutrosophic effect of the $j$th column, $\tau_{Nk}$ is the neutrosophic effect of the $k$th treatment, and $\varepsilon_{Nijk}$ is neutrosophic random error with zero mean and variance $\sigma_N^2 \in [\sigma_L^2, \sigma_U^2]$. Take $n$ as the total neutrosophic number of all assumed plots in blocks (rows and columns). It follows that $n = b^2$, where $b$ is number of neutrosophic repetitions, treatments, columns, and rows. Table 11.1 presents an example of the NLSD based on the neutrosophic statistical model (11.7), where $y_{Ni..}$ is the sum of all neutrosophic observations obtained in neutrosophic row $i$, $y_{N.j.}$ is the overall sum of every neutrosophic observations obtained under neutrosophic column $j$, $y_{N..k}$ is the total of all neutrosophic observations obtained under neutrosophic treatment $k$, and $y_{N...}$ is the grand sum of all neutrosophic observations. The following equations can be expressed mathematically:

$$y_{Ni..} = \sum_{j=1}^{b}\sum_{k=1}^{b} y_{Nijk}; i = 1, 2, \cdots, b$$

$$y_{N.j.} = \sum_{i=1}^{b}\sum_{k=1}^{b} y_{Nijk}; j = 1, 2, \cdots, b$$

$$y_{N..k} = \sum_{i=1}^{b}\sum_{j=1}^{b} y_{Nijk}; k = 1, 2, \cdots, b$$

$$y_{N...} = \sum_{i=1}^{b}\sum_{j=1}^{b}\sum_{k=1}^{b} y_{Nijk} = \sum_{i=1}^{b} y_{Ni..} = \sum_{j=1}^{b} y_{N.j.} = \sum_{k=1}^{b} y_{N..k}$$

Similarly, $\bar{y}_{Ni..}$ is the mean of neutrosophic observations taken in the neutrosophic row $i$, $\bar{y}_{N.j.}$ is the mean of neutrosophic observations taken in the neutrosophic column $j$, $\bar{y}_{N..k}$ is the mean of neutrosophic observations taken in treatment $k$, $\bar{y}_{N...}$ and is the overall mean of all neutrosophic observations. That is,

$$\bar{y}_{Ni..} = \frac{y_{Ni..}}{b}, \bar{y}_{N.j.} = \frac{y_{N.j.}}{b}, \bar{y}_{N..k} = \frac{y_{N..k}}{b}, \quad \text{and} \quad \bar{y}_{N...} = \frac{y_{N...}}{b^2}$$

**TABLE 11.1**: The neutrosophic complete LSD of $b \times b$ order.

| | **Column 1** | **Column 2** | **Column 3** | $\cdots$ | **Column b** | **Total rows** |
|---|---|---|---|---|---|---|
| **Row 1** | $A = y_{N111}$ | $B = y_{N122}$ | $C = y_{N133}$ | $\cdots$ | $F = y_{N1b6}$ | $y_{N1..}$ |
| **Row 2** | $D = y_{N214}$ | $E = y_{N225}$ | $A = y_{N231}$ | $\cdots$ | $B = y_{N2b2}$ | $y_{N2..}$ |
| **Row 3** | $B = y_{N312}$ | $C = y_{N323}$ | $F = y_{N336}$ | $\cdots$ | $A = y_{N3b1}$ | $y_{N3..}$ |
| $\vdots$ | $\vdots$ | $\vdots$ | $\vdots$ | $\vdots$ | $\vdots$ | $\vdots$ |
| **Row b** | $E = y_{Nb15}$ | $D = y_{Nb24}$ | $B = y_{Nb32}$ | $\cdots$ | $C = y_{Nbb3}$ | $y_{Nb..}$ |
| **Total columns** | $y_{N.1.}$ | $y_{N.2.}$ | $y_{N.3.}$ | $\cdots$ | $y_{N.b.}$ | |
| **Total treatments** | $y_{N..1}$ | $y_{N..2}$ | $y_{N..3}$ | $\cdots$ | $y_{N..b}$ | $y_{N...}$ |

It is possible to express the total neutrosophic sum of squares as follows:

$$\sum_{i=1}^{b}\sum_{j=1}^{b}\sum_{k=1}^{b}(y_{Nijk}-\bar{y}_{N...})^2 =$$

$$\sum_{i=1}^{b}\sum_{j=1}^{b}\sum_{k=1}^{b}\Bigg[(\bar{y}_{Ni..}-\bar{y}_{N...})+(\bar{y}_{N.j.}-\bar{y}_{N...})+(\bar{y}_{N..k}-\bar{y}_{N...})$$

$$+(y_{Nijk}-\bar{y}_{Ni..}-\bar{y}_{N.j.}-\bar{y}_{N..k}+2\bar{y}_{N...})\Bigg]^2$$

$$=b\sum_{i=1}^{b}(\bar{y}_{Ni..}-\bar{y}_{N...})^2+b\sum_{j=1}^{b}(\bar{y}_{N.j.}-\bar{y}_{N...})^2+b\sum_{k=1}^{b}(\bar{y}_{N..k}-\bar{y}_{N...})^2$$

$$+\sum_{i=1}^{b}\sum_{j=1}^{b}\sum_{k=1}^{b}(y_{Nijk}-\bar{y}_{Ni..}-\bar{y}_{N.j.}-\bar{y}_{N..k}+2\bar{y}_{N...})^2$$

and the final decomposition can be rewritten as

$$SS_{NT}=SS_{NR}+SS_{NC}+SS_{NTr}+SS_{NE} \tag{11.8}$$

Table 11.2 presents the analysis of variance for NLSD.

As an alternative, the computing formulas for NSSs can also be expressed as follows:

$$SS_{NR}=\frac{1}{b}\sum_{i=1}^{b}y_{Ni..}^2-\frac{y_{N...}^2}{n};\ SS_{NR}\in[SS_{LR},SS_{UR}]$$

$$SS_{NC}=\frac{1}{b}\sum_{j=1}^{b}y_{N.j.}^2-\frac{y_{N...}^2}{n};\ SS_{NC}\in[SS_{LC},SS_{UC}]$$

$$SS_{NTr}=\frac{1}{b}\sum_{k=1}^{b}y_{N..k}^2-\frac{y_{N...}^2}{n};\ SS_{NTr}\in[SS_{LTr},SS_{UTr}]$$

$$SS_{NE}=\sum_{i=1}^{b}\sum_{j=1}^{b}\sum_{k=1}^{b}(y_{Nijk}-\bar{y}_{Ni..}-\bar{y}_{N.j.}-\bar{y}_{N..k}+2\bar{y}_{N...})^2;$$

$$SS_{NE}\in[SS_{LE},SS_{UE}]$$

$$SS_{NT}=\sum_{i=1}^{b}\sum_{j=1}^{b}\sum_{k=1}^{b}y_{Nijk}^2-\frac{y_{N...}^2}{n};\ SS_{NT}\in[SS_{LT},SS_{UT}]$$

Neutrosophic mean squares (NMS) and neutrosophic proper test statistics $F_N$ are listed in Table 11.2, where the neutrosophic form of $F_N$ is:

$$F_N=F_L+F_UI_{F_N};I_{F_N}\in[I_{F_L},I_{F_U}] \tag{11.9}$$

**TABLE 11.2**: ANOVA table for NLSD.

| Sources of variation | NSS | Ndf | NMS | $F_N$ |
|---|---|---|---|---|
| **Rows** | $SS_{NR}$ | $b-1$ | $MS_{NR} = \dfrac{SS_{NR}}{b-1}$ | $\dfrac{MS_{NR}}{MS_{NE}}$ |
| **Columns** | $SS_{NC}$ | $b-1$ | $MS_{NC} = \dfrac{SS_{NC}}{b-1}$ | $\dfrac{MS_{NC}}{MS_{NE}}$ |
| **Treatments** | $SS_{NTr}$ | $b-1$ | $MS_{NTr} = \dfrac{SS_{NTr}}{b-1}$ | $\dfrac{MS_{NTr}}{MS_{NE}}$ |
| **Error** | $SS_{NE}$ | $(b-1)(b-2)$ | $MS_{NE} = \dfrac{SS_{NE}}{(b-1)(b-2)}$ | |
| **Total** | $SS_{NT}$ | $n-1$ | | |

where $F_L$ and $F_U I_{F_N}$ represent determinate part and indeterminate one, respectively. And when $F_U I_{F_N} = 0$, $F_N$ coincides with classical $F$ statistics.

---

## 11.4 Neutrosophic Incomplete Latin Square Design (NILSD)

The missing values cause disarrangement and incompleteness and hence non-orthogonality in blocks (rows and columns); that is, all treatments do not appear in each block. This can occur due to negligence or circumstances beyond our control. In this subsection, the authors discuss the existence of missing values for LSD under neutrosophic statistics.

### 11.4.1 Missing a treatment in NLSD

Now, consider NLSD with $b$ rows, columns, and treatments. Let one treatment missing; the statistical model (11.7) is also valid for NLSD with one value missing from treatments. There are two methods for the neutrosophic missing observation problem (neutrosophic approximate analysis and neutrosophic exact analysis).

#### 11.4.1.1 Neutrosophic approximate analysis (NAA)

In this subsection, we will discuss the neutrosophic approximate analysis for NLSD with one neutrosophic observation missing from treatments. To do this, we should follow these steps:

**Step 1:** Estimating the neutrosophic missing observation and placing it in the neutrosophic observations table.

**Step 2:** The neutrosophic error degrees of freedom have been reduced by 1. Consider that the neutrosophic observation of $u$th treatment in $r$th row and $m$th column is missed. In other words, the cell $y_{Nrmu}$ is missed. The mathematical formula to estimate the neutrosophic missing observation $y_{Nrmu}$ can be determined as follows:

$$\widehat{y}_{Nrmu} = \frac{b\,(y_{Nr..} + y_{N.m.} + y_{N..u}) - 2y_{N...}}{(b-2)(b-1)} \tag{11.10}$$

where $y_{Nr..}$ is a sum of $r$th neutrosophic row in which one observation is missed, $y_{N.m.}$ is a sum of $m$th neutrosophic column in which one observation is missed, and $y_{N..u}$ is a sum of $m$th neutrosophic treatment in which one observation is missed. Table 11.3 presents the ANOVA for NLSD with one value missing from treatments (NAA).

**TABLE 11.3**: ANOVA for NLSD with value missing from treatments (NAA).

| Sources of variation | NSS | Ndf | NMS | $F_N$ |
|---|---|---|---|---|
| **Rows** | $SS_{NR}$ | $b-1$ | $MS_{NR}=\frac{SS_{NR}}{b-1}$ | $\frac{MS_{NR}}{MS_{NE}}$ |
| **Columns** | $SS_{NC}$ | $b-1$ | $MS_{NC}=\frac{SS_{NC}}{b-1}$ | $\frac{MS_{NC}}{MS_{NE}}$ |
| **Treatments(C)** | $SS_{NTr(C)}$ | $b-1$ | $MS_{NTr(C)}=\frac{SS_{NTr(C)}}{b-1}$ | $\frac{MS_{NTr(C)}}{MS_{NE}}$ |
| **Error** | $SS_{NE}$ | $b^2-3b+1$ | $MS_{NE}=\frac{SS_{NE}}{b^2-3b+1}$ | |
| **Total** | $SS_{NT}$ | $n-1$ | | |

Neutrosophic sums of squares for NLSD with one value missing from treatments (NAA) are calculated as following:

$$SS_{NR} = \frac{1}{b}\sum_{i=1}^{b} y_{Ni..}^2 - \frac{y_{N...}^2}{n}; SS_{NR} \in [SS_{LR}, SS_{UR}]$$

$$SS_{NC} = \frac{1}{b}\sum_{j=1}^{b} y_{N.j.}^2 - \frac{y_{N...}^2}{n}; SS_{NC} \in [SS_{LC}, SS_{UC}]$$

$$SS_{NTr} = \frac{1}{b}\sum_{k=1}^{b} y_{N..k}^2 - \frac{y_{N...}^2}{n}; SS_{NTr} \in [SS_{LTr}, SS_{UTr}]\,, \text{where } n = b^2 - 1$$

The neutrosophic corrected treatment sum of squars $SS_{NTr(C)} = SS_{NTr} - \text{Bias}_N$ with

$$SS_{NTr(C)} \in \left[SS_{LTr(C)}, SS_{UTr(C)}\right] \text{ and}$$

$$\text{Bias}_N = \left(\frac{y_{N...} - y_{Nr..} - y_{N.m.} - (b-1)y_{N..u}}{(b-1)(b-2)}\right)^2$$

Also, we have

$$\text{Bias}_N \in [\text{Bias}_L, \text{Bias}_U]$$

$$SS_{NT} = \sum_{i=1}^{b}\sum_{j=1}^{b}\sum_{k=1}^{b} y_{Nijk}^2 - \frac{y_{N...}^2}{n}; SS_{NT} \in [SS_{LT}, SS_{UT}]$$

$$SS_{NE} = \text{By subtraction }; SS_{NE} \in [SS_{LE}, SS_{UE}]\,.$$

#### 11.4.1.2 Neutrosophic exact analysis (NEA)

In this subsection, we will discuss the neutrosophic exact analysis for NLSD with neutrosophic value missing from treatments. Table 11.4 presents the ANOVA for NLSD with one value missing from treatments.

Neutrosophic adjusted sums of squares for NLSD with one value missing from treatments are calculated as following:

$$SS_{NR(adj.)} =$$

$$\frac{1}{b}\sum_{i=1}^{b} y_{Ni..}^2 + \frac{1}{b-1}\left[\frac{(y_{Nr..} + y_{N.m.} + y_{N..c} - y_{N...})^2}{b-2} - \frac{(y_{N.m.} + y_{N..c} - y_{N...})^2}{b-1} + \frac{y_{N...}\,(2y_{Nr..} - y_{N...})}{b}\right];$$

$$SS_{NR(adj.)} \in \left[SS_{LR(adj.)}, SS_{UR(adj.)}\right]$$

**TABLE 11.4**: ANOVA table for NLSD with one value missing from treatments.

| **Sources of variation** | **NSS** | **Ndf** | **NMS** | $F_N$ |
|---|---|---|---|---|
| **Rows** | $SS_{NR}$ | $b-1$ | $MS_{NR} = \frac{SS_{NR}}{b-1}$ | $\frac{MS_{NR}}{MS_{NE}}$ |
| **Columns** | $SS_{NC}$ | $b-1$ | $MS_{NC} = \frac{SS_{NC}}{b-1}$ | $\frac{MS_{NC}}{MS_{NE}}$ |
| **Treatments(adj.)** | $SS_{NTr(adj.)}$ | $b-1$ | $MS_{NTr(adj.)} = \frac{SS_{NTr(adj.)}}{b-1}$ | $\frac{MS_{NTr(adj.)}}{MS_{NE}}$ |
| **Error** | $SS_{NE}$ | $b^2-3b+1$ | $MS_{NE} = \frac{SS_{NE}}{b^2-3b+1}$ | |
| **Total** | $SS_{NT}$ | $n-1$ | | |

$$SS_{NC(adj.)} =$$
$$\frac{1}{b}\sum_{i=1}^{b} y_{N.j.}^2 + \frac{1}{b-1}\left[\frac{\left(y_{Nr..} + y_{N.m.} + y_{N..c} - y_{N...}\right)^2}{b-2} - \frac{\left(y_{Nr..} + y_{N..c} - y_{N...}\right)^2}{b-1} + \frac{y_{N...}\left(2y_{N.m.} - y_{N...}\right)}{b}\right];$$
$$SS_{NC(adj.)} \in \left[SS_{LC(adj.)}, SS_{UC(adj.)}\right]$$
$$SS_{NTr(adj.)} =$$
$$\frac{1}{b}\sum_{i=1}^{b} y_{N..k}^2 + \frac{1}{b-1}\left[\frac{\left(y_{Nr..} + y_{N.m.} + y_{N..c} - y_{N...}\right)^2}{b-2} - \frac{\left(y_{Nr..} + y_{N.m.} - y_{N...}\right)^2}{b-1} + \frac{y_{N...}\left(2y_{N..c} - y_{N...}\right)}{b}\right];$$
$$SS_{NTr(adj.)} \in \left[SS_{LTr(adj.)}, SS_{UTr(adj.)}\right]$$
$$SS_{NT} = \sum_{i=1}^{b}\sum_{j=1}^{b}\sum_{k=1}^{b} y_{Nijk}^2 - \frac{y_{N...}^2}{n}; SS_{NT} \in \left[SS_{LT}, SS_{UT}\right]$$
$$SS_{NE} = \text{By subtraction }; SS_{NE} \in \left[SS_{LE}, SS_{UE}\right],$$
$$\text{where } n = b^2 - 1$$

where $y_{Nr..}$ is a sum of $r$th neutrosophic row in which one observation is missed, $y_{N.m.}$ is a sum of $m$th neutrosophic column in which one observation is missed, and $y_{N..c}$ is a sum of $r$th neutrosophic treatment in which one observation is missed.

### 11.4.2 NBILSD

When studying an NLSD with $b$ rows, columns, and treatments, but knowing that $b$ treatments are missing in a way that the remaining cells satisfy the condition of balanced occurrence of symbols, we say that this design is balanced incomplete, and can be called the neutrosophic balanced incomplete LSD (NBILSD). Remarkably, the model given in equation (11.7) is still valid to represent NBILSD. Table 11.5 is the ANOVA table for NBILSD.

To calculate neutrosophic sums of squares (NSS), we recalculate and present the neutrosophic adjusted sums of squares for rows, columns, and treatments using the neutrosophic full effect model and the neutrosophic reduced effect model.

**TABLE 11.5**: ANOVA table for NBILSD.

| Sources of variation | NSS | Ndf | NMS | $F_N$ |
|---|---|---|---|---|
| **Rows** | $SS_{NR}$ | $b-1$ | $MS_{NR} = \frac{SS_{NR}}{b-1}$ | $\frac{MS_{NR}}{MS_{NE}}$ |
| **Columns** | $SS_{NC}$ | $b-1$ | $MS_{NC} = \frac{SS_{NC}}{b-1}$ | $\frac{MS_{NC}}{MS_{NE}}$ |
| **Treatments(adj.)** | $SS_{NTr(adj.)}$ | $b-1$ | $MS_{NTr(adj.)} = \frac{SS_{NTr(adj.)}}{b-1}$ | $\frac{MS_{NTr(adj.)}}{MS_{NE}}$ |
| **Error** | $SS_{NE}$ | $b^2-4b+2$ | $MS_{NE} = \frac{SS_{NE}}{b^2-4b+2}$ | |
| **Total** | $SS_{NT}$ | $n-1$ | | |

Next, the estimated neutrosophic full effect model parameters are shown:

$$\widehat{\mu}_N = \frac{y_{N\ldots}}{b(b-1)}; \widehat{\mu}_N \in [\widehat{\mu}_L, \widehat{\mu}_U]$$

$$\widehat{\alpha}_{Ni_k} = \frac{(b-2)y_{Ni_k..} + y_{N.j_k.} + y_{N..k} - y_{N\ldots}}{b(b-3)}; \widehat{\alpha}_{Ni_k} \in [\widehat{\alpha}_{Li_k}, \widehat{\alpha}_{Ui_k}]$$

$$\widehat{\beta}_{Nj_k} = \frac{(b-2)y_{N.j_k.} + y_{Ni_k..} + y_{N..k} - y_{N\ldots}}{b(b-3)}; \widehat{\beta}_{Nj_k} \in \left[\widehat{\beta}_{Lj_k}, \widehat{\beta}_{Uj_k}\right]$$

$$\widehat{\tau}_{Nk} = \frac{(b-2)y_{N..k} + y_{Ni_k..} + y_{N.j_k.} - y_{N\ldots}}{b(b-3)}; \widehat{\tau}_{Nk} \in [\widehat{\tau}_{Lk}, \widehat{\tau}_{Uk}]$$

Estimation of parameters of the neutrosophic reduced-row (RR) effect model:

$$\widehat{\mu}_N^{RR} = \frac{y_{N\ldots}}{b(b-1)}; \widehat{\mu}_N^{RR} \in \left[\widehat{\mu}_L^{RR}, \widehat{\mu}_U^{RR}\right]$$

$$\widehat{\beta}_{Nj_k}^{RR} = \frac{(b-1)y_{N.j_k.} + y_{N..k} - y_{N\ldots}}{b(b-2)}; \widehat{\beta}_{Nj_k}^{RR} \in \left[\widehat{\beta}_{Lj_k}^{RR}, \widehat{\beta}_{Uj_k}^{RR}\right]$$

$$\widehat{\tau}_{Nk}^{RR} = \frac{(b-1)y_{N..k} + y_{N.j_k.} - y_{N\ldots}}{b(b-2)}; \widehat{\tau}_{Nk}^{RR} \in \left[\widehat{\tau}_{Lk}^{RR}, \widehat{\tau}_{Uk}^{RR}\right]$$

Estimation of parameters of the neutrosophic reduced-column (RC) effect model:

$$\widehat{\mu}_N^{RC} = \frac{y_{N\ldots}}{b(b-1)}; \widehat{\mu}_N^{RC} \in \left[\widehat{\mu}_L^{RC}, \widehat{\mu}_U^{RC}\right]$$

$$\widehat{\alpha}_{Ni_k}^{RC} = \frac{(b-1)y_{Ni_k..} + y_{N..k} - y_{N\ldots}}{b(b-2)}; \widehat{\alpha}_{Ni_k}^{RC} \in \left[\widehat{\alpha}_{Li_k}^{RC}, \widehat{\alpha}_{Ui_k}^{RC}\right]$$

$$\widehat{\tau}_{Nk}^{RC} = \frac{(b-1)y_{N..k} + y_{Ni_k..} - y_{N\ldots}}{b(b-2)}; \widehat{\tau}_{Nk}^{RC} \in \left[\widehat{\tau}_{Lk}^{RC}, \widehat{\tau}_{Uk}^{RC}\right]$$

Estimation of parameters of the neutrosophic reduced-treatment (RT) effect model:

$$\widehat{\mu}_N^{RT} = \frac{y_{N\ldots}}{b(b-1)}; \widehat{\mu}_N^{RT} \in \left[\widehat{\mu}_L^{RT}, \widehat{\mu}_U^{RT}\right]$$

$$\widehat{\alpha}_{Ni_k}^{RT} = \frac{(b-1)y_{Ni_k..} + y_{N.j_k.} - y_{N\ldots}}{b(b-2)}; \widehat{\alpha}_{Ni_k}^{RT} \in \left[\widehat{\alpha}_{Li_k}^{RT}, \widehat{\alpha}_{Ui_k}^{RT}\right]$$

$$\widehat{\beta}_{Nj_k}^{RT} = \frac{(b-1)y_{N.j_k.} + y_{Ni_k..} - y_{N\ldots}}{b(b-2)}; \widehat{\beta}_{Nj_k}^{RT} \in \left[\widehat{\beta}_{Lj_k}^{RT}, \widehat{\beta}_{Uj_k}^{RT}\right]$$

We follow [2] in defining suitable formulas for neutrosophic adjusted sums of squares of rows, columns, treatments, and errors, which turn out to be as follows:

$$SS_{NR(adj.)} = R(\mu_N, \alpha_N, \beta_N, \tau_N) - R(\mu_N, \beta_N, \tau_N); SS_{NR(adj.)} \in \left[SS_{LR(adj.)}, SS_{UR(adj.)}\right]$$

$$SS_{NC(adj.)} = R(\mu_N, \alpha_N, \beta_N, \tau_N) - R(\mu_N, \alpha_N, \tau_N); SS_{NC(adj.)} \in \left[SS_{LC(adj.)}, SS_{UC(adj.)}\right]$$

$$SS_{NTr(adj.)} = R(\mu_N, \alpha_N, \beta_N, \tau_N) - R(\mu_N, \alpha_N, \beta_N); SS_{NTr(adj.)} \in \left[SS_{LTr(adj.)}, SS_{UTr(adj.)}\right]$$

$$SS_{NE} = \text{By subtraction}; SS_{NE} \in [SS_{LE}, SS_{UE}]$$

where

$$R(\mu_N, \alpha_N, \beta_N, \tau_N) = \widehat{\mu}_N y_{N\ldots} + \sum_{i=1}^{b} \widehat{\alpha}_{Ni_k} y_{Ni_k..} + \sum_{i=1}^{b} \widehat{\beta}_{Nj_k} y_{N.j_k.} + \sum_{i=1}^{b} \widehat{\tau}_{Nk} y_{N..k}$$

$$R(\mu_N, \beta_N, \tau_N) = \widehat{\mu}_N^{RR} y_{N\ldots} + \sum_{i=1}^{b} \widehat{\beta}_{Nj_k}^{RR} y_{N.j_k.} + \sum_{i=1}^{b} \widehat{\tau}_{Nk}^{RR} y_{N..k}$$

$$R(\mu_N, \alpha_N, \tau_N) = \widehat{\mu}_N^{RC} y_{N\ldots} + \sum_{i=1}^{b} \widehat{\alpha}_{Ni_k}^{RC} y_{Ni_k..} + \sum_{i=1}^{b} \widehat{\tau}_{Nk}^{RC} y_{N..k}$$

$$R(\mu_N, \alpha_N, \beta_N) = \widehat{\mu}_N^{RT} y_{N\ldots} + \sum_{i=1}^{b} \widehat{\alpha}_{Ni_k}^{RT} y_{Ni_k..} + \sum_{i=1}^{b} \widehat{\beta}_{Nj_k}^{RT} y_{N.j_k.}$$

---

## 11.5 Decision Rule for Neutrosophic Hypotheses

It is worthy to remember that the neutrosophic null and alternative hypotheses are

$$H_{N0} : \alpha_{N1} = \alpha_{N2} = \cdots = \alpha_{Nb} = 0 \text{ v.s. } H_{N1} : \text{ at least one } \alpha_{Ni} \neq 0$$
$$H_{N0} : \beta_{N1} = \beta_{N2} = \cdots = \beta_{Nb} = 0 \text{ v.s. } H_{N1} : \text{ at least one } \beta_{Nj} \neq 0$$
$$H_{N0} : \tau_{N1} = \tau_{N2} = \cdots = \tau_{Nb} = 0 \text{ v.s. } H_{N1} : \text{ at least one } \tau_{Nk} \neq 0$$

In [8], the authors outlined the following decision rules, which are also neutrosophic in nature:
If $\min\{p_N\text{-value}\} > \alpha$, accept $H_{N0}$ at a nominal level of $\alpha$.
If $\max\{p_N\text{-value}\} \leqslant \alpha$, reject $H_{N0}$ at a nominal level of $\alpha$.
If $\min\{p_N\text{-value}\} < \alpha < \max\{p_N\text{-value}\}$, then there is indeterminacy and no decisive decision could be done. For this specific indecisive conclusion, take the ratio of

$\dfrac{\alpha - \min\{p_N\text{-value}\}}{\max\{p_N\text{-value}\} - \min\{p_N\text{-value}\}}$ as a chance to reject $H_{N0}$ at $\alpha$, andtake the ratio of $\dfrac{\max\{p_N\text{-value}\} - \alpha}{\max\{p_N\text{-value}\} - \min\{p_N\text{-value}\}}$ as a chance to accept it.

---

## 11.6 Numerical Examples

**Example 11.1:** *The following is a case study from the yields (in kgs.) of paddy with imprecise observations, where each row and column contains four*

**TABLE 11.6**: Data for the NLSD.

| | Columns | | | |
|---|---|---|---|---|
| **Rows** | 1 | 2 | 3 | 4 |
| 1 | $A_3[22, 27]$ | $A_2[20, 25]$ | $A_1[15, 20]$ | $A_4[17, 22]$ |
| 2 | $A_1[16, 21]$ | $A_4[15, 19]$ | $A_3[19, 24]$ | $A_2[15, 20]$ |
| 3 | $A_2[17, 22]$ | $A_1[10, 15]$ | $A_4[14, 19]$ | $A_3[18, 24]$ |
| 4 | $A_4[13, 17]$ | $A_3[15, 20]$ | $A_2[19, 24]$ | $A_1[10, 16]$ |

*different treatments. For more information, see [30]. Relying on this dataset, we discuss two cases, namely:*

- *NLSD.*
- *NLSD with one missing value of treatments (e.g., omit $A_2$ from the second row and fourth column).*

*Table 11.6 demonstrates how data are aligned in the assumed Latin square.*
*Where* $SS_{NTr(adj.)} = SS_{NTr(C)} = [97.56, 105.17]$.

The ANOVA summary for the NLSD is provided in Table 11.7.
Table 11.8 presents the ANOVA results for the NLSD when one treatment value is missing.

**Example 11.2:** *In this example, in order to get a balanced incomplete Latin square, we rearrange the data in the previous example with another Latin square and omit one treatment of each type, such that remaining cells satisfy the familiar balanced occurrence of symbols condition. Accounting on this dataset, we proceed to discuss the third scenario, namely:*

*NBILSD (e.g., omit $A_1, A_3, A_4, A_2$ from the rows $1 - 4$, respectively). For more information, see [5]. Dataset is distributed in the demonstrated Latin square of Table 11.9.*

**Remark 11.1:** *In the ANOVA Tables 11.7, 11.8, and 11.10, in many cases the null hypothesis was rejected and the alternative hypothesis was accepted. Therefore, to find out the significant differences between groups, we use post hoc multiple comparison tests in Equations* (11.1)–(11.6). *For more information, see [15].*

## 11.7 Simulation Study

The main objective of simulation study section is to compare the efficiency of the proposed test $F_N \in [F_L, F_U]$ for NLSD, NLSD with one missing value of

**TABLE 11.7**: ANOVA table for the NLSD

| Sources of variation | NSS | Ndf | NMS | $F_N$ | N form of F | $p_N$-value |
|---|---|---|---|---|---|---|
| **Rows** | $[43.69, 41.19]$ | 3 | $[14.56, 13.73]$ | $[4.09, 3.25]$ | $4.09 - 3.25I_N; I_N \in [0, 0.258]$ | $[0.07, 0.10]$ |
| **Columns** | $[14.19, 11.69]$ | 3 | $[4.73, 3.90]$ | $[1.33, 0.92]$ | $1.33 - 0.92I_N; I_N \in [0, 0.446]$ | $[0.35, 0.49]$ |
| **Treatments** | $[85.69, 90.69]$ | 3 | $[28.56, 30.23]$ | $[8.02, 7.15]$ | $8.02 - 7.15I_N; I_N \in [0, 0.122]$ | $[0.02, 0.02]$ |
| **Error** | $[21.37, 25.37]$ | 6 | $[3.56, 4.23]$ | | | |
| **Total** | $[164.94, 168.94]$ | 15 | | | | |

**TABLE 11.8**: ANOVA table for the NLSD with one missing value of treatments.

| **Sources of variation** | **NSS** | **Ndf** | **NMS** | $F_N$ | **N form of F** | $p_N$**-value** |
|---|---|---|---|---|---|---|
| **Rows** | $[44.83, 41.58]$ | 3 | $[14.94, 13.86]$ | $[8.01, 6.71]$ | $8.01 - 6.71I_N; I_N \in [0, 0.194]$ | $[0.02, 0.03]$ |
| **Columns** | $[12.28, 10.92]$ | 3 | $[4.09, 3.64]$ | $[2.19, 1.76]$ | $2.19 - 1.76I_N; I_N \in [0, 0.244]$ | $[0.21, 0.27]$ |
| **Treats(adj.)** | $[97.56, 105.17]$ | 3 | $[32.52, 35.06]$ | $[17.42, 16.96]$ | $17.42 - 16.96I_N; I_N \in [0, 0.027]$ | $[0.00, 0.00]$ |
| **Error** | $[9.33, 10.33]$ | 5 | $[1.87, 2.07]$ | | | |
| **Total** | $[164.00, 168.00]$ | 14 | | | | |

**TABLE 11.9**: Data for the NBILSD.

| | Columns | | | |
|---|---|---|---|---|
| **Rows** | 1 | 2 | 3 | 4 |
| 1 | $A_1[15, 20]$ | $A_2[20, 25]$ | $A_3[22, 27]$ | $A_4[17, 22]$ |
| 2 | $A_2[15, 20]$ | $A_1[16, 21]$ | $A_4[15, 19]$ | $A_3[19, 24]$ |
| 3 | $A_3[18, 24]$ | $A_4[14, 19]$ | $A_1[10, 15]$ | $A_2[17, 22]$ |
| 4 | $A_4[13, 17]$ | $A_3[15, 20]$ | $A_2[19, 24]$ | $A_1[10, 16]$ |

treatments and NBILSD with the ordinary previous classical approaches, in terms of uncertainty measuring.

To reveal the efficiency of our results, we compare the proposed test and the previous one, under NS settings, in terms of computing the empirical Type I error rate, and also the test power $(1-\beta)$. Moreover, the two values of $\alpha$ to be considered are $(0.05, 0.01)$.

As for neutrosophic observations total count, we relied on different previous examples, like [2, 3]. The underlying probabilistic distribution is NND, and we run MC simulations for 10000 times (replications).

A suitable algorithm that gives both neutrosophic empirical Type I error rate $\alpha_{\text{Empirical}}$, and empirical statistical test power $Power_{\text{Empirical}}$ is as follows:

Steps for computing $\alpha_{\text{Empirical}}$:

Step 1: Generate, at random, a sample of $x_{N1}^{(i)}, x_{N2}^{(i)}, \cdots, x_{Nn}^{(i)}$ from the NND under $H_{N0}$, where $i = 1, 2, \cdots, 10000$.
Step 2: Under $H_{N0}$, compute $F_{Ni}$-test statistics.
Step 3: Encode results by $I_{Ni} = 1$ when $H_{N0}$ is rejected, and $I_{Ni} = 0$ elsewhere.
Step 4: Compute the ratio of $\frac{1}{10000} \sum_{i=1}^{10000} I_{Ni}$, which stands for $\alpha_{\text{Empirical}}$.

Steps for computing $Power_{\text{Empirical}}$:

Step 1: Generate, at random, a sample of $x_{N1}^{(i)}, x_{N2}^{(i)}, \cdots, x_{Nn}^{(i)}$ from the NND under $H_{N1}$, for $i = 1, 2, \cdots, 10000$. As an example, take $(\mu_{N1}, \mu_{N2}, \mu_{N3}, \mu_{N4}) = (0, 1, 3, 4)$.
Step 2: Under $H_{N1}$, compute $F_{Ni}$-test statistics.
Step 3: Encode results by $I_{Ni} = 1$ when $H_{N1}$ is rejected, and $I_{Ni} = 0$ elsewhere.
Step 4: Compute the ratio of $\frac{1}{10000} \sum_{i=1}^{10000} I_{Ni}$, which stands for $Power_{\text{Empirical}}$.

Figures 11.1 and 11.2 illustrate the Monte Carlo procedures used in the analysis.

TABLE 11.10: ANOVA table for the NBILSD

| Sources of variation | NSS | Ndf | NMS | $F_N$ | N form of F | $p_N$-value |
|---|---|---|---|---|---|---|
| **Rows** | $[76.67, 76.67]$ | 3 | $[25.56, 25.56]$ | $[3.37, 3.88]$ | $3.37 + 3.88I_N; I_N \in [0, 0.131]$ | $[0.24, 0.21]$ |
| **Columns** | $[15.08, 17.08]$ | 3 | $[5.03, 5.69]$ | $[0.66, 0.86]$ | $0.66 + 0.86I_N; I_N \in [0, 0.233]$ | $[0.65, 0.58]$ |
| **Treats(adj.)** | $[33.75, 37.75]$ | 3 | $[11.25, 12.58]$ | $[1.48, 1.91]$ | $1.48 + 1.91I_N; I_N \in [0, 0.225]$ | $[0.43, 0.36]$ |
| **Error** | $[15.17, 13.17]$ | 2 | $[7.58, 6.58]$ | | | |
| **Total** | $[140.67, 144.67]$ | 11 | | | | |

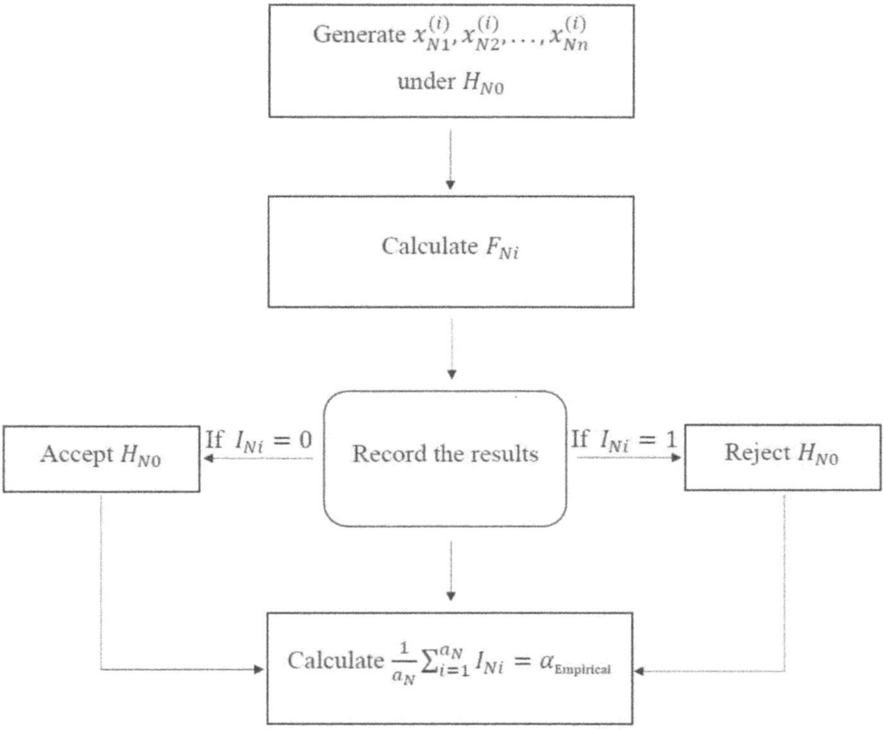


**FIGURE 11.1**: MC steps for $\alpha_{\text{Empirical}}$.

Now, consider the following alternative hypotheses that are planned to evaluate statistical tests power under NS:

$$(\mu_{N1}, \mu_{N2}, \mu_{N3}, \mu_{N4}) = (0, 1, 1, 1)$$
$$(\mu_{N1}, \mu_{N2}, \mu_{N3}, \mu_{N4}) = (0, 1, 2, 2)$$
$$(\mu_{N1}, \mu_{N2}, \mu_{N3}, \mu_{N4}) = (1, 1, 3, 3)$$
$$(\mu_{N1}, \mu_{N2}, \mu_{N3}, \mu_{N4}) = (0, 1, 2, 3)$$
$$(\mu_{N1}, \mu_{N2}, \mu_{N3}, \mu_{N4}) = (0, 2, 3, 3)$$
$$(\mu_{N1}, \mu_{N2}, \mu_{N3}, \mu_{N4}) = (0, 1, 3, 4)$$

Here $\bar{\alpha}_{\text{Empirical}}$ stands for mean empirical type I error.

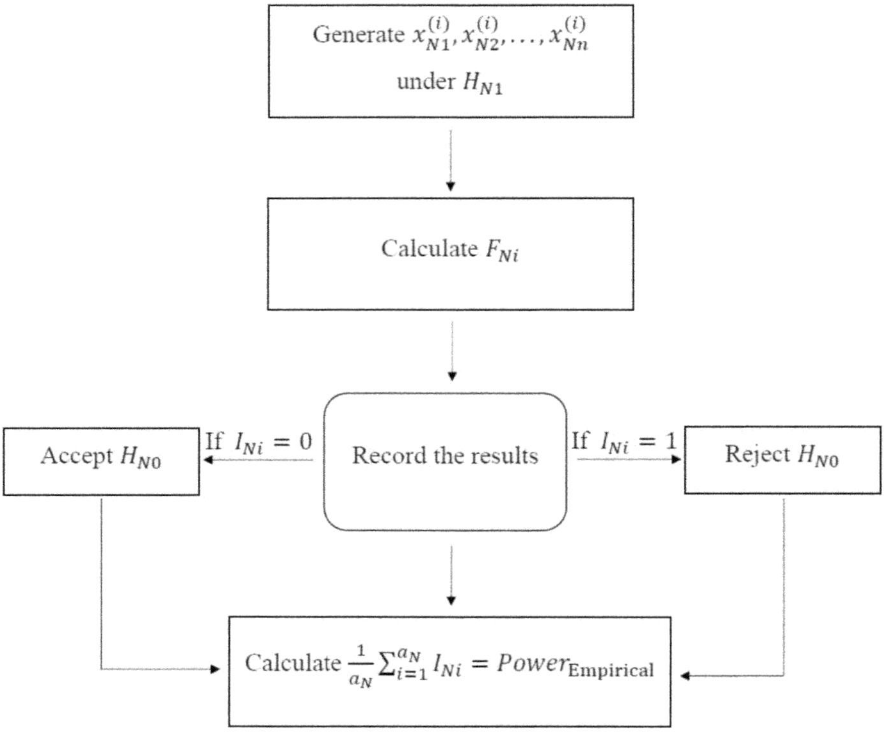


**FIGURE 11.2**: MC steps for $Power_{\text{Empirical}}$.

## 11.8 Comparison Analysis

This section is designated to compare the proposed neutrosophic approach with well-known previous classical designs.

### 11.8.1 In sense of neutrosophic formula of $F_N$-test

As stated earlier, the writing of $F_N = F_L + F_U I_{F_N}; I_{F_N} \in [I_{F_L}, I_{F_U}]$ is reduced to the classical $F$-test by putting $I_{F_N} = 0$. This reflects the fact that the NS thinking gives $F_N$-test in more general form than classical counter parts, since $F_N$ values are arranged in intervals. Take for example the neutrosophic form $1.48 + 1.91 I_N; I_N \in [0, 0.225]$ of $F_N$-test for treatments seen in Table 11.10. This means that the proposed $F_N$-test statistics for NBILSD ranges between 1.48 and 1.91 with the measure of indeterminacy equals to 0.225. To be more

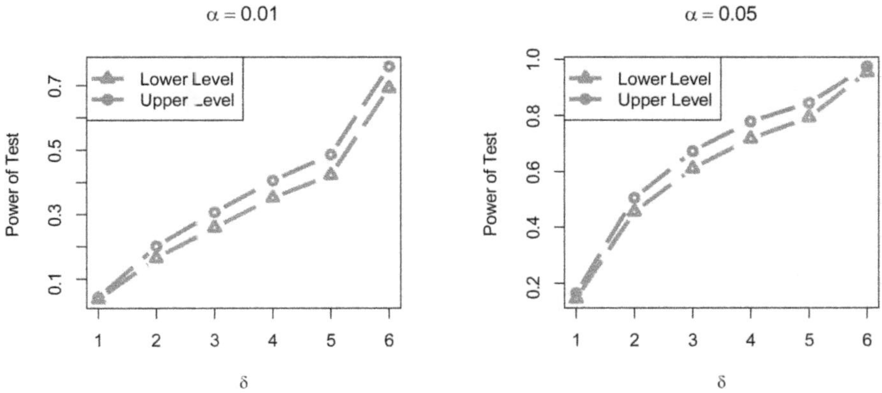


**FIGURE 11.3**: Test power for NLSD.

precise, this indicates that the proposed NS designs are more flexible and informative than classical designs.

### 11.8.2 Comparisons made by observing $\alpha_{\textbf{Empirical}}$ and $power_{\textbf{Empirical}}$

Using $\alpha_{\text{Empirical}}$ and $power_{\text{Empirical}}$, we illustrate the relative performance of the proposed neutrosophic analysis approach to analyze NLSDs, NLSDs with a missing value, and NILSDs in presence of indeterminacy. As the results in the numerical section, for example, Table 11.11, indicate the $\alpha$ and power of the test $1-\beta$ for NLSDs, NLSDs with a missing value, and NBILSDs are more general and flexible than the LSDs, LSDs with a missing value, and BILSDs under classic statistics. For example, in Figure 11.3, it is clear that NLSD for treatments has a higher test power and also $\alpha_{\text{Empirical}}$ lies in $[0.0517, 0.0529]$, which is so close to the nominal significance level of $\alpha = 0.05$. Similarly, under NS, all other $\alpha_{\text{Empirical}}$ are fairly close to its nominal level (0.05 or 0.01). Further, the significant differences between $power_{\text{Empirical}}$ of the determinate and indeterminate parts ensure both flexibility and superiority of the proposed method. Specifically, Figures 11.3–11.5 which are related to plotting power values of Table 11.11, show that the $power_{\text{Empirical}}$ of treatments for indeterminate part uniformly outperforms the determinate one (which corresponds to classical designs). All previous discussions approve our previous statements considering the superiority of neutrosophic approaches in handling LSDs, LSDs with one missing, and lastly BILSDs.

**TABLE 11.11**: Simulation results are shown for row = 4, column = 4, treatment = 4.

| | | | Mean empirical power | | | | | |
|---|---|---|---|---|---|---|---|---|
| **Test** | $\alpha$ | $\bar{\alpha}_{\text{Empirical}}$ | $\delta_1 =$ (0, 1, 1, 1) | $\delta_2 =$ (0, 1, 2, 2) | $\delta_3 =$ (1, 1, 3, 3) | $\delta_4 =$ (0, 1, 2, 3) | $\delta_5 =$ (0, 2, 3, 3) | $\delta_6 =$ (0, 1, 3, 4) |
| **NLSD** | 0.01 | [0.0102, 0.0106] | [0.0382, 0.0437] | [0.1646, 0.2012] | [0.2584, 0.3065] | [0.3515, 0.4057] | [0.4228, 0.4860] | [0.6913, 0.7578] |
| | 0.05 | [0.0517, 0.0529] | [0.1447, 0.1650] | [0.4549, 0.5049] | [0.6111, 0.6723] | [0.7170, 0.7782] | [0.7926, 0.8447] | [0.9540, 0.9746] |
| **NLSD with** | 0.01 | [0.0101, 0.0105] | [0.0321, 0.0391] | [0.1216, 0.1389] | [0.1841, 0.2181] | [0.2646, 0.3084] | [0.3175, 0.3664] | [0.5565, 0.6209] |
| **one missing** | 0.05 | [0.0473, 0.0516] | [0.1331, 0.1438] | [0.3822, 0.4253] | [0.5071, 0.5651] | [0.6478, 0.7125] | [0.7018, 0.7581] | [0.9089, 0.9366] |
| **NBILSD** | 0.01 | [0.0018, 0.0024] | [0.0039, 0.0055] | [0.0102, 0.0128] | [0.0178, 0.0215] | [0.0238, 0.0298] | [0.0253, 0.0289] | [0.0565, 0.0719] |
| | 0.05 | [0.0200, 0.0203] | [0.0365, 0.0394] | [0.1148, 0.1285] | [0.1746, 0.1942] | [0.2101, 0.2377] | [0.2134, 0.2429] | [0.4088, 0.4539] |

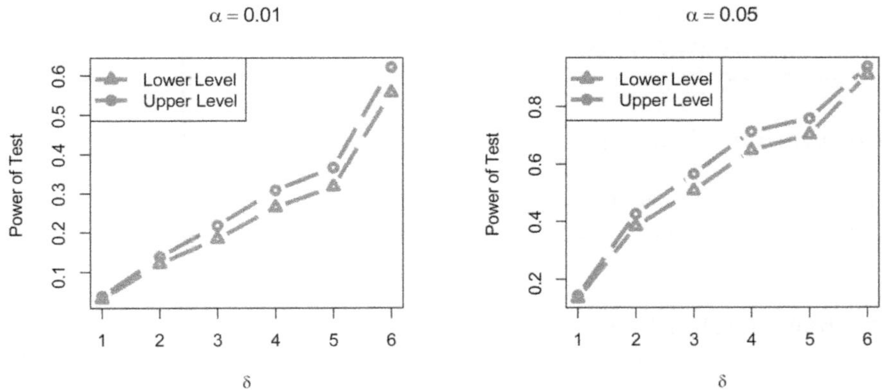


**FIGURE 11.4**: Test power for NLSD with one missing from treatments.

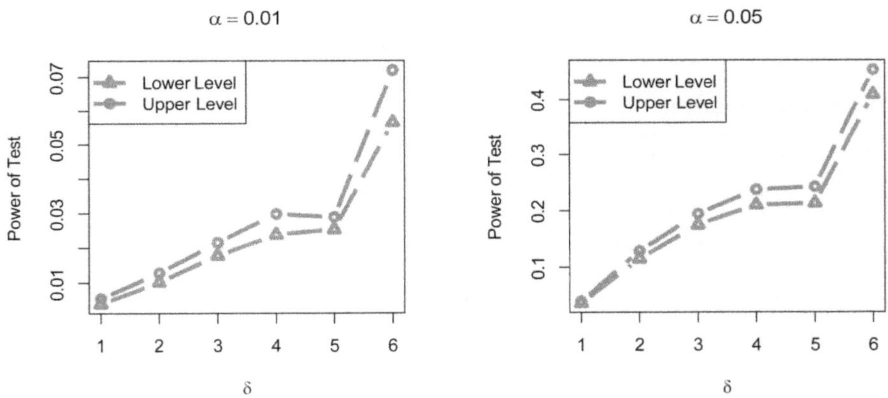


**FIGURE 11.5**: Test power of NBILSD.

---

## 11.9 Conclusions

This chapter presents an analysis of complete and incomplete LSD under neutrosophic statistics. The problem of ANOVA for these designs with indeterminate observations was resolved by determining the exact mathematical formulas for sums of squares. The neutrosophic test statistics $F_N \in [F_L, F_U]$ of the proposed designs have been calculated and have been compared with existing test statistics, which show that the proposed designs are efficient designs for analysis of the data under uncertainty. The MC simulation study has also been made to evaluate the proposed designs. Also, real-world examples have been included to compare the proposed designs with existing designs. It has been observed that the proposed designs are a valuable addition to the experimental designs under NS. The other neutrosophic experimental designs can be studied as future researches.

**Declarations:** Ethics approval and consent to participate: Not applicable
Consent for publication: Not applicable.
Availability of data and material: The data is given in the chapter.
Competing interests: No conflict of interest regarding the chapter.
Funding: None.

**Author contributions:**
AA and YAZ wrote the paper and did the statistics and the relevant table. MA critically and linguistically revised and reviewed the whole literature. All authors read and approved the final manuscript.

---

## Abbreviations:

| | |
|---|---|
| ANOVA | Analysis of variance |
| LSD | Latin square design |
| NANOVA | Neutrosophic analysis of variance |
| MC | Monte Carlo |
| NAA | Neutrosophic approximate analysis |
| NBILSD | Neutrosophic balanced incomplete Latin square design |
| NBT | Neutrosophic Bonferroni test |
| NDF | Neutrosophic degrees of freedom |
| NEA | Neutrosophic exact analysis |
| NILSD | Neutrosophic incomplete Latin square design |
| NLSD | Neutrosophic Latin square design |
| NLSDT | Neutrosophic LSD test |
| NMS | Neutrosophic mean square |
| NND | Neutrosophic normal distribution |
| NRV | Neutrosophic random variable |
| NSS | Neutrosophic sum of square |
| NS | Neutrosophic statistics |
| NST | Neutrosophic Scheffe test |
| RC | Reduced-column |
| RR | Reduced-row |
| RT | Reduced-treatment |

## Bibliography

[1] R. A. Fisher. *Statistical methods for research workers.* Oliver and Boyd, 1925.

[2] D. C. Montgomery. *Design and analysis of experiments.* John Wiley & Sons, 2017.

[3] R. L. Ott and M. T. Longnecker. *An introduction to statistical methods and data analysis.* Cengage Learning, 2015.

[4] K. Sirikasemsuk and K. Leerojanaprapa. One missing value problem in Latin square design of any order: Exact analysis of variance. *Cogent Engineering*, 4(1):1411222, 2017.

[5] M. Ai, K. Li, S. Liu, and D. Lin. Balanced incomplete Latin square designs. *Journal of Statistical Planning and Inference*, 143(9):1575–1582, 2013.

[6] B. Mandal and S. Dash. On balanced incomplete Latin square designs. *Communications in Statistics–Theory and Methods*, 46(13):6258–6263, 2017.

[7] F. Smarandache. *Neutrosophic logic–A generalization of the intuitionistic fuzzy logic.* Multispace & multistructure. Neutrosophic transdisciplinarity (100 collected papers of science), 4, p. 396, 2010.

[8] F. Smarandache. *Introduction to neutrosophic statistics: Infinite study.* Educational Publisher, 2014.

[9] J. Chen, J. Ye, and S. Du. Scale effect and anisotropy analyzed for neutrosophic numbers of rock joint roughness coefficient based on neutrosophic statistics. *Symmetry*, 9(10):208, 2017.

[10] J. Chen, J. Ye, S. Du, and R. Yong. Expressions of rock joint roughness coefficient using neutrosophic interval statistical numbers. *Symmetry*, 9 (7):123, 2017.

[11] M. Aslam. A new sampling plan using neutrosophic process loss consideration. *Symmetry*, 10(5):132, 2018.

[12] M. Aslam. A new attribute sampling plan using neutrosophic statistical interval method. *Complex & Intelligent Systems*, 5(4):365–370, 2019.

[13] M. Aslam. Neutrosophic analysis of variance: Application to university students. *Complex & Intelligent Systems*, 5(4):403–407, 2019.

[14] A. AlAita and A. Aslam. Analysis of covariance under neutrosophic statistics. *Journal of Statistical Computation and Simulation*, 93(3):1–19, 2022.

[15] M. Aslam and M. Albassam. Presenting post hoc multiple comparison tests under neutrosophic statistics. *Journal of King Saud University–Science*, 32(6):2728–2732, 2020.

[16] H. L. Huang. New distance measure of single-valued neutrosophic sets and its application. *International Journal of Intelligent Systems*, 31(10): 1021–1032, 2016.

[17] R. Chutia, M. Krishna, M. Gogoi, M. Firozja, and F. Smarandache. Ordering single-valued neutrosophic numbers based on flexibility parameters and its reasonable properties. *International Journal of Intelligent Systems*, 36(4):1831–1850, 2021.

[18] A. Nafei, A. Javadpour, H. Nasseri, and W. Yuan. Optimized score function and its application in group multiattribute decision making based on fuzzy neutrosophic sets. *International Journal of Intelligent Systems*, 36(12):7522–7543, 2021.

[19] T. Garai and H. Garg. Possibilistic multiattribute decision making for water resource management problem under single-valued bipolar neutrosophic environment. *International Journal of Intelligent Systems*, 2021.

[20] D. Pamucar, M. Yazdani, R. Obradovic, A. Kumar, and M. Torres-Jimenez. A novel fuzzy hybrid neutrosophic decision-making approach for the resilient supplier selection problem. *International Journal of Intelligent Systems*, 35(12):1934–1986, 2020.

[21] M. Abdel-Basset, A. Atef, and F. Smarandache. A hybrid neutrosophic multiple criteria group decision making approach for project selection. *Cognitive Systems Research*, 57:216–227, 2019.

[22] M. Abdel-Basset, M. Gunasekaran, M. Mohamed, and F. Smarandache. A novel method for solving the fully neutrosophic linear programming problems. *Neural Computing and Applications*, 31(5):1595–1605, 2019.

[23] M. Abdel-Basset, G. Manogaran, A. Gamal, and F. Smarandache. A hybrid approach of neutrosophic sets and dematel method for developing supplier selection criteria. *Design Automation for Embedded Systems*, 22 (3):257–278, 2018.

[24] M. Abdel-Basset, M. Mohamed, and V. Chang. Nmcda: A framework for evaluating cloud computing services. *Future Generation Computer Systems*, 86:12–29, 2018.

[25] M. Abdel-Basset, M. Mohamed, and F. Smarandache. An extension of neutrosophic AHP–SWOT analysis for strategic planning and decision-making. *Symmetry*, 10(4):116, 2018.

[26] M. Abdel-Basset, Y. Zhou, M. Mohamed, and V. Chang. A group decision making framework based on neutrosophic VIKOR approach for e-government website evaluation. *Journal of Intelligent & Fuzzy Systems*, 34(6):4213–4224, 2018.

[27] I. Sumathi and C. A. C. Sweety. New approach on differential equation via trapezoidal neutrosophic number. *Complex & Intelligent Systems*, 5 (4):417–424, 2019.

[28] S. Broumi, D. Nagarajan, A. Bakali, M. Talea, F. Smarandache, and M. Lathamaheswari. The shortest path problem in interval valued trapezoidal and triangular neutrosophic environment. *Complex & Intelligent Systems*, 5(4):391–402, 2019.

[29] M. Miari, M. T. Anan, and M. B. Zeina. Neutrosophic two way ANOVA. *International Journal of Neutrosophic Science*, 18(3):73–83, 2022.

[30] S. Parthiban and P. Gajivaradhan. A comparative study of Latin square design under fuzzy environments using trapezoidal fuzzy numbers. *Mathematical Theory and Modeling*, 1(6):57–77, 2016.

# Chapter 12

## *Neutrosophic Programming Approach for Mixed Allocations in Multivariate-Multiobjective Nonlinear Stratified Sampling Problem with Parabolic Fuzzy Numbers*

**Abdul Quddoos, Nisha Khanam, A. H. Ansari, and Md. Izhar Khan**

### 12.1 Introduction

In stratified random sampling (SRS), usually a particular type of allocation procedure is adopted according to the information available about the population. But, due to practical complications the adoption of only one type of allocation to all strata is not always feasible.

Considering the above scenario, Ahsan et al. [1] introduced the concept of "Mixed allocation" based on the work of Clark and Steel [2] for the univariate SRS. Further, its multivariate extension was considered by Varshney and Ahsan [3]. Khanam et al. [4] also considered the multivariate case of the stratified sampling problem to find the compromise mixed allocations by using fuzzy programming approach (FPA). Several other authors modified existing criteria and proposed new approaches for obtaining compromise allocations such as Yates [5], Aoyama [6], Chatterjee [7, 8], Ahsan and Khan [9, 10], Melaku and Sadasivan [11], Bethel [12], Kozak [13], Kreienbrock [14], Díaz-García and Cortez [15, 16], Khan et al. [17], Ansari et al. [18, 19], and many others.

Moreover, in many real-life situations, it is hardly possible for the decision-maker to always determine the exact or precise values of uncertain parameters of sample survey problems. Such type of inexact or imprecise parameters may be well represented using fuzzy numbers.

Fuzzy sets (FSs) were originally proposed by Zadeh [20] to handle the inexact or imprecise data in real-world problems. Later on, Atanassov [21]

DOI: 10.1201/9781003683339-12 

introduced the concept of intuitionistic fuzzy sets (IFSs), which is the extended version of FS. In an FS, a grade of belongingness (membership function) is associated with each element of the set, while in an IFS, a grade of non-belongingness (non-membership) is also associated with each element along with grade of belongingness. The concept of IFS has been further generalized by Smarandache [22] in the form of a neutrosophic sets (NSs). According to him, one more grade of indeterminacy is also associated with each element along with membership and non-membership grades.

Based on the concepts of FS, Zimmermann [23] developed an FPA for solving multi-objective programming problems (MOPP), which was later appreciated by Dyson [24]. FPA is considered as an efficient solution approach for MOPP. Another efficient approach named neutrosophic programming approach (NPA) has been proposed by Smarandache [22] for solving MOPP based on the concepts of NS. Ample contribution of NPA can be seen in the literature in various fields of research; for example, Ahmad and Adhami [25] suggested the utility of NPA to multi-objective non-linear transportation problem with fuzzy parameters; Das et al. [26] solved the multi-objective non-linear programming problem based on neutrosophic optimization technique and suggested its application in Riser design problem; and Sarkar et al. [27] proposed multi-objective neutrosophic optimization technique and its application to structural design. Ahmed et al. [28] developed a new method for solving linear programming problems (LPPs) in bi-polar single-valued neutrosophic environment. Jamil Ahmed [29] presented a method to solve linear regression (LR)-type single-valued LPPs and proposed ranking function for solving real-life problems.

Of late, Mona Mohamed and Abdullah Gamal [30] presented a study on appraising the obstacles facing Industry 5.0 in emerging economies under neutrosophic theory. Monem et al. [31] presented an integrated neutrosophic regional management ranking method for agricultural water management. Gamal et al. [32] suggested a decision-making framework to examine a responsive supply chain based on Industry 5.0 for healthcare systems.

Despite the application of NPA in various areas, we could not encounter any application of NPA in the area of SRS. Gupta and Bari [33] formulated an MMNSSP in which they represented the quadratic cost, stratum standard deviations, measurement cost, travel cost, and the total budget of the survey as parabolic fuzzy numbers and adopted optimum allocation procedure to find the compromise allocations. Their use of parabolic fuzzy numbers to handle uncertain parameters is an innovative and effective contribution to SRS. However, their assumption of an uncertain total survey budget is quite unconvincing, since, according to Neyman [34], the sampling variance of the population mean is minimized under a fixed survey budget. To overcome the issue, we removed the condition of uncertain total survey budget and formulated an MMNSSP with parabolic fuzzy parameters (stratum standard deviation and per unit measurement cost) to obtain the compromise mixed allocations under the fixed total survey budget. The grouping scheme of the proposed model is same as considered by Ahsan et al. [1], which depends upon a compromise criterion for working out a mixed allocation. Therefore the proposed allocation is called "Compromise Mixed Allocation". To obtain the compromise mixed allocations, an NPA has been discussed in detail for the formulated MMNSSP. Finally, a numerical problem is taken to demonstrate the practical utility of the proposed approach and the compromise mixed allocations are obtained using NPA and FPA both. The results thus obtained from both approaches have also been compared to check the efficiency of NPA over FPA.

This chapter is organized as follows: following the "Introduction", the "Preliminaries" section presents the basic concepts of fuzzy sets, parabolic fuzzy numbers, intuitionistic fuzzy sets, and single-valued neutrosophic sets. The next section provides the formulation of the MMNSSP under parabolic fuzzy parameters. In the next section, the NPA and the corresponding stepwise algorithm are discussed for obtaining the compromise mixed allocations. This is followed by a numerical illustration to demonstrate the practical applicability of the proposed approach, along with the estimation of relative efficiency. Finally, a comparison, Conclusion, and future scope are presented, and the advantages and limitations of the proposed methodology are listed.

## 12.2 Preliminaries

**Definition 12.1:** Fuzzy sets

Let $W$ be the universe whose generic element be denoted by $w$. A fuzzy set $X$ in $W$ is a function $X : W \longrightarrow [0, 1]$.

We frequently use $\mu_X(w)$ for the function $X$ and say that the fuzzy set $X$ is characterized by its membership function $\mu_X(w) : W \longrightarrow [0, 1]$ which associates with each $w$ in $W$, a real number $\mu_X(w)$ in $[0, 1]$. The value of $\mu_X(w)$ at $w$ represents the grade of membership of $W$ in $X$ and is interpreted as the degree to which $w$ belongs to $X$ (Bellman and Zadeh [35]).

**Definition 12.2:** Parabolic fuzzy number

Let $\tilde{X}$ be a parabolic fuzzy number which can be represented by the triplets $(a, b, c)$ denoting the lower, middle, and upper value of a membership function and is defined by,

$$\mu_{\tilde{X}}(w) = \begin{cases} (\frac{w-a}{b-a})^2, & \text{if } a \leq w \leq b \\ 1, & \text{if } w = b; \\ (\frac{c-w}{c-b})^2, & \text{if } b \leq w \leq c \\ 0, & \text{if otherwise} \end{cases}$$

**Definition 12.3:** Defuzzification of parabolic fuzzy number

The process of obtaining the single crisp/deterministic values of fuzzy numbers is called defuzzification or ranking. Among various methods of defuzzification, the centroid and the maxima methods are mostly used (Ross [36]). Therefore, the defuzzified value ($d$) of the parabolic fuzzy numbers obtained by using centroid method can be shown as follows:

$$d(\tilde{X}) = \frac{\int_w w\mu_{\tilde{X}}(w)dw}{\int_w \mu_{\tilde{X}}(w)dw} \tag{12.1}$$

where $w$ is the output variable and $\mu_{\tilde{X}}(w)$ is the membership function.

$$\int_w \mu_{\tilde{X}}(w)dw = \int_a^b \left(\frac{w-a}{b-a}\right)^2 dw + \int_b^c \left(\frac{c-w}{c-b}\right)^2 dw = \frac{(c-a)}{3} \tag{12.2}$$

$$\int_w w\mu_{\tilde{X}}(w)dw = \int_a^b w\left(\frac{w-a}{b-a}\right)^2 dw + \int_b^c w\left(\frac{c-w}{c-b}\right)^2 dw$$

$$= \frac{(c-a)(a+2b+c)}{12} \tag{12.3}$$

Hence by using equations (12.2) and (12.3), the defuzzified value or ranking function ($d$) of the parabolic fuzzy number $\tilde{X}(a,b,c)$ is

$$d(\tilde{X}) = \frac{(a+2b+c)}{4}. \tag{12.4}$$

**Definition 12.4:** Intuitionistic fuzzy sets (IFS)

Let, an IFS $X$ in the universal discourse $W$ is given by the ordered triplet as follows (Atanassov [21]):

The $\alpha_X(w)$ and $\beta_X(w)$ denote the membership and non-membership function of the element $w$ in to the set $X$.

$$\left.\begin{array}{l} \mathrm{X} = \left[w, \alpha_X(w), \beta_X(w) : w \in W\right] \\ \text{where} \\ \alpha_X(w) : W \longrightarrow [0,1]; \beta_X(w) : W \longrightarrow [0,1] \\ \text{with conditions} \\ 0 \leq \alpha_X(w) + \beta_X(w) \leq 1. \end{array}\right\}$$

**Definition 12.5:** Neutrosophic set (NS)

Let a neutrosophic set $A$ in the universal discourse $X$ s.t. $x \in X$ is denoted as

$$A = [\langle x, T_A(x), I_A(x), F_A(x)\rangle | x \in X]$$

$T_A(x) : X \longrightarrow ]0^-, 1^+[$ is called truth, $I_A(x) : X \longrightarrow ]0^-, 1^+[$ is called indeterminacy, and $F_A(x) : X \longrightarrow ]0^-, 1^+[$ is called falsity membership functions. $T_A(x)$, $I_A(x)$, $F_A(x)$ show the following relation (Smarandache [22]):

$$0^- \leq sup\ T_A(x) + I_A(x) + sup\ F_A(x) \leq 3^+$$

**Definition 12.6:**
Let $A$ be a single-valued neutrosophic set (SVNS) over universal discourse $X$ and is defined as

$$A = [\langle x, T_A(x), I_A(x), F_A(x)\rangle | x \in X]$$

where $T_A(x)$, $I_A(x)$, $F_A(x) \in [0,1]$ and $0^- \leq sup\, T_A(x) + I_A(x) + sup\, F_A(x) \leq 3^+$ for each $x \in X$ (Smarandache [22]).

**Definition 12.6.1:**
The union of two SVNSs $A$ and $B$ is also a single-valued neutrosophic set $C$, denoted by $C = (A \cup B)$, and is defined by three membership functions: truth $T_C(x)$, indeterminacy $I_C(x)$, and falsity $F_C(x)$ as

$$T_C(x) = \max\,(T_A(x), T_B(x))$$

$$I_C(x) = \max\,(I_A(x), I_B(x))$$

$$F_C(x) = \min\,(F_A(x), F_B(x))$$

for each $x \in X$ (Smarandache [22]).

**Definition 12.6.2:** The intersection of two SVNSs $A$ and $B$ is also a single-valued neutrosophic set $C$, denoted by $C = (A \cap B)$, and is defined by three membership functions: truth $T_C(x)$, indeterminacy $I_C(x)$, and falsity $F_C(x)$ as

$$T_C(x) = \min\,(T_A(x), T_B(x))$$

$$I_C(x) = \min\,(I_A(x), I_B(x))$$

$$F_C(x) = \max\,(F_A(x), F_B(x))$$

for each $x \in X$ (Smarandache [22]).

## 12.3 Formulation of Multi-objective Multivariate Non-Linear Stratified Sampling Problem with Parabolic Fuzzy Parameters

Let the population of sizes $N_1$,$N_2$,...,$N_L$ units has been divided into $L$ sub-populations. These sub-populations are exhaustive and non-overlapping with $\sum_{h=1}^{L} N_h = N$. The unbiased estimator $\bar{y}_{st} = \sum_{h=1}^{L} W_h \bar{y}_h$ of the population mean $\bar{Y}$ is given with variance $V(\bar{y}_{st}) = \sum_{h=1}^{L} \frac{W_h^2 S_h^2}{n_h}$ (ignoring finite population correction (fpc)), $\bar{y}_{st}$ is the stratified sample mean, $W_h^2$ are the weights, and $S_h^2$ are the stratum standard deviations of stratum $h$, respectively.

Let a linear cost function be $c_0 + \sum_{h=1}^{L} c_h n_h \leq C$ or $\sum_{h=1}^{L} c_h n_h \leq C_0$, where $C$ is the total cost of survey and $C_0 = C - c_o$ is the total available survey budget excluding overhead cost $c_0$ for the measurement of all units selected in the stratified samples and $c_h$ is the per unit measurement cost of stratum $h$.

Ansari et al. [19] suggested an individual mixed allocations problem of MMNSSP for $l$ characteristics as given in (12.5):

$$\left.\begin{array}{l} \text{Minimize } f_l(\alpha_{lj}) \\ \text{Subject to;} \\ \sum_{j=1}^{k}\sum_{h\in I_j} c_h \alpha_{lj}\beta_h \leq C_0 \\ \text{and} \\ \alpha_{lj} \geq 0;\ j = 1, 2, ..., k \end{array}\right\} \tag{12.5}$$

where $f_l(\alpha_{lj}) = \sum_{j=1}^{k}\sum_{h\in I_j} \frac{W_h^2 s_{lh}^2}{\alpha_{lj}\beta_h};\ l = 1, 2, ..., p.$
Therefore, the mixed allocations for (12.5) are $n_{lh(mixed)} = \alpha_{lj}\beta_h;\ h = 1, 2, ..., L;\ j = 1, 2..., k;\ l = 1, 2, ..., p$, with an estimated sampling variance, $v_l^* = \sum_{h=1}^{L} \frac{W_h^2 s_{lh}^2}{n_{lh(mixed)}}$.
Further, let us suppose that the stratum standard deviations and per unit measurement costs of stratum $h$ in (12.5) reflect some uncertainty. It may be dealt well by representing such parameters in the form of parabolic fuzzy numbers. The problem shown in (12.5) may be reformulated as (12.6),

$$\left.\begin{array}{l} f_k(n_{kh}) = \sum_{h=1}^{L} \frac{W_h^2 \tilde{s}_{kh}^2}{n_{kh(mixed)}} \\ \text{Subject to;} \\ \sum_{h=1}^{L} \tilde{c_h} n_{kh} \leq C_o \\ \text{and} \\ 0 \leq n_{kh} \leq N_h;\ h = 1, 2, ..., L;\ k = 1, 2, ..., p. \end{array}\right\} \tag{12.6}$$

where $\tilde{s}_{kh}^2$ and $\tilde{c_h}$ denote the parabolic fuzzy values of stratum standard deviations and per unit measurement costs and $f_k(n_{kh})$ denotes the variance of the estimate of the population mean $\bar{Y}_k$ (ignoring fpc) for characteristic $k$. (Here, we take characteristic $k$ instead of $l$.)

To avoid the problem of oversampling, the usual non-negativity restrictions $\alpha_{kj} \geq 0$ have been modified as $0 \leq n_{kh} \leq N_h$ as in (12.6).

Equation (12.6) can easily result in an equivalent crisp problem shown in (12.7) by defuzzifying the fuzzy quantities using (12.4),

$$\left.\begin{array}{l} f_k(n_{kh}) = \sum_{h=1}^{L} \frac{W_h^2 d(\tilde{s}_{kh}^2)}{n_{kh(mixed)}} \\ \text{Subject to;} \\ \sum_{h=1}^{L} d(\tilde{c_h}) n_{kh} \leq C_o \\ \text{and} \\ 0 \leq n_{kh} \leq N_h;\ h = 1, 2, ..., L;\ k = 1, 2, ..., p. \end{array}\right\} \tag{12.7}$$

where $d(\tilde{s}_{kh}^2)$ and $d(\tilde{c_h})$ are the defuzzified values of stratum standard deviations and per unit measurement costs, respectively.

Let $(n_{kh}^*)$ be the optimum allocations for characteristic $k$ and corresponding optimal value of objective function is denoted by $f_k(n_{kh}^*)$ for each $k = 1, 2..., p$. Further, let $n_h$ be the compromise allocations and $f_k(n_h)$ be the compromise value of objective function for characteristic $k$ $(k = 1, 2..., p)$.

Then, obviously we may write,

$$f_k(n_{kh}^*) \leq f_k(n_h), k = 1, 2, ...p;$$

or

$$f_k(n_h) - f_k(n_{kh}^*) \geq 0 \tag{12.8}$$

where $n_h$; $h = 1, 2, ..., L$ denotes the compromise allocation. Equation(12.8) shows the increase in sampling variances if compromise allocation is used in place of individual optimal mixed allocation for characteristic $k$.

## 12.4 Neutrosophic Programming Approach (NPA)

Let us consider a non-linear multi-objective minimization problem as given in (12.9),

$$\left.\begin{array}{l}\text{Minimize } \{f_1(x), f_2(x), ..., f_p(x)\} \\ \text{Subject to;} \\ g_j(x) \leq b_j, j = 1, 2, ..., q. \\ x_j \geq 0\end{array}\right\} \tag{12.9}$$

A neutrosophic decision set $D_N$, a conjunction of neutrosophic objectives and constraints, may be defined as,

$D_N = (\cap_{k=1}^{p} G_k) \cap (\cap_{j=1}^{q} C_j) = (x, T_D(x), I_D(x), F_D(x))$, where

$$T_D(x) = \min \begin{Bmatrix} T_{G1}(x), T_{G2}(x), ..., T_{Gp}(x) \\ T_{C1}(x), T_{C2}(x), ..., T_{Cq}(x) \end{Bmatrix} \forall x \in X.$$

$$I_D(x) = \min \begin{Bmatrix} I_{G1}(x), I_{G2}(x), ..., I_{Gp}(x) \\ I_{C1}(x), I_{C2}(x), ..., I_{Cq}(x) \end{Bmatrix} \forall x \in X.$$

$$F_D(x) = \max \begin{Bmatrix} F_{G1}(x), F_{G2}(x), ..., F_{Gp}(x) \\ F_{C1}(x), F_{C2}(x), ..., F_{Cq}(x) \end{Bmatrix} \forall x \in X.$$

$T_D(x)$, $I_D(x)$, and $F_D(x)$ are the truth, indeterminacy, and falsity membership functions of a neutrosophic decision set $D_N$, respectively.

The problem shown in (12.9) can now be transformed to a vector-minimum non-linear multiobjective programming problem as

$$\left.\begin{array}{l} \text{Max } \lambda \\ \text{Max } \theta \\ \text{Min } \eta \\ \text{Subject to;} \\ g_j(x) \leq b_j, j = 1, 2, ..., q. \\ x_j \geq 0 \\ T_{G_k}(x) \geq \lambda, T_{C_j}(x) \geq \lambda \\ I_{G_k}(x) \geq \theta, I_{C_j}(x) \geq \theta \\ F_{G_k}(x) \leq \eta, F_{C_j}(x) \leq \eta \\ \text{and} \\ \lambda \geq \theta,\ \lambda \geq \eta,\ \lambda + \theta + \eta \leq 3 \\ \lambda, \theta, \eta \in (0, 1)) \end{array}\right\} \quad (12.10)$$

The problem shown in (12.10) may further be written as single objective non-linear programming problem (NLPP) (12.11) as

$$\left.\begin{array}{l} \text{Max } \lambda + \theta - \eta \\ \text{Subject to constraints of (12.10)} \end{array}\right\} \quad (12.11)$$

---

## 12.5 Step-Wise Algorithm for Solving MMNSSP in (12.7)

This section discusses the procedure of NPA for solving (12.7). The step-wise algorithm of NPA may be described as follows:

**Step 1:** Obtain the values of each objective function by considering only one objective at a time and ignoring all others. After solving $k$ objectives individually, record all the individual optimal mixed allocations $(n^*_{1h})$, $(n^*_{2h})$, ..., $(n^*_{ph})$.

**Step 2:** Find all $f_k(n_{kh})$ using each $(n^*_{kh})$, $k = 1, 2, ..., p$ and construct the pay-off matrix as follows:

$$\begin{array}{c} \\ n_{1h} \\ n_{2h} \\ \cdots \\ n_{ph} \end{array} \begin{array}{c} \begin{array}{cccc} f_1 & f_2 & \cdots & f_p \end{array} \\ \begin{bmatrix} f_1(n_{1h}) & f_2(n_{1h}) & \cdots & f_p(n_{1h}) \\ f_1(n_{2h}) & f_2(n_{2h}) & \cdots & f_p(n_{2h}) \\ \cdots & \cdots & \cdots & \cdots \\ f_1(n_{ph}) & f_2(n_{ph}) & \cdots & f_p(n_{ph}) \end{bmatrix} \end{array}$$

**Step 3:** Determine the upper and lower bounds for each objective function as: $U_k = max\ [f_k(n_{kh})]$ and $L_k = min\ [f_k(n_{kh})]$, respectively.

**Step 4:** Represent the upper and lower bounds $U_k$ and $L_k$ $(k = 1, 2, ..., p)$ for truth, indeterminacy, and falsity membership function for each objective as

$$U_k^T = max\ [f_k(n_{kh})],\ L_k^T = min\ [f_k(n_{kh})] \quad \text{for truth membership}$$
$$U_k^I = L_k^T + s_k(U_k^T - L_k^T),\ L_k^I = L_k^T \quad \text{for indeterminacy membership}$$
$$U_k^F = U_k^T,\ L_k^F = L_k^T + t_k(U_k^T - L_k^T) \quad \text{for falsity membership}$$

where $s_k,\ t_k \in (0, 1)$ are predetermined real-numbers assigned by decision-maker.

**Step 5:** The linear membership function for truth, indeterminacy, and falsity can be defined by using the values of upper and lower bounds as

$$T_k[f_k(n_{kh})] = \begin{Bmatrix} 1, & \text{if } [f_k(n_{kh})] \leq L_k^T \\ \dfrac{U_k^T - [f_k(n_{kh})]}{U_k^T - L_k^T}, & \text{if } L_k^T \leq [f_k(n_{kh})] \leq U_k^T \\ 0, & \text{if } [f_k(n_{kh})] \geq U_k^T \end{Bmatrix} \tag{12.12}$$

$$I_k[f_k(n_{kh})] = \begin{Bmatrix} 1, & \text{if } [f_k(n_{kh})] \leq L_k^I \\ \dfrac{U_k^I - [f_k(n_{kh})]}{U_k^I - L_k^I}, & \text{if } L_k^I \leq [f_k(n_{kh})] \leq U_k^I \\ 0, & \text{if } [f_k(n_{kh})] \geq U_k^I \end{Bmatrix} \tag{12.13}$$

$$F_k[f_k(n_{kh})] = \begin{Bmatrix} 1, & \text{if } [f_k(n_{kh})] \leq L_k^F \\ \dfrac{[f_k(n_{kh})] - L_k^F}{U_k^F - L_k^F}, & \text{if } L_k^F \leq [f_k(n_{kh})] \leq U_k^F \\ 0, & \text{if } [f_k(n_{kh})] \geq U_k^F \end{Bmatrix} \tag{12.14}$$

**Step 6:** Let $\lambda = min\ T_k[f_k(n_{kh})]$, $\theta = min\ I_k[f_k(n_{kh})]$, and $\eta = max\ F_k[f_k(n_{kh})]$. An equivalent vector-minimum problem of the problem shown in (12.7) can be written as

$$\left.\begin{array}{l} \text{Max } \lambda + \theta - \eta \\ \text{Subject to;} \\ T_k[f_k(n_{kh})] \geq \lambda,\ I_k[f_k(n_{kh})] \geq \theta,\ F_k[f_k(n_{kh})]) \leq \eta \\ \sum_{h=1}^{L} d(\tilde{c_h})n_{kh} \leq d(\tilde{C_o}) \\ \text{and} \\ 0 \leq n_{kh} \leq N_h;\ h = 1, 2, ..., L;\ k = 1, 2, ..., p. \\ \lambda \geq \theta,\ \lambda \geq \eta,\ \lambda + \theta + \eta \leq 3 \\ \lambda, \theta, \eta \in (0, 1) \end{array}\right\} \tag{12.15}$$

where $d(.)$ represents the defuzzified values of parabolic fuzzy numbers.

TABLE 12.1: Data for seven strata and two characteristics.

| h | $N_h$ | $W_h$ | $s_{1h}$ | $s_{2h}$ | $c_h$ |
|---|---|---|---|---|---|
| 1 | 472 | 0.1888 | (5.2, 5.4, 5.6) | (7.5, 7.7, 7.9) | (6, 7, 8 ) |
| 2 | 559 | 0.2236 | (5.8, 5.10, 5.12) | (30.1, 31.2, 32.3) | (8, 10, 12) |
| 3 | 425 | 0.1700 | (5.1, 5.2, 5.3) | (7.8, 7.9, 7.10) | (6, 8, 10) |
| 4 | 218 | 0.0872 | (22.2, 24.2, 26.2) | (32.2, 34.2, 36.2) | (7, 9, 11) |
| 5 | 233 | 0.0932 | (21.2, 23.4, 25.6) | (16.2, 18.6, 20.2) | (11, 12, 13) |
| 6 | 328 | 0.1312 | (15.1, 16.2, 17.3) | (32.4, 33.6, 34.8) | (10, 11, 12) |
| 7 | 265 | 0.1060 | (39.2, 40.3, 41.4) | (44.2, 45.3, 46.4) | (12, 14, 16) |

**Step 7:** The problem shown in (12.15) reduces to an equivalent NLPP with the help of (12.12)–(12.14) as,

$$\left.\begin{array}{l}\text{Max } \lambda+\theta-\eta \\ \text{Subject to;} \\ \sum_{h=1}^{L} d(\tilde{c_h})n_{kh} \leq d(\tilde{C_o}) \\ \text{and} \\ 0 \leq n_{kh} \leq N_h;\ h=1,2,...,L;\ k=1,2,...,p. \\ [f_k(n_{kh})]+(U_k^T-L_k^T)\lambda \leq U_k^T, \\ [f_k(n_{kh})]+(U_k^I-L_k^I)\theta \leq U_k^I, \\ [f_k(n_{kh})]-(U_k^F-L_k^F)\eta \leq L_k^F, \\ \lambda \geq \theta,\ \lambda \geq \eta,\ \lambda+\theta+\eta \leq 3 \\ \lambda,\theta,\eta \in (0,1)\end{array}\right\} \quad (12.16)$$

**Step 8:** Solve the problem shown in (12.16) using Lingo 18.0 or any other software and obtain the required compromise mixed allocations.

---

## 12.6 A Numerical Illustration

The data for the following numerical are partly from Ansari et al. [19]. In stratification with seven strata and two characteristics, the values of $N_h$, $s_1h$, $s_2h$, $c_h$ are given in Table 12.1. The values of stratum standard deviations for both characteristics and per unit measurement costs are assumed to be fuzzy and the total available budget of the survey $C_0 = 4000$ (excluding overhead cost $c_0$) units are available for the measurement within strata.

The grouping scheme for the above data of both characteristics in this chapter is based on the same grouping scheme as considered by Ahsan et al.[1]

**For the first characteristic:**

1. Equal allocation is to be used for group $G_1$, which comprises strata 1,

2, and 3.
i.e, $\beta_h = 1$ ; $h \in I_1 = \{1, 2, 3\}$.

2. Proportional allocation is to be used for group $G_2$ which constitutes strata 4 and 5,
i.e, $\beta_h = W_h$; $h \in I_2 = \{4, 5\}$.

3. Optimum allocation used for group $G_3$ constitutes strata 6 and 7,
i.e, $\beta_h = \frac{W_h s_h}{\sqrt{c_h}}$ ; $h \in I_3 = \{6, 7\}$.

Thus $I_1 = \{1, 2, 3\}$, $I_2 = \{4, 5\}$ and $I_3 = \{6, 7\}$.
**For the second characteristic:** Equal allocation is to be used in group $G_1$ while proportional and optimum allocation is to be used for groups $G_2$ and $G_3$, respectively.
Therefore, $I_1 = \{1, 3\}$, $I_2 = \{2, 4, 6\}$ and $I_3 = \{5, 7\}$.
It is assumed that $I_j$; $j = 1, 2, 3$ are mutually exclusive and exhaustive.
Table 12.2 shows the values of $\frac{W_h^2 s_{1h}^2}{\beta_h}$ and $c_h\beta_h$ for the first characteristic. From the computations of Table 12.2, the mixed allocations can be obtained for the first characteristic as

$$
\begin{aligned}
n_{11(mixed)} &= n_{12(mixed)} = n_{13(mixed)} = 31.22397 \approx 31 \\
n_{14(mixed)} &= 55.6136 \approx 56, \ n_{15(mixed)} = 59.4402 \approx 59 \\
n_{16(mixed)} &= 55.8012 \approx 56, \ n_{17(mixed)} = 99.4129 \approx 99
\end{aligned}
$$

with an estimated sampling variance $v_1^* = \sum_{h=1}^{L} \frac{W_h^2 s_{1h}^2}{n_{1h(mixed)}} = 0.528786$.

Using Table 12.3, the mixed allocations for the second characteristic can be obtained as

$$
\begin{aligned}
n_{21(mixed)} &= n_{23(mixed)} = 26.15039 \approx 26 \\
n_{22(mixed)} &= 118.6405 \approx 119, \ n_{24(mixed)} = 46.2677 \approx 46, \\
n_{26(mixed)} &= 69.6137 \approx 70 \\
n_{25(mixed)} &= 25.6598 \approx 26, \ n_{27(mixed)} = 66.5185 \approx 67
\end{aligned}
$$

with an estimated sampling variance $v_2^* = \sum_{h=1}^{L} \frac{W_h^2 s_{2h}^2}{n_{2h(mixed)}} = 1.484$.
Following Step 2, the pay-off matrix for the problem shown in (12.7) may be constructed as follows:

$$
\begin{array}{c} \\ n_{1h} \\ n_{2h} \end{array}
\begin{array}{c}
\begin{array}{cc} f_1 & \quad f_2 \end{array} \\
\begin{bmatrix} 0.528786 & 2.481681 \\ 0.69835 & 1.484 \end{bmatrix}
\end{array}
$$

The upper and lower bounds of each objective function can be obtained by following Step 3 as : $U_1 = 0.69835$, $U_2 = 2.481681$, $L_1 = 0.528786$, $L_2 = 1.484$.

**TABLE 12.2**: Computations regarding the first characteristic.

| $\mathbf{h}$ | $\mathbf{N_h}$ | $\mathbf{W_h}$ | $\mathbf{d(s_{1h})}$ | $\mathbf{W_h^2 s_{1h}^2}$ | $\mathbf{d(c_h)}$ | $\beta_\mathbf{h}$ | $\mathbf{c_h}\beta_\mathbf{h}$ | $\frac{\mathbf{W_h^2 s_{1h}^2}}{\beta_\mathbf{h}}$ |
|---|---|---|---|---|---|---|---|---|
| 1 | 472 | 0.1888 | 5.4 | 1.0394 | 7 | 1 | 7 | 1.0394 |
| 2 | 559 | 0.2236 | 5.28 | 1.3938 | 10 | 1 | 10 | 1.3938 |
| 3 | 425 | 0.1700 | 5.2 | 0.7814 | 8 | 1 | 8 | 0.7814 |
| | | | $h \in I_1$ | | | | 25 | 3.2146 |
| 4 | 218 | 0.0872 | 24.2 | 4.4531 | 9 | 0.0872 | 0.7848 | 51.0676 |
| 5 | 233 | 0.0932 | 23.4 | 4.7562 | 12 | 0.0932 | 1.1184 | 51.0321 |
| | | | $h \in I_2$ | | | | 1.9032 | 102.0997 |
| 6 | 328 | 0.1312 | 16.2 | 4.5175 | 11 | 0.64084 | 7.04924 | 7.0493 |
| 7 | 265 | 0.1060 | 40.3 | 18.2483 | 14 | 1.14169 | 15.98366 | 15.9836 |
| | | | $h \in I_3$ | | | | 23.0329 | 23.0329 |

**TABLE 12.3**: Computations regarding the second characteristic.

| $\mathbf{h}$ | $\mathbf{N_h}$ | $\mathbf{W_h}$ | $\mathbf{d(s_{2h})}$ | $\mathbf{W_h^2 s_{2h}^2}$ | $\mathbf{d(c_h)}$ | $\beta_\mathbf{h}$ | $\mathbf{c_h}\beta_\mathbf{h}$ | $\frac{\mathbf{W_h^2 s_{2h}^2}}{\beta_\mathbf{h}}$ |
|---|---|---|---|---|---|---|---|---|
| 1 | 472 | 0.1888 | 7.7 | 2.1134 | 7 | 1 | 7 | 2.1134 |
| 3 | 425 | 0.1700 | 7.68 | 1.7046 | 8 | 1 | 8 | 1.7046 |
| | | | $h \in I_1$ | | | | 15 | 3.818 |
| 2 | 559 | 0.2236 | 31.2 | 48.6690 | 10 | 0.2236 | 2.236 | 217.6610 |
| 4 | 218 | 0.0872 | 34.2 | 8.8938 | 9 | 0.0872 | 0.7848 | 101.9931 |
| 6 | 328 | 0.1312 | 33.6 | 19.4332 | 11 | 0.1312 | 1.4432 | 148.1189 |
| | | | $h \in I_2$ | | | | 4.464 | 467.773 |
| 5 | 233 | 0.0932 | 18.4 | 2.9408 | 12 | 0.49504 | 5.94048 | 5.9405 |
| 7 | 265 | 0.1060 | 45.3 | 23.0572 | 14 | 1.2833 | 17.9662 | 17.9671 |
| | | | $h \in I_3$ | | | | 23.90668 | 23.9076 |

Following Steps 4 and 5, the upper and lower bounds for each objective are obtained separately for truth, indeterminacy, and falsity membership function,

For $[f_1(n_{1h})]$:
$U_1^T = U_1 = 0.69835$, $U_1^I = L_1^T + s_1(U_1^T - L_1^T) = 0.528786 + s_1(0.169564)$, $U_1^F = U_1^T = 0.69835$
$L_1^T = L_1 = 0.528786$, $L_1^I = L_1^T = 0.528786$, $L_1^F = L_1^T + t_1(U_1^T - L_1^T) = 0.528786 + t_1(0.169564)$

$$T_1[f_1(n_{1h})] = \begin{cases} 1, & \text{if } [f_1(n_{1h})] \leq 0.528786 \\ \frac{0.69835-[f_1(n_{1h})]}{0.169564}, & \text{if } 0.528786 \leq [f_1(n_{1h})] \leq 0.69835 \\ 0, & \text{if } [f_1(n_{1h})] \geq 0.69835 \end{cases}$$

$$I_1[f_1(n_{1h})]) = \begin{cases} 1, & \text{if } [f_1(n_{1h})] \leq 0.528786 \\ \frac{0.528786+s_1(0.169564)-[f_1(n_{1h})]}{s_1(0.169564)}, & \text{if } 0.528786 \leq [f_1(n_{1h})] \\ & \quad \leq [f_1(n_{1h})] \leq 0.528786 \\ & \quad +s_1(0.169564) \\ 0, & \text{if } [f_1(n_{1h})] \geq 0.528786 \\ & \quad +s_1(0.169564) \end{cases}$$

$$F_1[f_1(n_{1h})] = \begin{cases} 1, & +t_1(0.169564) \\ & \quad +t_1(0.169564) \\ \frac{[f_1(n_{1h})]-(0.528786+t_1(0.169564))}{0.69835-0.528786-t_1(0.169564)}, & \text{if } 0.528786 + t_1(0.169564) \\ & \quad \leq [f_1(n_{1h})] \leq 0.69835 \\ 0, & \text{if } [f_1(n_{1h})] \geq 0.69835 \end{cases}$$

For $[f_2(n_{2h})]$:
$U_2^T = U_2 = 2.481681$, $U_2^I = L_2^T + s_2(U_2^T - L_2^T) = 1.484 + s_2(0.997681)$, $U_2^F = U_2^T = 2.481681$
$L_2^T = L_2 = 1.484$, $L_2^I = L_2^T = 1.484$, $L_2^F = L_2^T + t_2(U_2^T - L_2^T) = 1.484 + t_2(0.997681)$

$$T_2[f_2(n_{2h})] = \begin{cases} 1, & \text{if } [f_2(n_{2h})] \leq 1.484 \\ \frac{2.481681-[f_2(n_{2h})]}{0.997681}, & \text{if } 1.484 \leq [f_2(n_{2h})] \leq 2.481681 \\ 0, & \text{if } [f_2(n_{2h})] \geq 2.481681 \end{cases}$$

$$I_2[f_2(n_{2h})] = \begin{cases} 1, & \text{if } [f_2(n_{2h})] \leq 1.484 \\ \frac{1.484+s_2(0.997681)-[f_2(n_{2h})]}{s_2(0.997681)}, & \text{if } 1.484 \leq [f_2(n_{2h})] \leq 1.484 \\ & \quad + s_2(0.997681) \\ 0, & \text{if } [f_2(n_{2h})] \geq 1.484 \\ & \quad +s_2(0.997681) \end{cases}$$

$$F_2[f_2(n_{2h})] = \begin{cases} 1, & \text{if } [f_2(n_{2h})] \leq 1.484 + t_2(0.997681) \\ \frac{[f_2(n_{2h})] - (1.484 + t_2(0.997681))}{2.481681 - 1.484 - t_2(0.997681)}, & \text{if } 1.484 + t_2(0.997681) \leq [f_2(n_{2h})] \\ & \leq 2.481681 \\ 0, & \text{if } [f_2(n_{2h})] \geq 2.481681 \end{cases}$$

Following Steps 6 and 7, we obtained the single objective crisp NLPP as follows:

$$\begin{aligned}
&\text{Max } \lambda + \theta - \eta \\
&\text{subject to;} \\
&7\ n_1 + 10\ n_2 + 8\ n_3 + 9\ n_4 + 12\ n_5 + 11\ n_6 + 14\ n_7 \leq 4000 \\
&\text{and} \\
&0 \leq n_h \leq N_h;\ h = 1, 2, ..., 7 \\
&\left[\frac{1.0394}{n_1} + \frac{1.3938}{n_2} + \frac{0.7814}{n_3} + \frac{4.4531}{n_4} + \frac{4.7562}{n_5} + \frac{4.5175}{n_6} + \frac{18.2483}{n_7}\right] \\
&+(0.69835 - 0.528786)\lambda \leq 0.69835 \\
&\left[\frac{2.1134}{n_1} + \frac{48.6690}{n_2} + \frac{1.7046}{n_3} + \frac{8.8938}{n_4} + \frac{2.9408}{n_5} + \frac{19.4332}{n_6} + \frac{23.0572}{n_7}\right] \\
&+ (2.481681 - 1.484)\lambda \leq 2.481681 \\
&\left[\frac{1.0394}{n_1} + \frac{1.3938}{n_2} + \frac{0.7814}{n_3} + \frac{4.4531}{n_4} + \frac{4.7562}{n_5} + \frac{4.5175}{n_6} + \frac{18.2483}{n_7}\right] \\
&+ s_1\theta(0.169564) \leq 0.528786 + s_1(0.169564) \\
&\left[\frac{2.1134}{n_1} + \frac{48.6690}{n_2} + \frac{1.7046}{n_3} + \frac{8.8938}{n_4} + \frac{2.9408}{n_5} + \frac{19.4332}{n_6} + \frac{23.0572}{n_7}\right] \\
&+ s_2\theta(0.997681) \leq 1.484 + s_2(0.997681) \\
&\left[\frac{1.0394}{n_1} + \frac{1.3938}{n_2} + \frac{0.7814}{n_3} + \frac{4.4531}{n_4} + \frac{4.7562}{n_5} + \frac{4.5175}{n_6} + \frac{18.2483}{n_7}\right] \\
&- (0.69835 - 0.528786 - t_1(0.169564))\eta \leq 0.528786 + t_1(0.169564) \\
&\left[\frac{2.1134}{n_1} + \frac{48.6690}{n_2} + \frac{1.7046}{n_3} + \frac{8.8938}{n_4} + \frac{2.9408}{n_5} + \frac{19.4332}{n_6} + \frac{23.0572}{n_7}\right] \\
&- (2.481681 - 1.484 - t_2(0.997681))\eta \leq 1.484 + t_2(0.997681) \\
&\lambda \geq \theta,\ \lambda \geq \eta,\ \lambda + \theta + \eta \leq 3,\ \lambda, \theta, \eta \in (0, 1) \\
&0 \leq s_k,\ t_k \leq 1, \forall\ k = 1, 2.
\end{aligned}$$

The above problem has been coded in Lingo 18.0 and the following solutions are obtained as $n_1 = 30.69 \approx 31$, $n_2 = 69.24 \approx 69$, $n_3 = 25.14 \approx 25$, $n_4 = 55.89 \approx 56$, $n_5 = 45.12 \approx 45$, $n_6 = 58.31 \approx 58$, $n_7 = 86.10 \approx 86$ with $\sum_{h=1}^{7} c_h n_h = 3993$.

### 12.6.1 Estimation of relative efficiency

Let the estimated relative efficiency (ERE) of the estimate $(\bar{y}_{kst})$ of the population mean $\bar{Y}_k$ of the $kth$ characteristic be defined as

$$\text{ERE} = \frac{v_{comp}(\bar{y}_{kst})}{v_k^*}$$

where $v_{comp}(\bar{y}_{1st}) = 0.560236$, $v_{comp}(\bar{y}_{2st}) = 1.669004$, and $v_1^* = 0.528786$, $v_2^* = 1.484$.
The mean estimated relative efficiency (MERE) is given by (*see* Semiz [37])

$$\text{MERE} = \frac{1}{p}\sum_{k=1}^{p}\frac{v_{comp}(\bar{y}_{kst})}{v_k^*} \tag{12.17}$$

$$= \frac{2.18413}{2} \tag{12.18}$$

$$= 1.092065. \tag{12.19}$$

Therefore, the MERE for compromise allocations is 1.092065.

## 12.7 Comparison, Conclusion, and Future Research Work

In this chapter, a new variant of MMNSSP has been presented by considering the uncertain behavior of stratum standard deviations and per unit measurement costs. Parabolic fuzzy numbers are used to deal with uncertain parameters. A solution procedure based on the concept of neutrosophic programming has been proposed to solve the formulated MMNSSP. The proposed approach has been demonstrated and compared with FPA with the help of a numerical example. The results obtained from NPA and FPA are tabulated in Table 12.4, which show that the maximum MERE can be obtained with the help of proposed NPA as compared to FPA (see Figure 12.1).

Some of the advantages of NPA along with their limitations have been summarized in Table 12.5. NPA, based on neutrosophic sets is more flexible in nature in a decision-making process as it consists of three membership functions, namely truth membership function, indeterminacy membership function, and falsity membership function; these provide a realistic framework for solving multi-objective NLPPs under uncertainty. Thus, we may conclude that the uncertain nature of parameters can be dealt well by using parabolic fuzzy numbers and best compromise mixed allocations may be obtained using proposed NPA.

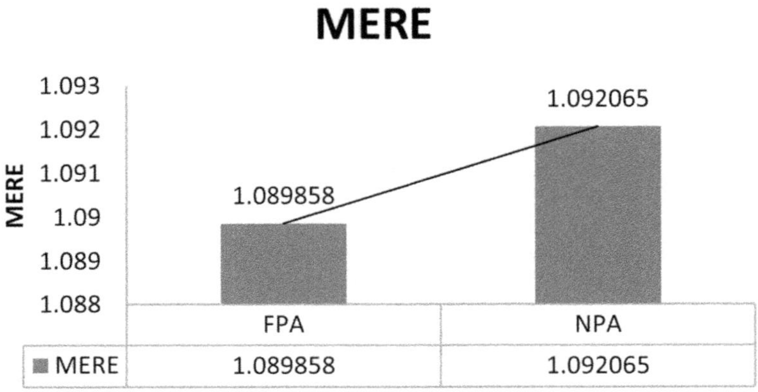


**FIGURE 12.1**: Comparison of MERE of NPA and FPA.

**TABLE 12.4**: Mean estimated relative efficiencies (MERE) of NPA and FPA.

| **S.No.** | **Allocations** | $n_1$ | $n_2$ | $n_3$ | $n_4$ | $n_5$ | $n_6$ | $n_7$ | **MERE** |
|---|---|---|---|---|---|---|---|---|---|
| 1 | FPA | 31 | 68 | 25 | 56 | 46 | 58 | 87 | 1.089858 |
| 2 | Proposed approach | 31 | 69 | 25 | 56 | 45 | 58 | 86 | 1.092065 |

In future, NPA can be applied to different variants of sample surveys namely two-stage sampling, cluster sampling, the problem of non-response, etc. Furthermore, an extension of MMNSSP using pentagonal neutrosophic numbers may be proposed in future research work.

**TABLE 12.5**: Advantages and limitations.

| **Advantages** | **Limitations** |
|---|---|
| Works well in reducing the total survey cost and providing better results under uncertainty. | A large-scale study is needed for the practical utility of the problem. |
| Useful approach for solving nonlinear multiobjective programming problems in stratified sampling. | Multiobjective programming problems require more computational effort. |

## Bibliography

[1] M.J. Ahsan, Najmussehar, and M.G.M. Khan. Mixed allocation in stratified sampling. *Aligarh Journal of Statistics*, 25:87–97, 2005.

[2] R.G. Clark and D.G. Steel. Optimum allocation of sample to strata and stages with simple additional constraints. *The Statistician*, 49(2):197–207, 2000.

[3] R. Varshney and M.J. Ahsan. Compromise mixed allocation in multivariate stratified sampling. *Journal of the Indian Society of Agricultural Statistics*, 65(3):291–296, 2011.

[4] N. Khanam, A.H. Ansari, and A. Quddoos. Compromise mixed allocation using fuzzy programming approach. *Journal of Statistical Computation and Simulation*, 91(3):435–444, 2020.

[5] F. Yates. *Sampling Methods for Censuses and Surveys.* Charles Griffin and Co. Ltd., London, 3rd edition, 1960.

[6] H. Aoyama. Stratified random sampling with optimum allocation for multivariate populations. *Annals of the Institute of Statistical Mathematics*, 14:251–258, 1963.

[7] S. Chatterjee. A note on optimum allocation. *Scandinavian Actuarial Journal*, 50:40–44, 1967. `https://api.semanticscholar.org/CorpusID:121496514`.

[8] S. Chatterjee. Multivariate stratified surveys. *Journal of the American Statistical Association*, 63(322):530–534, 1968.

[9] M.J. Ahsan and S.U. Khan. Optimum allocation in multivariate stratified random sampling using prior information. *Journal of Indian Statistical Association*, 15:57–67, 1977.

[10] M.J. Ahsan and S.U. Khan. Optimum allocation in multivariate stratified random sampling with overhead cost. *Metrika*, 29(1):71–78, 1982.

[11] A. Melaku and G. Sadasivan. L1-norm and other methods for sample allocation in multivariate stratified surveys. *Computational Statistics and Data Analysis*, 5(4):415–423, 1987.

[12] J. Bethel. Sample allocation in multivariate surveys. *Survey Methodology*, 15(1):47–57, 1989.

[13] M. Kozak. Multivariate sample allocation: Application of random search method. *Statistics in Transition*, 7(4):889–900, 2006.

[14] L. Kreienbrock. Generalized measures of dispersion to solve the allocation problem in multivariate stratified random sampling. *Communications in Statistics – Theory and Methods*, 22(1):219–239, 1993.

[15] J.A. Díaz-García and L.U. Cortez. Optimum allocation in multivariate stratified sampling: Multi-objective programming. Technical Report Comunicación Técnica No. I-06-07/28-03-2006, PE/CIMAT, 2006.

[16] J.A. Díaz-García and L.U. Cortez. Multi-objective optimisation for optimum allocation in multivariate stratified sampling. *Survey Methodology*, 34(2):215–222, 2008.

[17] M.G.M. Khan, M.J. Ahsan, and N. Jahan. Compromise allocation in multivariate stratified sampling: An integer solution. *Naval Research Logistics*, 44:69–79, 1997.

[18] A.H. Ansari, R. Varshney, and M.J. Ahsan. Compromise mixed allocation in multivariate stratified sampling using dynamic programming technique. *Journal of Advanced Statistics*, 3(4):71–79, 2018.

[19] A.H. Ansari, R. Varshney, and M.J. Ahsan. A compromise allocation based on the individual mixed allocations in multivariate stratified sampling: A goal programming approach. *Calcutta Statistical Association Bulletin*, 64:253–254, 2012.

[20] L.A. Zadeh. Fuzzy sets. *Information and Control*, 8(3):338–353, 1965.

[21] K.T. Atanassov. Intuitionistic fuzzy sets. *Fuzzy Sets and Systems*, 20(1):87–96, 1986.

[22] F. Smarandache. *A Unifying Field in Logics: Neutrosophic Logic in Philosophy.* American Research Press, Rehoboth, 1999.

[23] H.J. Zimmermann. Fuzzy programming and linear programming with several objectives. *Fuzzy Sets and Systems*, 1(1):45–55, 1978.

[24] R.G. Dyson. Maximin programming, fuzzy linear programming and multicriteria decision making. *Journal of the Operational Research Society*, 31:263–267, 1980.

[25] F. Ahmad and A.Y. Adhami. Neutrosophic programming approach to multi-objective non-linear transportation problem with fuzzy parameters. *International Journal of Management Science and Engineering Management*, 14(3):218–229, 2018.

[26] P. Das and T.K. Roy. Multi-objective non-linear programming problem based on neutrosophic optimization technique and its application in riser design problem. *Neutrosophic Sets and Systems*, 9(1):88–95, 2015.

[27] M. Sarkar, S. Dey, and T.K. Roy. Multi-objective neutrosophic optimization technique and its application to structural design. *International Journal of Computer Applications*, 148(12):31–37, 2016.

[28] J. Ahmed, M.G. Alharbi, M. Akram, and S. Bashir. A new method to evaluate linear programming problem in bipolar single-valued neutrosophic environment. *Computer Modeling in Engineering and Sciences*, 129(2):881–906, 2021.

[29] J. Ahmed. LR-type fully single-valued neutrosophic linear programming problems. *Neutrosophic Sets and Systems*, 46:416–444, 2021.

[30] M. Mohamed and A. Gamal. Toward sustainable emerging economics based on industry 5.0: Leveraging neutrosophic theory in appraisal decision framework. *Neutrosophic Systems with Applications*, 1(1):14–21, 2023.

[31] A. Abdel-Monem, N. Nabeeh, and M. Abouhawwash. An integrated neutrosophic regional management ranking method for agricultural water management. *Neutrosophic Systems with Applications*, 1:22–28, 2023.

[32] A. Gamal, A.F.A. Amalie, and M. Abouhawwash. Towards a responsive resilient supply chain Industry 5.0: A case study in healthcare systems. *Neutrosophic Systems with Applications*, 2:8–24, 2023.

[33] N. Gupta and A. Bari. Fuzzy multi-objective optimization for optimum allocation in multivariate stratified sampling with quadratic cost and parabolic fuzzy numbers. *Journal of Statistical Computation and Simulation*, 87(12):2372–2383, 2017.

[34] J. Neyman. On the two different aspects of the representative method: The method of stratified sampling and the method of purposive selection. *Journal of the Royal Statistical Society*, 97(4):558–625, 1934.

[35] R.E. Bellman and L.A. Zadeh. Decision-making in a fuzzy environment. *Management Science*, 17(4):B–141–B–164, 1970.

[36] T.J. Ross. *Fuzzy Logic with Engineering Applications.* John Wiley and Sons, Chichester, 2nd edition, 2005.

[37] M. Semiz. Determination of compromise integer strata sample sizes using goal programming. *Hacettepe Journal of Mathematics and Statistics*, 33:91–96, 2004.

# Chapter 13

## A Neutrosophic Control Chart for the Conway-Maxwell-Poisson Distribution Using Multiple Dependent State Sampling

**G. Srinivasa Rao, Ambreen Shafqat, Muhammad Saleem, Liaquat Ahmad, Jean-Claude Malela-Majika, Sandile C. Shongwe, Muhammad Aslam, Abdulrahman AlAita, and Yasser Al Zaim**

### 13.1 Introduction

The improvement in speed and responsiveness of monitoring schemes in the production process is the basic problem of the quality control experts since the inception of statistical process control (SPC) production processes and services providing agencies. The solution to this problem was suggested by [1] in the form of developing the upper control limit (UCL) and the lower control limit (LCL) within which the process is defined as in-control, and if any observation lies outside these limits, then the process is declared as out-of-control. Since the inception of this idea by [1], quality control experts are endeavoring to develop robust monitoring schemes and significant contributions have been observed under the assumption of perfect measurements and certainty. However, little has been done under the assumption of uncertainty of the observed quality characteristic.

A successful process monitoring procedure entails the choice of a right (or appropriate) control chart. The later depend on the type of distribution of the quality characteristic and charting statistic. COM-Poisson distribution is the most commonly used probability distribution for modeling the counts data when the mean and the variance of the underlying distribution are not equal. In many situations, the assumption of identical mean and variance is often violated when the mean is smaller than the variance (i.e. this is said to be over-dispersion) or when the mean is greater than the variance (this is said to be under-dispersion). Ignoring over- and under-dispersion may lead to erroneous

DOI: 10.1201/9781003683339-13

results, see for instance [2]. Conway and Maxwell [3] developed the COM-Poisson distribution, which can be viewed as a Poisson distribution with an additional parameter (i.e. two-parameter Poisson distribution), in the context of a queuing system for over-dispersed, under-dispersed, and equally dispersed observations. The additional parameter governs the rate of decay using the ratio of probabilities as $\mathbb{P}(X = x - 1) / \mathbb{P}(X = x) = x^{\nu}/\lambda$. In the context of practical applications of the COM-Poisson distribution, it is considered as well-performed distribution for several settings in addition to being the most efficient, flexible, and easy to use distribution, and provides a stronger base for fitting the under-dispersed and the over-dispersed data. In [4], the authors explored the COM-Poisson distribution for modeling the counts data related to the motor vehicle crashes. In [5], the authors developed the methodology for conjugate analysis of the COM-Poisson distribution as an exponential family of distributions in the context of the Bayesian statistics. In [2], the authors proposed an efficient monitoring of the production process using the COM-Poisson distribution. Many researchers of the statistical process control used the COM-Poisson distribution for its nice properties; see also [6–10].

In [11], the author introduced the neutrosophic statistic which is the generalization of the fuzzy logic. The fuzzy logic is used to measure the percentage of the truth and the falsehood of any phenomena. The neutrosophic logic was developed by considering the percentage of indeterminacy in addition to fuzzy logic (see [12]). The neutrosophic statistics is used when the observations under consideration are vague, incomplete, confusing, and undecided [13]. In SPC, several authors have explored and applied the neutrosophic statistics to address the non-crispy observations [14–23].

In [24], the authors initiated the concept of MDSS which is based on the conditional sampling procedure. The decision of the in-control or the out-of-control process is not only based on the current sampling inspection but also considers the past behavior of the process. This sampling scheme is known to be efficient with respect to the sample size and has attracted the attention of several researchers in SPC due to its interesting properties. In [25], the authors proposed the MDSS plan for the measurement data for deciding the lot sentencing. In [26], the authors developed an MDSS control chart using exponential distribution. In [16], the authors designed an efficient monitoring for t-chart using the MDSS scheme. In [27], the authors developed the variance control chart under the MDSS and the repetitive sampling schemes. Currently, in the SPC literature, there is no study on MDSS under neutrosophic statistics; the latter studies are based on classical statistics. Therefore, in this chapter, a new control chart for monitoring observations from a neutrosophic COM-Poisson (NCOM-Poisson) distribution using the MDSS method is proposed.

The rest of the chapter is laid out as follows: the methodology of the NCOM-Poisson chart is explained in Section 13.2. The performance analysis of the proposed chart is given in Section 13.3. The simulation study of the proposed methodology is provided in Section 13.4. A practical application is given in Section 13.5 and finally, the concluding remarks and recommendations are given in Section 13.6.

## 13.2 Design of the NCOM-Poisson Distribution Chart Using MDSS

### 13.2.1 Neutrosophic framework

The design of the proposed chart is based on the conventional distribution of the COM-Poisson model developed by [28] known as neutrosophic COM-Poisson distribution. Let $y_N = [y, y']$, where $y = y' = 0, 1, 2, \cdots$. Thus, the neutrosophic mass function of the COM-Poisson distribution is described as

$$\mathbb{P}(Y_N = y_N; \mu_N, \nu_N) = \frac{\mu_N^{y_N}}{(Y_N!)^{\nu} Z_N(\mu_N, \nu_N)};$$
$$Y_N \in \{Y_L, Y_U\}, \text{for } y_N = [0,0], [1,1], [2,2] \cdots \quad (13.1)$$

The neutrosophic scale and dispersion parameters are presented as $\mu_N \in \{\mu_L, \mu_U\}$ and $\nu_N \in \{\nu_L, \nu_U\}$, respectively. Further, the neutrosophic normalization constant is defined by

$$z_N(\mu_N, \nu_N) = \sum_{j_N}^{\infty} \frac{\mu_N^{y_N}}{(j_N!)^{\nu_N}}; \quad \text{where } j_N = [0,0], [1,1], [2,2] \cdots \quad (13.2)$$

Note that the NCOM-Poisson reduces to the conventional COM-Poisson distribution if the sample or the population under study has no uncertainty. The NCOM-Poisson distribution has numerous interesting approximations as if $\nu_N \in \{\infty, \infty\}$, then it is the Bernouli distribution; if $\nu_N \in \{0, 0\}$, then it is neutrosophic geometric distribution; and if $\nu_N \in \{1, 1\}$, then it tends to form the neutrosophic Poisson distribution. Then the mean and variance of the proposed distribution can be respectively written as

$$\mu_{Y_N} = \mu_N^{1/\nu_N} - \frac{\nu_N - 1}{2\nu_N}; \quad \mu_N \in \{\mu_L, \mu_U\}, \nu_N \in \{\nu_L, \nu_U\} \quad (13.3)$$

$$\sigma_{Y_N}^2 = \frac{\mu_N^{1/\nu_N}}{\nu_N}; \quad \mu_N \in \{\mu_L, \mu_U\}, \nu_N \in \{\nu_L, \nu_U\} \quad (13.4)$$

### 13.2.2 NCOM-Poisson control chart using MDSS

Using the neutrosophic statistics under the MDSS, the NCOM-Poisson control chart is constructed using the following steps:

**Steps for constructing NCOM-Poisson control chart:**

**Step 1:** Choose an item randomly from the manufacturing process with the quality characteristic $Y_N \in \{Y_L, Y_U\}$ and record the number of non-conformities say $Y_{tN} \in \{Y_{tL}, Y_{tU}\}$.

**Step 2:** Process is said to be in-control if $LCL_{2N} \leqslant Y_{tN} \leqslant UCL_{2N}; Y_{tN} \in \{Y_{tL}, Y_{tU}\}$. Assert the process as out-of-control if $Y_{tN} \geqslant UCL_{1N}$ or $Y_{tN} \leqslant LCL_{1N}$. Otherwise, go to Step 3.

**Step 3:** The process declared as in-control if $i_N; i_N \in \{i_L, i_U\}$ preceding subgroups have been declared as in-control. Otherwise, declare the process to be out-of-control.

Here $LCL_{1N}$ and $UCL_{1N}$ and $LCL_{2N}$ and $UCL_{2N}$ are the outer and inner lower and upper limits, respectively, under the neutrosophic statistical interval.

Thus, the outer control limits are given by

$$LCL_{1N} = \mu_{Y_N} - k_{1N}\sigma_{Y_N} = \left(\mu_N^{1/\nu_N} - \frac{\nu_N - 1}{2\nu_N}\right) - k_{1N}\left(\frac{\mu_N^{1/\nu_N}}{\nu_N}\right)^{1/2};$$
$$k_{1N} \in \{k_{1L}, k_{1U}\} \tag{13.5}$$

$$UCL_{1N} = \mu_{Y_N} + k_{1N}\sigma_{Y_N} = \left(\mu_N^{1/\nu_N} - \frac{\nu_N - 1}{2\nu_N}\right) + k_{1N}\left(\frac{\mu_N^{1/\nu_N}}{\nu_N}\right)^{1/2};$$
$$k_{1N} \in \{k_{1L}, k_{1U}\} \tag{13.6}$$

Also,

$$LCL_{2N} = \mu_{Y_N} - k_{2N}\sigma_{Y_N} = \left(\mu_N^{1/\nu_N} - \frac{\nu_N - 1}{2\nu_N}\right) - k_{2N}\left(\frac{\mu_N^{1/\nu_N}}{\nu_N}\right)^{1/2};$$
$$k_{2N} \in \{k_{2L}, k_{2U}\} \tag{13.7}$$

$$UCL_{2N} = \mu_{Y_N} + k_{2N}\sigma_{Y_N} = \left(\mu_N^{1/\nu_N} - \frac{\nu_N - 1}{2\nu_N}\right) + k_{2N}\left(\frac{\mu_N^{1/\nu_N}}{\nu_N}\right)^{1/2};$$
$$k_{2N} \in \{k_{2L}, k_{2U}\} \tag{13.8}$$

where $k_{1N} \in \{k_{1L}, k_{1U}\}$ and $k_{2N} \in \{k_{2L}, k_{2U}\}$ are the control limits coefficients of the neutrosophic chart to be determined such that the attained in-control average run length ($ARL_0$) is equal to a pre-specified $ARL_0$. The proposed chart is reduced to the single sampling chart if $k_{1N} = k_{2N} = k_N$ and $i_N = 1$. Suppose that $\mu_{0N} \in \{\mu_{0L}, \mu_{0U}\}$ is the mean of the neutrosophic statistics of the in-control process. Then, the probability of the in-control process is given by

$$\begin{aligned}\mathbb{P}_{0N}^{\text{in}} &= \mathbb{P}\left(LCL_{2N} \leqslant Y_{tN} \leqslant UCL_{2N} | \mu = \mu_{0N}\right) + \\ &\quad \{\mathbb{P}\left(LCL_{1N} \leqslant Y_{tN} \leqslant LCL_{2N} | \mu = \mu_{0N}\right) \\ &\quad + \mathbb{P}\left(UCL_{2N} \leqslant Y_{tN} \leqslant UCL_{1N} | \mu = \mu_{0N}\right)\} \\ &\quad \{\mathbb{P}\left(LCL_{2N} \leqslant Y_{tN} \leqslant UCL_{2N} | \mu = \mu_{0N}\right)\}^{i_N}\end{aligned} \tag{13.9}$$

where

$$\begin{aligned}&\mathbb{P}\left(LCL_{2N} \leqslant Y_{tN} \leqslant UCL_{2N} | \mu = \mu_{0N}\right) \\ &= \mathbb{P}\left(Y_{tN} \leqslant UCL_{2N} | \mu = \mu_{0N}\right) - \mathbb{P}\left(Y_{tN} \leqslant LCL_{2N} | \mu = \mu_{0N}\right) \\ &= \mathbb{P}\left(\frac{Y_{tN} - \mu_{Y_N}}{\sigma_{Y_N}} \leqslant \frac{UCL_{2N} - \mu_{Y_N}}{\sigma_{Y_N}}\right) - \mathbb{P}\left(\frac{Y_{tN} - \mu_{Y_N}}{\sigma_{Y_N}} \leqslant \frac{LCL_{2N} - \mu_{Y_N}}{\sigma_{Y_N}}\right) \\ &= \mathbb{P}\left(Z_{tN} \leqslant k_{2N}\right) - \mathbb{P}\left(Z_{tN} \leqslant -k_{2N}\right) \\ &= 2\Phi\left(k_{2N}\right) - 1\end{aligned} \tag{13.10}$$

also

$$\begin{aligned}&\mathbb{P}\left(LCL_{1N} \leqslant Y_{tN} \leqslant LCL_{2N} | \mu = \mu_{0N}\right) \\ &= \mathbb{P}\left(Y_{tN} \leqslant LCL_{2N} | \mu = \mu_{0N}\right) - \mathbb{P}\left(Y_{tN} \leqslant LCL_{1N} | \mu = \mu_{0N}\right) \\ &= \mathbb{P}\left(\frac{Y_{tN} - \mu_{Y_N}}{\sigma_{Y_N}} \leqslant \frac{LCL_{2N} - \mu_{Y_N}}{\sigma_{Y_N}}\right) - \mathbb{P}\left(\frac{Y_{tN} - \mu_{Y_N}}{\sigma_{Y_N}} \leqslant \frac{LCL_{1N} - \mu_{Y_N}}{\sigma_{Y_N}}\right) \\ &= \mathbb{P}\left(Z_{tN} \leqslant -k_{2N}\right) - \mathbb{P}\left(Z_{tN} \leqslant -k_{1N}\right) \\ &= \Phi\left(k_{1N}\right) - \Phi\left(k_{2N}\right)\end{aligned} \tag{13.11}$$

and

$$\begin{aligned}&\mathbb{P}\left(UCL_{2N} \leqslant Y_{tN} \leqslant UCL_{1N} | \mu = \mu_{0N}\right) \\ &= \mathbb{P}\left(Y_{tN} \leqslant UCL_{1N} | \mu = \mu_{0N}\right) - \mathbb{P}\left(Y_{tN} \leqslant UCL_{2N} | \mu = \mu_{0N}\right) \\ &= \mathbb{P}\left(\frac{Y_{tN} - \mu_{Y_N}}{\sigma_{Y_N}} \leqslant \frac{UCL_{1N} - \mu_{Y_N}}{\sigma_{Y_N}}\right) - \mathbb{P}\left(\frac{Y_{tN} - \mu_{Y_N}}{\sigma_{Y_N}} \leqslant \frac{UCL_{2N} - \mu_{Y_N}}{\sigma_{Y_N}}\right) \\ &= \mathbb{P}\left(Z_{tN} \leqslant k_{1N}\right) - \mathbb{P}\left(Z_{tN} \leqslant k_{2N}\right) \\ &= \Phi\left(k_{1N}\right) - \Phi\left(k_{2N}\right)\end{aligned} \tag{13.12}$$

where $\Phi(\cdot)$ is the cumulative standard normal distribution for the neutrosophic statistics and $Z_{tN} = \left(Y_{tN} - \mu_{Y_N}\right)/\sigma_{Y_N}$ is the standard normal variable based on the neutrosophic statistics. Further information can be obtained from [29] and [30].

In the above expressions, the superscript 0 denotes in-control process. The performance of any chart can be examined by calculating its ARL, which may be defined as the average number of units that falls in the in-control region before the first out-of-control signal [31]. Therefore, an efficient chart will show larger value of ARL when it is in the state of in-control situation, while the value of ARL will be smaller when the process shows any deterioration [31]. The ARL is mathematically defined by

$$\mathrm{ARL}_{0N} = \frac{1}{1-\mathbb{P}_{0N}^{\mathrm{in}}}; \quad \mathrm{ARL}_{0N} \in \{\mathrm{ARL}_{0L}, \mathrm{ARL}_{0U}\} \tag{13.13}$$

For the out-of-control case, the out-of-control ARL (denoted as $\mathrm{ARL}_1$) value is expected to be smaller for any proposed chart. Suppose that the neutrosophic process mean parameter $\mu_{0N} \in \{\mu_{0L}, \mu_{0U}\}$ has shifted, $\mu_{1N} = \delta\mu_{0N}$ with a shift constant $\delta$ is $\mu_{1N} \in \{\mu_{1L}, \mu_{1U}\}$. Then, the process is declared out-of-control when a shift of size $\delta$ in standard deviation unit has occurred. Thus, the probability of the out-of-control process is given by

$$\begin{aligned}\mathbb{P}_{1N}^{\mathrm{in}} = {} & \mathbb{P}\left(LCL_{2N} \leqslant Y_{tN} \leqslant UCL_{2N} | \mu = \mu_{1N}\right) + \\ & \{\mathbb{P}\left(LCL_{1N} \leqslant Y_{tN} \leqslant LCL_{2N} | \mu = \mu_{1N}\right) \\ & + \mathbb{P}\left(UCL_{2N} \leqslant Y_{tN} \leqslant UCL_{1N} | \mu = \mu_{1N}\right)\} \\ & \{\mathbb{P}\left(LCL_{2N} \leqslant Y_{tN} \leqslant UCL_{2N} | \mu = \mu_{1N}\right)\}^{i_N}\end{aligned} \tag{13.14}$$

where

$$\begin{aligned}&\mathbb{P}\left(LCL_{2N} \leqslant Y_{tN} \leqslant UCL_{2N} | \mu = \mu_{1N}\right) = \\ &\mathbb{P}\left(Y_{tN} \leqslant UCL_{2N} | \mu = \mu_{1N}\right) - \mathbb{P}\left(Y_{tN} \leqslant LCL_{2N} | \mu = \mu_{1N}\right) = \\ &\Phi\left(\frac{\left(1-\delta^{1/\nu_N}\right)\mu_{0N}^{1/\nu_N} + k_{2N}\sqrt{\mu_{0N}^{1/\nu_N}/\nu_N}}{\sqrt{\delta^{1/\nu_N}\mu_{0N}^{1/\nu_N}/\nu_N}}\right) \\ &\quad - \Phi\left(\frac{\left(1-\delta^{1/\nu_N}\right)\mu_{0N}^{1/\nu_N} - k_{2N}\sqrt{\mu_{0N}^{1/\nu_N}/\nu_N}}{\sqrt{\delta^{1/\nu_N}\mu_{0N}^{1/\nu_N}/\nu_N}}\right)\end{aligned} \tag{13.15}$$

also

$$\begin{aligned}&\mathbb{P}\left(LCL_{1N} \leqslant Y_{tN} \leqslant LCL_{2N} | \mu = \mu_{1N}\right) = \\ &\Phi\left(\frac{\left(1-\delta^{1/\nu_N}\right)\mu_{0N}^{1/\nu_N} - k_{2N}\sqrt{\mu_{0N}^{1/\nu_N}/\nu_N}}{\sqrt{\delta^{1/\nu_N}\mu_{0N}^{1/\nu_N}/\nu_N}}\right) \\ &\quad - \Phi\left(\frac{\left(1-\delta^{1/\nu_N}\right)\mu_{0N}^{1/\nu_N} - k_{1N}\sqrt{\mu_{0N}^{1/\nu_N}/\nu_N}}{\sqrt{\delta^{1/\nu_N}\mu_{0N}^{1/\nu_N}/\nu_N}}\right)\end{aligned} \tag{13.16}$$

and

$$\mathbb{P}\left(UCL_{2N} \leqslant Y_{tN} \leqslant UCL_{1N} | \mu = \mu_{1N}\right) =$$
$$\Phi\left(\frac{\left(1-\delta^{1/\nu_N}\right)\mu_{0N}^{1/\nu_N} + k_{1N}\sqrt{\mu_{0N}^{1/\nu_N}/\nu_N}}{\sqrt{\delta^{1/\nu_N}\mu_{0N}^{1/\nu_N}/\nu_N}}\right)$$
$$-\,\Phi\left(\frac{\left(1-\delta^{1/\nu_N}\right)\mu_{0N}^{1/\nu_N} + k_{2N}\sqrt{\mu_{0N}^{1/\nu_N}/\nu_N}}{\sqrt{\delta^{1/\nu_N}\mu_{0N}^{1/\nu_N}/\nu_N}}\right) \tag{13.17}$$

As a result, the neutrosophic ARL (NARL) for the out-of-control process is given as follows:

$$\text{ARL}_{1N} = \frac{1}{1-\mathbb{P}_{1N}^{\text{in}}};\quad \text{ARL}_{1N} \in \{\text{ARL}_{1L}, \text{ARL}_{1U}\} \tag{13.18}$$

## 13.3 Performance Analysis

In this section, the performance of the NCOM-Poisson chart is investigated in terms of the NARL values when $\nu_N \in [0.4, 0.6]$, $\mu_N \in [2.5, 3.5]$, and $i_N \in [2, 3]$ for pre-specified $\text{ARL}_0$ values of 200, 300, and 370. The results in Table 13.1 can be summarized as follows:

- As the size of the shift increases, the sensitivity of the proposed chart increases as well.
- When both the mean and variance increase, the sensitivity of the proposed chart decreases.

In this section, the advantages of the proposed chart for neutrosophic statistic using MDSS have been described. Using the above-mentioned design and methodology, an *R*-code program is used to calculate the values of the parameters $\nu_N$, $\mu_N$, and $i_N$, along with control chart coefficients. As mentioned above, the performance of the proposed chart is investigated in terms of the ARL values. Thus, Tables 13.1–13.3 have been generated for the control chart coefficients and the in-control and out-of-control NARL using different process settings. For example, Table 13.1 has been generated for the parametric values as $\nu_N \in [0.4, 0.6]$, $\mu_N \in [2.5, 3.5]$, and $i_N \in [2, 3]$, and the control chart coefficients are estimated for the predefined in-control ARLs equal to 200, 300, and 370. The same values at shift level testify for the accuracy of the developed methodology as these values are [200.01, 200.00], [300.00, 300.03],

**TABLE 13.1**: The NARLs of proposed chart when $\nu_N \in [0.4, 0.6]$, $\mu_N \in [2.5, 3.5]$, and $i_N \in [2, 3]$.

| $k_{1N}$ | $[3.0787, 2.9014]$ | $[3.0102, 2.9847]$ | $[3.0739, 30589]$ |
|---|---|---|---|
| $k_{2N}$ | $[2.0573, 2.2754]$ | $[2.3177, 2.4464]$ | $[2.3586, 2.4595]$ |
| $\delta$ | | NARLs | |
| 1.0000 | $[200.01, 200.00]$ | $[300.00, 300.03]$ | $[370.00, 370.17]$ |
| 1.0125 | $[168.82, 178.72]$ | $[251.42, 267.44]$ | $[306.09, 328.27]$ |
| 1.0250 | $[138.81, 157.16]$ | $[205.03, 233.64]$ | $[245.94, 285.98]$ |
| 1.0375 | $[111.91, 136.84]$ | $[163.84, 202.15]$ | $[193.34, 245.15]$ |
| 1.0500 | $[89.04, 117.94]$ | $[129.19, 172.08]$ | $[149.83, 209.10]$ |
| 1.0625 | $[70.32, 100.89]$ | $[101.14, 146.04]$ | $[115.21, 176.58]$ |
| 1.0750 | $[55.40, 85.85]$ | $[79.01, 123.27]$ | $[88.38, 148.89]$ |
| 1.1250 | $[22.17, 43.28]$ | $[30.62, 62.41]$ | $[32.04, 72.81]$ |
| 1.1500 | $[14.71, 31.02]$ | $[19.97, 44.57]$ | $[20.34, 51.40]$ |
| 1.2000 | $[7.27, 17.8]$ | $[9.51, 24.63]$ | $[9.31, 27.48]$ |
| 1.2500 | $[4.19, 10.01]$ | $[5.26, 13.32]$ | $[5.04, 15.68]$ |
| 1.3000 | $[2.76, 6.10]$ | $[3.32, 8.97]$ | $[3.16, 9.22]$ |
| 1.3500 | $[2.03, 4.92]$ | $[2.35, 5.03]$ | $[2.23, 6.96]$ |
| 1.4000 | $[1.63, 3.63]$ | $[1.82, 4.32]$ | $[1.74, 4.39]$ |
| 1.4500 | $[1.39, 2.83]$ | $[1.51, 3.28]$ | $[1.46, 3.78]$ |
| 1.5000 | $[1.25, 2.31]$ | $[1.32, 2.62]$ | $[1.29, 2.67]$ |
| 1.6000 | $[1.10, 1.72]$ | $[1.13, 1.87]$ | $[1.11, 1.89]$ |
| 1.8000 | $[1.01, 1.24]$ | $[1.01, 1.29]$ | $[1.01, 1.30]$ |
| 1.9000 | $[1.00, 1.14]$ | $[1.00, 1.17]$ | $[1.00, 1.17]$ |
| 2.0000 | $[1.00, 1.08]$ | $[1.00, 1.10]$ | $[1.00, 1.10]$ |

and $[370.00, 370.17]$. The NARL values have been generated for different shift levels of 1.0125, 1.0250, 1.0375, 1.0500, 1.0625, 1.0750, 1.1250, 1.1500, 1.0200, 1.25000, 1.3000, 1.3500, 1.4000, 1.4500, 1.5000, 1.6000, 1.8000, 1.9000, and 2.0000. Then the out-of-control process is detected by calculating the $\text{NARL}_1$ given against each shift level; for instance, from Table 13.2, a shift of size 1.1250 is detected after $[83.11, 95.59]$, $[116.14, 134.32]$, and $[138.23, 161.03]$ for different $\text{NARL}_0$ values of 200, 300, and 370. From Tables 13.1 and 13.2, it can be observed that as the shift size increases, the values of $\text{NARL}_1$ decrease rapidly.

### 13.3.1 Comparison in NARL

In this section, the proposed chart is compared with the existing chart proposed by [2]. The values of the NARL of both control charts are shown in Table 13.4. From Table 13.4, it can be seen that the proposed control chart gives the smaller values of NARLs as compared to the chart proposed by [2]. For example, when $c = 1.05$, $\text{NARL}_0 = 370$, and $i_N \in [2, 3]$, the existing chart indicates that the first out-of-control is expected from 165th sample to 220th sample. On the other hand, at the same setting of the parameters, the

**TABLE 13.2**: The NARLs of proposed chart when $\nu_N \in [0.9, 1.1]$, $\mu_N \in [3.5, 4.5]$, and $i_N \in [3, 4]$.

| $k_{1N}$ | [3.0177, 3.0101] | [3.1637, 3.1936] | [3.2303, 3.2933] |
|---|---|---|---|
| $k_{2N}$ | [2.1656, 2.2223] | [2.2347, 2.2725] | [2.2735, 2.2999] |
| $\delta$ | | NARLs | |
| 1.0000 | [200.00, 200.00] | [300.01, 300.03] | [370.00, 371.73] |
| 1.0125 | [185.93, 188.22] | [277.24, 280.83] | [340.89, 347.00] |
| 1.0250 | [172.02, 176.43] | [254.88, 261.72] | [312.41, 322.46] |
| 1.0375 | [158.49, 164.79] | [233.27, 242.97] | [284.96, 298.43] |
| 1.0500 | [145.49, 153.43] | [212.65, 224.78] | [258.87, 275.21] |
| 1.0625 | [133.15, 142.45] | [193.21, 207.33] | [234.36, 253.00] |
| 1.0750 | [121.54, 131.95] | [175.08, 190.74] | [211.58, 231.97] |
| 1.1250 | [83.11, 95.59] | [116.14, 134.32] | [138.23, 161.03] |
| 1.1500 | [68.50, 80.95] | [94.27, 112.12] | [111.36, 133.44] |
| 1.2000 | [46.79, 58.06] | [62.52, 78.18] | [72.78, 91.70] |
| 1.2500 | [32.53, 42.04] | [42.28, 55.09] | [48.54, 63.73] |
| 1.3000 | [23.17, 30.93] | [29.36, 39.53] | [33.27, 45.13] |
| 1.3500 | [16.96, 23.21] | [20.98, 28.97] | [23.51, 32.68] |
| 1.4000 | [12.75, 17.78] | [15.45, 21.73] | [17.12, 24.24] |
| 1.4500 | [9.85, 13.92] | [11.70, 16.67] | [12.84, 18.41] |
| 1.5000 | [7.81, 11.11] | [9.11, 13.08] | [9.91, 14.31] |
| 1.6000 | [5.24, 7.51] | [5.94, 8.57] | [6.36, 9.22] |
| 1.8000 | [2.95, 4.15] | [3.20, 4.52] | [3.34, 4.75] |
| 1.9000 | [2.41, 3.33] | [2.57, 3.57] | [2.66, 3.71] |
| 2.0000 | [2.05, 2.77] | [2.15, 2.94] | [2.22, 3.03] |

proposed control chart indicates that the first out-of-control can be expected from 149th sample to 149th sample. From this chapter, it can be concluded that the proposed control chart has the ability to indicate a shift in the process quicker than the chart proposed by [2].

## 13.4 Simulation Study

The productivity of the proposed chart compared to that of [2] is discussed by generating the data from the NCOM-Poisson distribution. For the simulation dataset, let there be in-control NARL, $\text{NARL}_0 \in [370, 370]$, $\upsilon_N \in [0.4, 0.6]$, and $\mu_N \in [2.5, 3.5]$. The first half values are generated by stating the in-control process and remaining half values are generated when process has shifted with $c = 1.2$. For these specified parameters, the calculated values $\text{NARL}_0$, $\text{NARL}_0 \in [9.31, 27.48]$. Its means, the expected out-of-control values must be between 9th and 27th samples. The neutrosophic charting statistics of the proposed chart for simulated data are displayed in Figure 13.1. It can be

**TABLE 13.3**: The NARLs of proposed chart when $\nu_N \in [0.36, 0.37]$, $\mu_N \in [2.85, 2.90]$, and $i_N \in [2, 3]$.

| $k_{1N}$ | [2.9523, 2.8933] | [3.2171, 3.0991] | [3.2259, 3.1942] |
|---|---|---|---|
| $k_{2N}$ | [2.1413, 2.2883] | [2.1339, 2.2770] | [2.2011, 2.2945] |
| $\delta$ | | NARLs | |
| 1.0000 | [200.01, 200.01] | [300.00, 300.02] | [370.00, 370.04] |
| 1.0125 | [163.30, 164.62] | [240.51, 241.35] | [294.15, 294.79] |
| 1.0250 | [126.17, 128.61] | [181.56, 183.27] | [219.77, 221.14] |
| 1.0375 | [93.97, 96.99] | [131.66, 133.92] | [157.62, 159.40] |
| 1.0500 | [68.64, 71.74] | [93.47, 95.87] | [110.69, 112.50] |
| 1.0625 | [49.85, 52.71] | [65.94, 68.19] | [77.27, 78.90] |
| 1.0750 | [36.33, 38.81] | [46.70, 48.68] | [54.20, 55.56] |
| 1.1250 | [11.64, 12.77] | [13.56, 14.45] | [15.20, 15.74] |
| 1.1500 | [7.29, 8.04] | [8.18, 8.77] | [9.02, 9.38] |
| 1.2000 | [3.51, 3.88] | [3.72, 4.03] | [3.99, 4.19] |
| 1.2500 | [2.14, 2.35] | [2.19, 2.38] | [2.30, 2.43] |
| 1.3000 | [1.55, 1.69] | [1.57, 1.69] | [1.62, 1.71] |
| 1.3500 | [1.28, 1.37] | [1.28, 1.37] | [1.31, 1.38] |
| 1.4000 | [1.14, 1.20] | [[1.14, 1.20 | [1.15, 1.20] |
| 1.4500 | [1.07, 1.10] | [1.07, 1.10] | [1.07, 1.10] |
| 1.5000 | [1.03, 1.05] | [1.00, 1.05] | [1.03, 1.05] |
| 1.6000 | [1.00, 1.01] | [1.00, 1.01] | [1.00, 1.01] |
| 1.8000 | [1.00, 1.00] | [1.00, 1.00] | [1.00, 1.00] |
| 1.9000 | [1.00, 1.00] | [1.00, 1.00] | [1.00, 1.00] |
| 2.0000 | [1.00, 1.00] | [1.00, 1.00] | [1.00, 1.00] |

seen that the 22nd subgroup is out of control. From Figure 13.1, it can also be seen that the first indeterminacy interval point occurs at the 23rd subgroup. Figure 13.2 displays the plot of the charting statistics of the existing chart for the sample level of all specified parameters. From Figure 13.2, it can be noted that the first out-of-control point occurs at the 25th subgroup. Proposed chart spots four samples out-of-control; while the existing chart just shows one out-of-control point. By comparing both figures, it can be seen that the proposed chart under the neutrosophic statistic using the MDSS is more efficient as compared to its counterpart proposed by [2].

## 13.5 Real-life Applications

In this section, the application of the proposed control chart is provided using the well-reputable Saudi Arabia-based electric company data. This company manufactures printed circuits boards (PCB) which have been used in different computers, electronic goods, and electrical items. The PCB items are used to connect different features of electrical products with each other and give

**TABLE 13.4**: The comparison of proposed chart with existing when $\nu_N \in [0.4, 0.6]$ and $\mu_N \in [2.5, 3.5]$.

| $c$ | Existing Chart $k_N \in [3.289, 3.3982]$ NARLs | | Proposed Chart $i_N \in [2, 3]$ NARLs | |
|---|---|---|---|---|
| 1.0000 | [300.2, 302.0] | [370.1, 372.7] | [300.00, 300.03] | [370.00, 370.17] |
| 1.0125 | [255.2271.4] | [312.5, 333.5] | [251.42, 267.44] | [306.09, 328.27] |
| 1.0250 | [211.6, 240.4] | [257.2, 293.9] | [205.03, 233.64] | [245.94, 285.98] |
| 1.0375 | [172.3, 210.3] | [207.7, 255.8] | [163.84, 202.15] | [193.34, 245.15] |
| 1.0500 | [138.6, 182.3] | [165.6, 220.5] | [129.19, 172.08] | [149.83, 209.10] |
| 1.0625 | [110.8, 157.0] | [131.3, 188.8] | [101.14, 146.04] | [115.21, 176.58] |
| 1.0750 | [88.5, 134.6] | [104.0, 160.9] | [79.01, 123.27] | [88.38, 148.89] |
| 1.1250 | [37.4, 72.1] | [42.7, 84.3] | [30.62, 62.41] | [32.04, 72.81] |
| 1.1500 | [25.3, 53.4] | [28.5, 61.8] | [19.97, 44.57] | [20.34, 51.40] |
| 1.2000 | [12.8, 30.5] | [14.1, 34.7] | [9.51, 24.63] | [9.31, 27.48] |
| 1.2500 | [7.3, 18.6] | [7.9, 20.8] | [5.26, 13.32] | [5.04, 15.68] |
| 1.3000 | [4.6, 12.1] | [4.9, 13.3] | [3.32, 8.97] | [3.16, 9.22] |
| 1.3500 | [3.2, 8.3] | [3.3, 9.0] | [2.35, 5.03] | [2.23, 6.96] |
| 1.4000 | [2.3, 6.0] | [2.4, 6.4] | [1.82, 4.32] | [1.74, 4.39] |
| 1.4500 | [1.8, 4.5] | [1.9, 4.8] | [1.51, 3.28] | [1.46, 3.78] |
| 1.5000 | [1.5, 3.5] | [1.6, 3.7] | [1.32, 2.62] | [1.29, 2.67] |
| 1.6000 | [1.2, 2.4] | [1.2, 2.5] | [1.13, 1.87] | [1.11, 1.89] |
| 1.8000 | [1.0, 1.4] | [1.0, 1.5] | [1.01, 1.29] | [1.01, 1.30] |
| 1.9000 | [1.0, 1.2] | [1.0, 1.3] | [1.00, 1.17] | [1.00, 1.17] |
| 2.0000 | [1.0, 1.1] | [1.0, 1.18] | [1.00, 1.10] | [1.00, 1.10] |

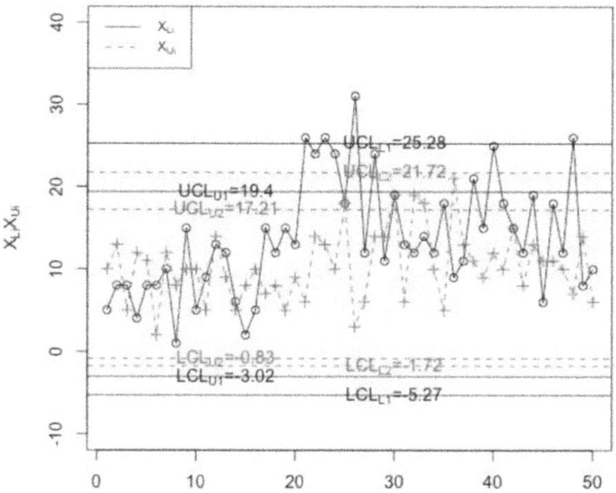


**FIGURE 13.1**: The proposed chart with $k_{1N} \in [3.0739, 30589]$ and $k_{2N} \in [2.3586, 2.4595]$.

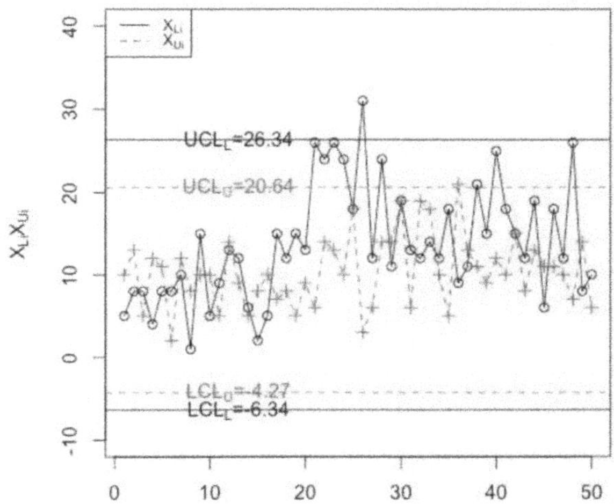


**FIGURE 13.2**: The existing chart with $k_N \in [3.289, 3.3982]$.

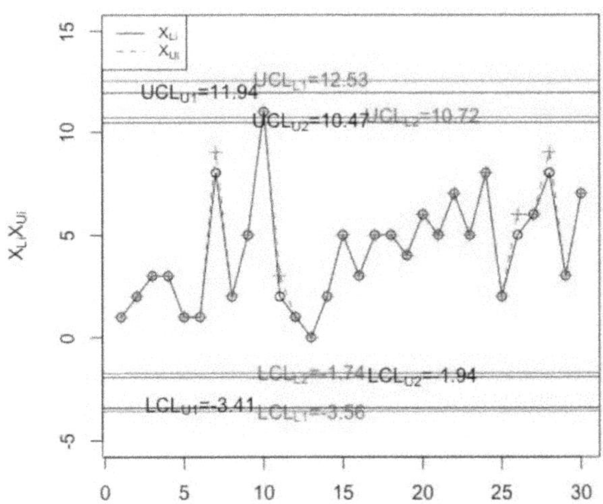


**FIGURE 13.3**: The proposed chart for real-life data.

mechanical support. The electric company is not sure about the proportion defective parameter for checking the state of the PCB products. Furthermore, due to the uncertainty in the amount defective parameter, it is not possible to apply the control chart designed under the classical statistics; hence, the neutrosophic statistics for checking the products manufacturing process status are used. The company is interested in the use of the proposed control chart for the monitoring of the number of defective products items. Thus, the company decides to use $\text{NARL}_0 \in [370, 370]$ with a sample size of 100. The data on PCB is shown in Table 13.5.

**TABLE 13.5**: Printed circuit boards dataset.

| Sample no. | Non-conformities | Sample no. | Non-conformities |
|---|---|---|---|
| 1 | [1, 1] | 16 | [3, 3] |
| 2 | [2, 2] | 17 | [5, 5] |
| 3 | [3, 3] | 18 | [5, 5] |
| 4 | [3, 3] | 19 | [4, 4] |
| 5 | [1, 1] | 20 | [6, 6] |
| 6 | [1, 1] | 21 | [5, 5] |
| 7 | [8, 9] | 22 | [7, 7] |
| 8 | [2, 2] | 23 | [5, 5] |
| 9 | [5, 5] | 24 | [8, 8] |
| 10 | [11, 11] | 25 | [2, 2] |
| 11 | [2, 3] | 26 | [5, 6] |
| 12 | [1, 1] | 27 | [6, 6] |
| 13 | [0, 0] | 28 | [8, 9] |
| 14 | [2, 2] | 29 | [3, 3] |
| 15 | [5, 5] | 30 | [7, 7] |

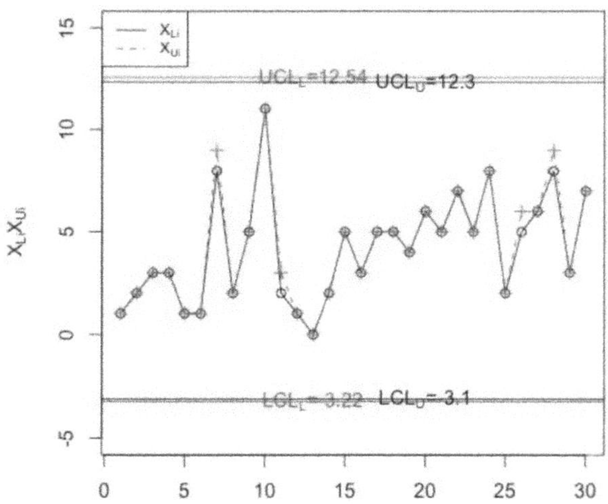


**FIGURE 13.4**: The existing chart for real data.

The data in Table 13.5 follows the NCOM-Poisson distribution with parameters $\nu_N \in [0.634222, 0.657101]$ and $\mu_N \in [2.5485, 2.5485]$ when $\text{NARL}_0 \in [370, 370]$. The non-conformities of the products are plotted in Figure 13.3. From Figure 13.3, it can be seen that no point is out of control or no shift is detected. But, it can be noted that the 10th point is in indeterminacy interval and this point is also near to the out-of-control limit; so it needs engineer's attention. The existing chart under the neutrosophic statistic by [2] is plotted in Figure 13.4. From Figure 13.4, it can be seen that the existing chart does not detect any shift. By comparing Figures 13.3 and 13.4, the existing control chart does not detect any point in the indeterminacy interval and shows that all points are in-control, but the proposed chart shows one indeterminacy interval point. Therefore, we can say that the proposed chart is more efficient and flexible as compared to its counterpart under the uncertainty environment.

## 13.6 Conclusions

In this chapter, a new control chart methodology for the neutrosophic statistics using the COM-Poisson distribution under the MDSS scheme is developed. Control chart coefficients are estimated to efficiently monitor a production process. The advantages of the proposed chart have been highlighted. Illustrative applications reveal the superiority of the proposed chart over the existing one. The smaller values of ARLs of the shifted process show the better detecting ability of the proposed chart. The proposed chart is a valuable addition to the toolkit of the quality control personnel when applied under appropriate conditions. The proposed chart can further be extended for other sampling techniques under multivariate statistics.

**Declarations:** Ethics approval and consent to participate: Not applicable
Consent for publication: Not applicable.
Availability of data and material: The data is given in the chapter.
Competing interests: No conflict of interest regarding the chapter.
Funding: None.

**Author contributions:**
SR, AS, MS, LA, JCM, SS, AA and YAZ had all contributed in writing the manuscript equally and done all statistics and relevant tables. MA critically and linguistically revised and reviewed the whole literature. All authors read and approved the final manuscript.

## Abbreviations

ARL: average run lengths
LCL: lower control limit
MDSS: multiple dependent state sampling
NARL: neutrosophic ARL
NCOM-Poisson: Neutrosophic Conway-Maxwell-Poisson
PCB: printed circuits boards
SPC: statistical process control
UCL: upper control limit

## Bibliography

[1] W. A. Shewhart. Quality control charts. *Bell System Technical Journal*, 5(4):593–603, 1926.

[2] M. Aslam and A. H. Al-Marshadi. Design of a control chart based on COM-Poisson distribution for the uncertainty environment. *Complexity*, 1:8178067, 2019.

[3] R. W. Conway and W. L. Maxwell. A queuing model with state dependent service rates. *Journal of Industrial Engineering*, 12:132–136, 1962.

[4] D. Lord, S. D. Guikema, and S. R. Geedipally. Application of the conway–maxwell–poisson generalized linear model for analyzing motor vehicle crashes. *Accident Analysis & Prevention*, 40(3):1123–1134, 2008. doi: http://dx.doi.org/10.1016/j.aap.2007.12.003.

[5] J. B. Kadane, G. Shmueli, T. P. Minka, S. Borle, and P. Boatwright. Conjugate analysis of the conway-maxwell-poisson distribution. *Bayesian Analysis*, 1(2):363–374, 2006.

[6] M. Aslam, L. Ahmad, C.-H. Jun, and O. H. Arif. A control chart for COM–Poisson distribution using multiple dependent state sampling. *Quality and Reliability Engineering International*, 32(8):2803–2812, 2016. doi: 10.1002/qre.1965.

[7] M. Aslam, A. Saghir, L. Ahmad, C.-H. Jun, and J. Hussain. A control chart for com-poisson distribution using a modified EWMA statistic. *Journal of Statistical Computation and Simulation*, 87(18):3491–3502, 2017.

[8] V. Alevizakos and C. Koukouvinos. A double exponentially weighted moving average control chart for monitoring com-poisson attributes. *Quality and Reliability Engineering International*, 35(7):2130–2151, 2019a.

[9] V. Alevizakos and C. Koukouvinos. A progressive mean control chart for com-poisson distribution. *Communications in Statistics-Simulation and Computation*, 51(3):1–19, 2019b.

[10] J.-H. Chen. A double generally weighted moving average chart for monitoring the COM-Poisson processes. *Symmetry*, 12(6):1014, 2020.

[11] F. Smarandache. *Neutrosophic Probability, Set, and Logic.* ProQuest Information & Learning; Infolearnquest, Ann Arbor, MI, 1998.

[12] F. Smarandache. *A Unifying Field in Logics. Neutrosophy: Neutrosophic Probability, Set and Logic.* American Research Press, Rehoboth, 1999.

[13] M. Aslam, M. Azam, L. Ahmad, and F. Smarandache. Product acceptance determination using repetitive sampling based on process loss consideration under neutrosophic numbers. In *Optimization Theory Based on Neutrosophic and Plithogenic Sets*. Academic Press, London, 2020.

[14] M. Albassam, N. Khan, and M. Aslam. The W/S test for data having neutrosophic numbers: An application to USA village population. *Complexity*, 3690879, 2020. doi: 10.1155/2020/3690879.

[15] M. Aslam and M. A. Raza. Design of new sampling plans for multiple manufacturing lines under uncertainty. *International Journal of Fuzzy Systems*, 21(3):978–992, 2019.

[16] M. Aslam, M. A. Raza, L. Ahmad, and C.-H. Jun. Design of a t-chart using generalized multiple dependent state sampling. *Quality and Reliability Engineering International*, 35(6):1789–1802, 2019. doi: 10.1002/qre.2475.

[17] M. Aslam, G. Srinivasa Rao, A. Shafqat, L. Ahmad, and R. A. Khan Sherwani. Monitoring circuit boards products in the presence of indeterminacy. *Measurement*, 168:108404, 2021. doi: https://doi.org/10.1016/j.measurement.2020.108404.

[18] M. Kashif, H. Nida, M. I. Khan, and M. Aslam. Decomposition of matrix under neutrosophic environment. *Neutrosophic Sets & Systems*, 30, 2019.

[19] Z. Khan, M. Gulistan, R. Hashim, N. Yaqoob, and W. Chammam. Design of S-control chart for neutrosophic data: An application to manufacturing industry. *Journal of Intelligent & Fuzzy Systems*, 38(4):4743–4751, 2020. doi: 10.3233/JIFS-191439.

[20] P. Muralikrishna and D. S. Kumar. Neutrosophic approach on normed linear space. *Neutrosophic Sets & Systems*, 30, 2019.

[21] M. Saad, T. Mahmood, K. Ullah, and N. Jan. Computing shortest path in a single valued neutrosophic hesitant fuzzy network. *The Nucleus*, 56 (3):123–130, 2020.

[22] C. M. Villamar, J. Suarez, L. D. L. Coloma, C. Vera, and M. Leyva. Analysis of technological innovation contribution to gross domestic product based on neutrosophic cognitive maps and neutrosophic numbers. *Neutrosophic Sets & Systems*, 30, 2019.

[23] X. Zhang, X. Wang, F. Smarandache, T. G. Jaíyéolá, and T. Lian. Singular neutrosophic extended triplet groups and generalized groups. *Cognitive Systems Research*, 57:32–40, 2019.

[24] A. W. Wortham and R. C. Baker. Multiple deferred state sampling inspection. *International Journal of Production Research*, 14(6):719–731, 1976. doi: 10.1080/00207547608956391.

[25] S. Balamurali and C.-H. Jun. Multiple dependent state sampling plans for lot acceptance based on measurement data. *European Journal of Operational Research*, 180(3):1221–1230, 2007.

[26] M. Aslam, M. Azam, N. Khan, and C.-H. Jun. A control chart for an exponential distribution using multiple dependent state sampling. *Quality & Quantity*, 49(2):455–462, 2014.

[27] M. S. Aldosari, M. Aslam, N. Khan, L. Ahmad, and C.-H. Jun. A new $S^2$ control chart using multiple dependent state repetitive sampling. *IEEE Access*, 6:49224–49236, 2018.

[28] K. F. Sellers, S. Borle, and G. Shmueli. The COM-Poisson model for count data: A survey of methods and applications. *Applied Stochastic Models in Business and Industry*, 28(2):104–116, 2012.

[29] F. Smarandache. Neutrosophic logic-A generalization of the intuitionistic fuzzy logic. In *Multispace & Multistructure. Neutrosophic Transdisciplinarity (100 Collected Papers of Science)*, volume 4, page 396, North-European Scientific Publishers, Hanko, 2010.

[30] M. Aslam. A variable acceptance sampling plan under neutrosophic statistical interval method. *Symmetry*, 11(1):114, 2019.

[31] C. D. Montgomery. *Introduction to Statistical Quality Control (6th edition)*. John Wiley & Sons, Inc, New York, 2009.

# Chapter 14

## A Hierarchical Framework for Assessing Groundwater Quality Using Evidential Neutrosophic Tree Soft Sets

**Devi M, Irene Hepzibah, and Priscilla Rajkumar**

### 14.1 Introduction

Water is a very essential resource; its existence is essential for supporting life of all living things. Groundwater is a very important source [1] amongst other sources and is of vital necessity in ensuring the demands of farming, consumption as drinking water [2], and even the industrial needs. Nevertheless, the increasing water pollution in the groundwater [3] is a major health and environmental issue to the population. As the world becomes more urbanized, industrialized, and more destructive in agricultural practices, groundwater quality [4] is fast declining. Thus, the necessity to monitor, analyse, and manage groundwater pollution has increased more urgently than ever before.

The intrusion of seawater is a major trigger of groundwater contamination in coastal regions due to the excessive pumping of groundwater [5] and the consequent imbalance in natural hydraulic gradients. In certain areas such as Pappakovil in the Nagapattinam district, groundwater has been grossly contaminated due to intrusion of saline water, making the water undrinkable and inadequate for agricultural practices [6, 7].

Here, researchers have examined the character and quality of groundwater [8] in certain vulnerable regions. The preliminary results review indicates that many of the parameters outlined in the water exceed accepted standards; thus, adequate research is needed to assess the various impacts on the human health and the environment. The parameters, such as pH, might seem merely harmless when considered [9] in isolation, but the concentration of ions, when combined with other parameters such as total dissolved solids (TDS) and electrical conductivity (EC), can have an incredible impact on the water quality. In order to evaluate groundwater contamination, researchers [10] usually

DOI: 10.1201/9781003683339-14

examine some chemical factors such as pH, TDS, EC, Cl, Na, Mg, and ratios of heavier metals. Hydro geochemical studies conducted in the coastal areas [11] indicate that high EC levels represent intrusion and deterioration of the water quality [12]. Spatial geochemistry and multivariate statistical modelling have been also heavily used in characterization of contamination sources and temporal trends [13, 14]. The standardized approaches to evaluating the quality of groundwater tend to fail when it comes to managing uncertainty, vagueness, and conflicting information, which are the features of the environment.

Machine learning (ML) has also shown ingenuity in modelling seawater intrusion dynamics through surrogate models in optimization schemes of modelling simulations. Nguyen et al., [15] have used XGBoost [16, 17] and Latin hypercube sampling to predict EC and chloride intrusion in coastal aquifers based on inputs in the form of satellite, soil, precipitation, transmissivity, and distance from the shoreline. In a recent study [18], an automatic model choice ML algorithm was trained to forecast the presence of intrusion of seawater in coastal aquifers, supporting the optimal groundwater mining plans [19].

It is against this backdrop that despite the developments in conventional ML models [20, 21], environmental systems that are characterized by data uncertainty, contradicting signals, and the absence of complete information are typically a challenge. In order to address this ambiguity, evidence theory (Dempster–Shafer evidence theory or DSET) [22, 23], has been integrated with ML algorithms in water quality forecasting [24] and the classification of hydro pollution [25].

Moreover, neutrosophic soft set theory has been explored more frequently in decision-making in indeterminacy. Incorporation of neutrosophic soft sets and bipolar neutrosophic sets [26] and proving their modelling capability of uncertainty, indeterminacy, and conflicting evidence in multi-criteria evaluation [27] have been accomplished by Karaaslan and Deli et al.[28]

Nevertheless, there are not many applications concerning the use of neutrosophic systems [29] in evaluating groundwater quality perse. Evidence fusion mechanisms based on interval neutrosophic logic can be used to rectify the differences between measured [30, 31] and computed values of conductivity, incomplete data on ionic concentrations, and soft information needs to integrate quality control thresholds [32]

To fill these gaps, the evidential neutrosophic tree soft set (ENTSS) method is proposed as a new tool of analysis. In managing uncertainty, quality control through conductivity coincidence checks, and relating total ion concentration to electrical conductivity, ENTSS combines the capabilities of evidence theory, neutrosophic soft set logic, and the hierarchical tree structured parameter relationship to examine groundwater quality. In such a way, ENTSS provides an adaptable, more realistic, and a powerful decision-support framework which can efficiently cope with the grounds of ambiguity and interdependence of data on groundwater, with decision-wise information to safeguard the lives of people and maintain ecological balance.

To conclude, this is the beginning of the further investigation of ENTSS as a crucially effective means of trying to solve the dilemma of groundwater pollution. It highlights the strong desire to consider multi-dimensional and uncertainty-aware frameworks that can lead sustainable water resource management in an environment that is under increasing stressors.

The remaining sections are structured in such a way that they provide the graded exploration of the offered ENTSS concept and its use in groundwater quality determination. Section 14.2 presents the preliminary concepts used in this chapter, such as basic probability assignments, Dempster rule of combination, soft sets, tree soft sets, and neutrosophic tree soft sets, which together form the theoretical background. The proposed model, ENTSS, is formally defined in Section 14.3, and its computational structure is introduced in the form of a detailed pseudo-code algorithm, step-by-step procedure outline, and the comparison table with the traditional methods. Section 14.4 utilizes the ENTSS model on a realistic dataset of groundwater and outlines the mechanism of sampling, the nature of the data, and the assignment of neutrosophic values to the hydrogeochemical parameters. Section 14.5, Results and Discussion, analyses the findings using three analytical profiles, highlighting the changes in the parameters, contamination patterns, and giving a viable recommendation to manage the groundwater. Lastly, Section 14.6 gives conclusion and a summary of the key findings, points to the efficiency and flexibility of ENTSS in handling the uncertainty in groundwater studies, and identifies prospects of future research on sustainable water quality assessment.

---

## 14.2 Preliminaries

DSET [33] is extensively utilized to handle uncertainties in various applications. Some fundamental concepts related to this theory are outlined here.

### 14.2.1 Basic probability assignments

Let $X$ be a collection of mutually exhaustive and exclusive events, commonly called frame of discernment (FOD). The power set of $X$, which is denoted as $2^X$, contains all empty set and $X$ itself.

For a given FOD $X = \theta_1, \theta_2, \theta_3, \cdots, \theta_X$, a basic probability assignment (BPA) (or mass function) $m$ is a mapping from $2^X$ to the interval [0,1], which is mathematically expressed as:

$$m : 2^X \to [0, 1]$$

This function satisfies the following conditions:

1. The mass of the empty set is to be zero.

$$m(\emptyset) = 0$$

2. Sum of all the mass must be equal to 1.

$$\sum_{A \in 2^X} [m(A)] = 1$$

BPA is also referred to as the mass function in DSET. Any subset A of $2^X$ that has a positive mass assignment ($m(A) > 0$) is called a focal element. If a BPA assigns all belief to the entire frame of discernment (i.e., $m(X) = 1$), it is known as a vacuous BPA, indicating a complete lack of specific information.

### 14.2.2 Dempster's combination rule

Consider two independent BPAs, $m_1$ and $m_2$. Their combination, represented as, $m = m_1 \bigoplus m_2$ is defined as follows:

$$m(A) = 1/(1 - K) \sum_{B \cap C = A} [m_1(B) \times m_2(C)], A \neq \emptyset \tag{14.1}$$

where, $K = \sum_{B \cap C = \emptyset} [m_1(B) \times m_2(C)]$

### 14.2.3 Soft sets

A soft set, proposed initially by Molodtsov (1999) [34], is a mathematical tool designed to deal with uncertainty without relying on traditional membership grades as in fuzzy sets. Suppose that there is a universal set $U$ and a set of attributes $E$. An order pair $(F, A)$ is referred to as a soft set, where:

- $A \subseteq D$,
- $F$ is any non-empty function such that $F : A \to \mathcal{P}(U)$ i.e., for each attribute $a \in A$,$F(a)$ is a subset of $U$ and $\mathcal{P}(U)$ is the power set.

### 14.2.4 Tree soft sets

A tree soft set is a variant of a soft set with a parameter space arranged in a hierarchical tree structure.

Suppose the set of attributes $D = D_1, D_2, \cdots$ is represented as nodes of a rooted tree, the next set of attributes $D_i = D_{i,1}, D_{i,2}, \cdots$ consists of first-level sub-attributes, and $D_{i,j} = D_{i,j,1}, D_{i,j,2}, \cdots$ is represented as nodes of second-level sub-attributes.

The final set of attributes $D_(i_1, i_2, \cdots, i_m)$ is represented as final nodes of $m$th attributes. A tree soft set over a universal set $U$ is defined as:

$$F : P(Tree(D)) \to \mathcal{P}(U), \tag{14.2}$$

where $Tree(D)$ represents the set of all nodes and leaves, and $P(Tree(D))$ denotes its power set. This structure is useful in situations where parameters are not independent but linked in a hierarchy.

### 14.2.5 Neutrosophic sets

A neutrosophic set, proposed by Smarandache (1998) [35], is an extension of the fuzzy set, introduced by Zadeh (1965) [36] in which three separate component are found: truth, indeterminacy, and falsity, which are treated as independent values rather than complementary. Suppose that $U$ is a universal set, and a neutrosophic set $A$ is defined as:

$$A = \{\langle x, T_A(x), I_A(x), F_A(x)\rangle : x \in U\} \tag{14.3}$$

where,

- $T_A(x)$ represents the truth degree,
- $I_A(x)$ represents the degree of indeterminacy,
- $F_A(x)$ represents the degree of falsity,
- Each of these values lies in the interval $[0, 1]$.

These values satisfy the following condition:

$$0 \leq \sum(T_A(x), I_A(x), F_A(x)) \leq 3$$

### 14.2.6 Neutrosophic soft set

A neutrosophic soft set integrates the soft set framework with neutrosophic theory to handle parameterized uncertainty involving three independent components: truth, indeterminacy, and falsity.

Formally, for a universal set $U$ and a parameter set $E$, a neutrosophic soft set is a pair $(F, A)$. If, $A \subseteq E$, then

$$F : A \to N(U)$$

The family of all neutrosophic sets on $U$ is denoted as $N(U)$. Thus, for each parameter $a \in A$,

$$F(a) = \{\langle x, T_{\mathcal{F}_a}(x), I_{\mathcal{F}_a}(x), F_{\mathcal{F}_a}(x)\rangle | \text{for all x} \in U\} \tag{14.4}$$

is a neutrosophic set on $U$, which satisfies

$$0 \leq T_{\mathcal{F}_a}(x) + I_{\mathcal{F}_a}(x) + F_{\mathcal{F}_a}(x) \leq 3.$$

This model allows more comprehensive decision-making by considering uncertainty in the evaluation of each parameter.

### 14.2.7 Neutrosophic tree soft sets

A neutrosophic tree soft set is a type of soft set in which the hierarchical structure of tree soft sets and the uncertainty of neutrosophic soft sets are combined. This model has the parameter space as a rooted tree, where a parameter is associated with a neutrosophic set rather than a traditional crisp or fuzzy set. Let $X \subseteq U$ be any non-empty subset of $U$, and $Tree(E)$ is a hierarchical structure of all parameters. Then, a neutrosophic tree soft set is represented as

$$\mathcal{F} : P(Tree(E)) \rightarrow \mathcal{N}(U)$$

For each parameter $\mathcal{F}(\sigma)$ is a neutrosophic set on $X$,

$$\mathcal{F}(\sigma) = \{\langle x, T_{\mathcal{F}(\sigma)}(x), I_{\mathcal{F}(\sigma)}(x), F_{\mathcal{F}(\sigma)}(x)\rangle | x \in U\} \tag{14.5}$$

with $T_{\mathcal{F}(\sigma)}, I_{\mathcal{F}(\sigma)}, F_{\mathcal{F}(\sigma)} : X \rightarrow [0,1]$ satisfying the following condition

$$0 \leq T_{\mathcal{F}(\sigma)}(x) + I_{\mathcal{F}(\sigma)}(x) + F_{\mathcal{F}(\sigma)}(x) \leq 3$$

This model is particularly useful for complex decision-making problems where attributes are interdependent and uncertain.

---

## 14.3 Evidential Neutrosophic Tree Soft Set (ENTSS)

This section proposes a new framework, namely ENTSS, to manage complex, hierarchical, and uncertain information in environmental decision-making. This is a model that combines three potent paradigms:

- DSET: Works with lack of information and uncertainty on how to approach this lack of information. It operates based on the belief structure rather than exact probabilities.
- Neutrosophic sets: Represent indeterminate and vague information as models to each element's truth value ($T$), indeterminacy value ($I$), and falsity value ($F$).
- Tree soft sets: Propose a hierarchical (tree-shaped) dependency between parameters, enabling multiple level decision-making that is contextually sensitive.

Collectively, ENTSS will facilitate multi-criteria analysis in areas of ambiguity, hierarchy, and insufficient knowledge, e.g., the assessment of groundwater quality, medical diagnosis, risk analysis, etc.

### 14.3.1 Definition of ENTSS

Let $X \subseteq U$ be a non-empty subset of a universal set $U$, and $Tree(E)$ be a hierarchical structure of parameters. Let $N_{\mathcal{F}_{D(\sigma)}}(m) : 2^U \rightarrow [0,1]$ be a neutrosophic basic probability assignment (NBPA), if

$$\sum_{A \in 2^U} N_{\mathcal{F}_{D(\sigma)}}(m(A)) = 1$$

and

$$N_{\mathcal{F}_{D(\sigma)}}(m(\emptyset)) = 0.$$

An ENTSS satisfies the following conditions:

- $F : P(Tree(E)) \rightarrow N_{\mathcal{F}_D}(m(X))$,
- $\mathcal{F}_{D(\sigma)} = \{T_{\mathcal{F}_{D(\sigma)}}, I_{\mathcal{F}_{D(\sigma)}}, F_{\mathcal{F}_{D(\sigma)}}\}$. For each parameter. $\mathcal{F}_{D(\sigma)}$ is an evidential neutrosophic set on $X$ and $\sigma \subseteq$ Tree(E).

Consider two independent NBPAs as $N_{\mathcal{F}_{D(\sigma)}}(m_1)$ and $N_{\mathcal{F}_{D(\sigma)}}(m_2)$. Their combination, represented as $N_{\mathcal{F}_{D(\sigma)}}(m) = N_{\mathcal{F}_{D(\sigma)}}(m_1) \bigoplus N_{\mathcal{F}_{D(\sigma)}}(m_2)$, is defined as follows:

$$\begin{aligned} N_{\mathcal{F}_{D(\sigma)}}(m(A)) = 1/(1-K_N) & \sum_{(B \cap C = A)} \\ & \times [T_{\mathcal{F}_{D(\sigma)}}(m_1(B)), I_{\mathcal{F}_{D(\sigma)}}(m_1(B)), F_{\mathcal{F}_{D(\sigma)}}(m_1(B))] \\ & \times [T_{\mathcal{F}_{D(\sigma)}}(m_2(C)), I_{\mathcal{F}_{D(\sigma)}}(m_2(C)), F_{\mathcal{F}_{D(\sigma)}}(m_2(C))], \end{aligned} \tag{14.6}$$

where

$$\begin{aligned} K_N = \sum_{B \cap C = \emptyset} & [T_{\mathcal{F}_{D(\sigma)}}(m_1(B)), I_{\mathcal{F}_{D(\sigma)}}(m_1(B)), F_{\mathcal{F}_{D(\sigma)}}(m_1(B))] \\ & \times [T_{\mathcal{F}_{D(\sigma)}}(m_2(C)), I_{\mathcal{F}_{D(\sigma)}}(m_2(C)), F_{\mathcal{F}_{D(\sigma)}}(m_2(C))]. \end{aligned}$$

Suppose $Bel_{N1}$ and $Bel_{N2}$ are belief functions over the same frame $\Theta$, with NBPAs $N_{\mathcal{F}_{D(\sigma)}}(m_1)$ and $N_{\mathcal{F}_{D(\sigma)}}(m_2)$ with the focal elements $1, 2, \cdots, n$. Suppose

$$\begin{aligned} \sum_{B \cap C = \emptyset} & [T_{\mathcal{F}_{D(\sigma)}}(m_1(B)), I_{\mathcal{F}_{D(\sigma)}}(m_1(B)), F_{\mathcal{F}_{D(\sigma)}}(m_1(B))] \\ & \times [T_{\mathcal{F}_{D(\sigma)}}(m_2(C)), I_{\mathcal{F}_{D(\sigma)}}(m_2(C)), F_{\mathcal{F}_{D(\sigma)}}(m_2(C))] < 1 \end{aligned} \tag{14.7}$$

Then the function $N_{\mathcal{F}_{D(\sigma)}}(m) : 2^U \rightarrow [0,1]$ defined by $N_{\mathcal{F}_{D(\sigma)}}(m(\emptyset)) = 0$ and

$$\begin{aligned} N_{\mathcal{F}_{D(\sigma)}}(m(A)) = 1/(1-K_N) & \sum_{(B \cap C = A)} \\ & \times [T_{\mathcal{F}_{D(\sigma)}}(m_1(B)), I_{\mathcal{F}_{D(\sigma)}}(m_1(B)), F_{\mathcal{F}_{D(\sigma)}}(m_1(B))] \\ & \times [T_{\mathcal{F}_{D(\sigma)}}(m_2(C)), I_{\mathcal{F}_{D(\sigma)}}(m_2(C)), F_{\mathcal{F}_{D(\sigma)}}(m_2(C))] \end{aligned}$$

where

$$K_N = \sum_{B\cap C=\emptyset} [T_{\mathcal{F}_{D(\sigma)}}(m_1(B)), I_{\mathcal{F}_{D(\sigma)}}(m_1(B)), F_{\mathcal{F}_{D(\sigma)}}(m_1(B))] \times [T_{\mathcal{F}_{D(\sigma)}}(m_2(C)), I_{\mathcal{F}_{D(\sigma)}}(m_2(C)), F_{\mathcal{F}_{D(\sigma)}}(m_2(C))].$$

For all non-empty $A \subset \Theta$, A is considered an NBPA. The core of the belief function given by $N_{\mathcal{F}_{D(\sigma)}}(m)$ is equal to the intersection of the cores of $Bel_{N1}$ and $Bel_{N2}$.

Since (14.7) is assumed, the function $N_{\mathcal{F}_{D(\sigma)}}(m)$ is certainly well-defined, which takes non-negative values.

By definition (14.6), $N_{\mathcal{F}_{D(\sigma)}}(m(\emptyset)) = 0$ and the NBPA $N_{\mathcal{F}_{D(\sigma)}}(m(A))$ defined such that the values sum to one (i.e., $\sum_{A\in 2^U} N_{\mathcal{F}_{D(\sigma)}}(m(A)) = 1$).

Now,

$$\sum_{A\subset\Theta} N_{\mathcal{F}_{D(\sigma)}}(m(A)) = N_{\mathcal{F}_{D(\sigma)}}(m(\emptyset)) + \sum_{\substack{A\subset\Theta\\ A\neq\emptyset}} N_{\mathcal{F}_{D(\sigma)}}(m(A))$$

$$= \sum_{\substack{A\subset\Theta\\ A\neq\emptyset}} \left( \frac{\sum\limits_{(B\cap C=A)} [T_{\mathcal{F}_{D(\sigma)}}(m_1(B)), I_{\mathcal{F}_{D(\sigma)}}(m_1(B)), F_{\mathcal{F}_{D(\sigma)}}(m_1(B))] \times [T_{\mathcal{F}_{D(\sigma)}}(m_2(C)), I_{\mathcal{F}_{D(\sigma)}}(m_2(C)), F_{\mathcal{F}_{D(\sigma)}}(m_2(C))]}{1 - \sum\limits_{B\cap C=\emptyset} [T_{\mathcal{F}_{D(\sigma)}}(m_1(B)), I_{\mathcal{F}_{D(\sigma)}}(m_1(B)), F_{\mathcal{F}_{D(\sigma)}}(m_1(B))] \times [T_{\mathcal{F}_{D(\sigma)}}(m_2(C)), I_{\mathcal{F}_{D(\sigma)}}(m_2(C)), F_{\mathcal{F}_{D(\sigma)}}(m_2(C))]} \right)$$

$$= \left( \frac{1}{1 - \sum\limits_{B\cap C=\emptyset} [T_{\mathcal{F}_{D(\sigma)}}(m_1(B)), I_{\mathcal{F}_{D(\sigma)}}(m_1(B)), F_{\mathcal{F}_{D(\sigma)}}(m_1(B))] \times [T_{\mathcal{F}_{D(\sigma)}}(m_2(C)), I_{\mathcal{F}_{D(\sigma)}}(m_2(C)), F_{\mathcal{F}_{D(\sigma)}}(m_2(C))]} \right) \times \left( \sum_{B\cap C\neq\emptyset} [T_{\mathcal{F}_{D(\sigma)}}(m_1(B)), I_{\mathcal{F}_{D(\sigma)}}(m_1(B)), F_{\mathcal{F}_{D(\sigma)}}(m_1(B))] \times [T_{\mathcal{F}_{D(\sigma)}}(m_2(C)), I_{\mathcal{F}_{D(\sigma)}}(m_2(C)), F_{\mathcal{F}_{D(\sigma)}}(m_2(C))] \right) = 1.$$

The core of the belief function given by $N_{\mathcal{F}_{D(\sigma)}}(m)$ will be

$$\bigcup (N_{\mathcal{F}_{D(\sigma)}}(m_1(B)) \cap N_{\mathcal{F}_{D(\sigma)}}(m_2(C)) = \left(\bigcup N_{\mathcal{F}_{D(\sigma)}}(m_1(B))\right) \cap \left(\bigcup N_{\mathcal{F}_{D(\sigma)}}(m_2(C))\right)$$

And $\bigcup N_{\mathcal{F}_{D(\sigma)}}(m_1(B))$ and $\bigcup N_{\mathcal{F}_{D(\sigma)}}(m_2(C))$ are the cores of $Bel_{N1}$ and $Bel_{N2}$, respectively.

Let $N_{\mathcal{F}_{D(\sigma)}}(m_1)$, $N_{\mathcal{F}_{D(\sigma)}}(m_2)$, and $N_{\mathcal{F}_{D(\sigma)}}(m_3)$ be three ENTSSs over the common universe $(U, E)$. Then, the following properties hold:

1. Commutative law:
$N_{\mathcal{F}_{D(\sigma)}}(m_1) \cup N_{\mathcal{F}_{D(\sigma)}}(m_2) = N_{\mathcal{F}_{D(\sigma)}}(m_2) \cup N_{\mathcal{F}_{D(\sigma)}}(m_1)$,
$N_{\mathcal{F}_{D(\sigma)}}(m_1) \cap N_{\mathcal{F}_{D(\sigma)}}(m_2) = N_{\mathcal{F}_{D(\sigma)}}(m_2) \cap N_{\mathcal{F}_{D(\sigma)}}(m_1)$.

2. Associative law:

$$\left(N_{\mathcal{F}_{D(\sigma)}}(m_1) \cup N_{\mathcal{F}_{D(\sigma)}}(m_2)\right) \cup N_{\mathcal{F}_{D(\sigma)}}(m_3) \\ = N_{\mathcal{F}_{D(\sigma)}}(m_1) \cup \left(N_{\mathcal{F}_{D(\sigma)}}(m_2) \cup N_{\mathcal{F}_{D(\sigma)}}(m_3)\right),$$

$$\left(N_{\mathcal{F}_{D(\sigma)}}(m_1) \cap N_{\mathcal{F}_{D(\sigma)}}(m_2)\right) \cap N_{\mathcal{F}_{D(\sigma)}}(m_3) \\ = N_{\mathcal{F}_{D(\sigma)}}(m_1) \cap \left(N_{\mathcal{F}_{D(\sigma)}}(m_2) \cap N_{\mathcal{F}_{D(\sigma)}}(m_3)\right).$$

3. Distributive law:

$$N_{\mathcal{F}_{D(\sigma)}}(m_1) \cup \left(N_{\mathcal{F}_{D(\sigma)}}(m_2) \cap N_{\mathcal{F}_{D(\sigma)}}(m_3)\right) \\ = \left(N_{\mathcal{F}_{D(\sigma)}}(m_1) \cup N_{\mathcal{F}_{D(\sigma)}}(M_2)\right) \cap \\ \left(N_{\mathcal{F}_{D(\sigma)}}(m_1) \cup N_{\mathcal{F}_{D(\sigma)}}(M_3)\right)$$

$$N_{\mathcal{F}_{D(\sigma)}}(m_1) \cap \left(N_{\mathcal{F}_{D(\sigma)}}(m_2) \cup N_{\mathcal{F}_{D(\sigma)}}(m_3)\right) \\ = \left(N_{\mathcal{F}_{D(\sigma)}}(m_1) \cap N_{\mathcal{F}_{D(\sigma)}}(m_2)\right) \cup \\ \left(N_{\mathcal{F}_{D(\sigma)}}(m_1) \cap N_{\mathcal{F}_{D(\sigma)}}(m_3)\right)$$

4. Involution law:
$(N_{\mathcal{F}_{D(\sigma)}}(m\prime)\prime = N_{\mathcal{F}_{D(\sigma)}}(m)$

5. Idempotent law:
$N_{\mathcal{F}_{D(\sigma)}}(m) \cup N_{\mathcal{F}_{D(\sigma)}}(m) = N_{\mathcal{F}_{D(\sigma)}}(m)$,
$N_{\mathcal{F}_{D(\sigma)}}(m) \cap N_{\mathcal{F}_{D(\sigma)}}(m) = N_{\mathcal{F}_{D(\sigma)}}(m)$.

6. Identity law:
$N_{\mathcal{F}_{D(\sigma)}}(m) \cup \emptyset = N_{\mathcal{F}_{D(\sigma)}}(m)$
$N_{\mathcal{F}_{D(\sigma)}}(m) \cap \emptyset = \emptyset$

7. De Morgan's law:
$\left(N_{\mathcal{F}_{D(\sigma)}}(m_1) \cup N_{\mathcal{F}_{D(\sigma)}}(m_2)\right)\prime = N_{\mathcal{F}_{D(\sigma)}}(m_1)\prime \cap N_{\mathcal{F}_{D(\sigma)}}(m_2)\prime$,
$\left(N_{\mathcal{F}_{D(\sigma)}}(m_1) \cap N_{\mathcal{F}_{D(\sigma)}}(m_2)\right)\prime = N_{\mathcal{F}_{D(\sigma)}}(m_1)\prime \cup N_{\mathcal{F}_{D(\sigma)}}(m_2)\prime$

**1. Commutative law**

Let $x \in \Theta$ and $N_{\mathcal{F}_{D(\sigma)}}(m_1)$ and $N_{\mathcal{F}_{D(\sigma)}}(m_2)$ be two ENTSSs. Since $\cdot$ & $*$ are commutative, then

$N_{\mathcal{F}_{D(\sigma)}}(m_1) \cup N_{\mathcal{F}_{D(\sigma)}}(m_2) \Rightarrow$

$T_{\mathcal{F}_{D(\sigma)}}(m_1(x)) \cdot T_{\mathcal{F}_{D(\sigma)}}(m_2(x)) = T_{\mathcal{F}_{D(\sigma)}}(m_2(x)) \cdot T_{\mathcal{F}_{D(\sigma)}}(m_1(x)),$

$I_{\mathcal{F}_{D(\sigma)}}(m_1(x)) * I_{\mathcal{F}_{D(\sigma)}}(m_2(x)) = I_{\mathcal{F}_{D(\sigma)}}(m_2(x)) * I_{\mathcal{F}_{D(\sigma)}}(m_1(x)),$

$F_{\mathcal{F}_{D(\sigma)}}(m_1(x)) * F_{\mathcal{F}_{D(\sigma)}}(m_2(x)) = F_{\mathcal{F}_{D(\sigma)}}(m_2(x)) * F_{\mathcal{F}_{D(\sigma)}}(m_1(x)),$

$= N_{\mathcal{F}_{D(\sigma)}}(m_2) \cup N_{\mathcal{F}_{D(\sigma)}}(m_1)$

$\Rightarrow N_{\mathcal{F}_{D(\sigma)}}(m_1) \cup N_{\mathcal{F}_{D(\sigma)}}(m_2) = N_{\mathcal{F}_{D(\sigma)}}(m_2) \cup N_{\mathcal{F}_{D(\sigma)}}(m_1).$

Therefore, commutative law is proved.

Similarly, we can prove

$N_{\mathcal{F}_{D(\sigma)}}(m_1) \cap N_{\mathcal{F}_{D(\sigma)}}(m_2) = N_{\mathcal{F}_{D(\sigma)}}(m_2) \cap N_{\mathcal{F}_{D(\sigma)}}(m_1).$

**2. Associative law**

Let $x \in \Theta and N_{\mathcal{F}_{D(\sigma)}}(m_1)$, $N_{\mathcal{F}_{D(\sigma)}}(m_2)$, and $N_{\mathcal{F}_{D(\sigma)}}(m_3)$ be three ENTSSs. Since $\cdot$ & $*$ are commutative and associative, then

$N_{\mathcal{F}_{D(\sigma)}}(m_1) \cup N_{\mathcal{F}_{D(\sigma)}}(m_2) \cup N_{\mathcal{F}_{D(\sigma)}}(m_3) \Rightarrow$

$$[T_{\mathcal{F}_{D(\sigma)}}(m_1(x)) \cdot T_{\mathcal{F}_{D(\sigma)}}(m_2(x))] \cdot T_{\mathcal{F}_{D(\sigma)}}(m_3(x)) = T_{\mathcal{F}_{D(\sigma)}}(m_1(x)) \cdot [T_{\mathcal{F}_{D(\sigma)}}(m_2(x)) \cdot T_{\mathcal{F}_{D(\sigma)}}(m_3(x))],$$

$$[I_{\mathcal{F}_{D(\sigma)}}(m_1(x)) * I_{\mathcal{F}_{D(\sigma)}}(m_2(x))] * I_{\mathcal{F}_{D(\sigma)}}(m_3(x)) = I_{\mathcal{F}_{D(\sigma)}}(m_1(x)) * [I_{\mathcal{F}_{D(\sigma)}}(m_2(x)) * I_{\mathcal{F}_{D(\sigma)}}(m_3(x))],$$

$$[F_{\mathcal{F}_{D(\sigma)}}(m_1(x)) * F_{\mathcal{F}_{D(\sigma)}}(m_2(x))] * F_{\mathcal{F}_{D(\sigma)}}(m_3(x)) = F_{\mathcal{F}_{D(\sigma)}}(m_1(x)) * [F_{\mathcal{F}_{D(\sigma)}}(m_2(x)) * F_{\mathcal{F}_{D(\sigma)}}(m_3(x))]$$

($\because *$ is associative)

$$\begin{aligned} &= N_{\mathcal{F}_{D(\sigma)}}(m_1) \cup \left(N_{\mathcal{F}_{D(\sigma)}}(m_2) \cup N_{\mathcal{F}_{D(\sigma)}}(m_3)\right) \\ &\times \left(N_{\mathcal{F}_{D(\sigma)}}(m_1) \cup N_{\mathcal{F}_{D(\sigma)}}(m_2)\right) \cup N_{\mathcal{F}_{D(\sigma)}}(m_3) \\ &= N_{\mathcal{F}_{D(\sigma)}}(m_1) \cup \left(N_{\mathcal{F}_{D(\sigma)}}(m_2) \cup N_{\mathcal{F}_{D(\sigma)}}(m_3)\right) \end{aligned}$$

Hence, associativity has been proved.

Similarly,

$$\left(N_{\mathcal{F}_{D(\sigma)}}(m_1) \cap N_{\mathcal{F}_{D(\sigma)}}(m_2)\right) \cap N_{\mathcal{F}_{D(\sigma)}}(m_3)$$
$$= N_{\mathcal{F}_{D(\sigma)}}(m_1) \cap \left(N_{\mathcal{F}_{D(\sigma)}}(m_2) \cap N_{\mathcal{F}_{D(\sigma)}}(m_3)\right).$$

**3. Distributive law**

Let $x \in \Theta$ and $N_{\mathcal{F}_{D(\sigma)}}(m_1)$, $N_{\mathcal{F}_{D(\sigma)}}(m_2)$, and $N_{\mathcal{F}_{D(\sigma)}}(m_3)$ be three ENTSSs. Since $\cdot$ & $*$ are commutative and associative, then

$$N_{\mathcal{F}_{D(\sigma)}}(m_1) \cup (N_{\mathcal{F}_{D(\sigma)}}(m_2) \cap N_{\mathcal{F}_{D(\sigma)}}(m_3)$$
$$= T_{\mathcal{F}_{D(\sigma)}}(m_1(x)) \cdot [T_{\mathcal{F}_{D(\sigma)}}(m_2(x)) * T_{\mathcal{F}_{D(\sigma)}}(m_3(x))]$$
$$= [T_{\mathcal{F}_{D(\sigma)}}(m_1(x)) \cdot T_{\mathcal{F}_{D(\sigma)}}(m_2(x))] * [T_{\mathcal{F}_{D(\sigma)}}(m_1(x)) \cdot T_{\mathcal{F}_{D(\sigma)}}(m_3(x))]$$

$$I_{\mathcal{F}_{D(\sigma)}}(m_1(x)) \cdot [I_{\mathcal{F}_{D(\sigma)}}(m_2(x)) * I_{\mathcal{F}_{D(\sigma)}}(m_3(x))]$$
$$= [I_{\mathcal{F}_{D(\sigma)}}(m_1(x)) \cdot I_{\mathcal{F}_{D(\sigma)}}(m_2(x))] * [I_{\mathcal{F}_{D(\sigma)}}(m_1(x)) \cdot I_{\mathcal{F}_{D(\sigma)}}(m_3(x))],$$

$$F_{\mathcal{F}_{D(\sigma)}}(m_1(x)) \cdot [F_{\mathcal{F}_{D(\sigma)}}(m_2(x)) * F_{\mathcal{F}_{D(\sigma)}}(m_3(x))]$$
$$= [F_{\mathcal{F}_{D(\sigma)}}(m_1(x)) \cdot F_{\mathcal{F}_{D(\sigma)}}(m_2(x))] * [F_{\mathcal{F}_{D(\sigma)}}(m_1(x)) \cdot F_{\mathcal{F}_{D(\sigma)}}(m_3(x))]$$
$$= \left(N_{\mathcal{F}_{D(\sigma)}}(m_1) \cup N_{\mathcal{F}_{D(\sigma)}}(m_2)\right) \cap \left(N_{\mathcal{F}_{D(\sigma)}}(m_1) \cup N_{\mathcal{F}_{D(\sigma)}}(m_3)\right)$$
$$\therefore N_{\mathcal{F}_{D(\sigma)}}(m_1) \cup \left(N_{\mathcal{F}_{D(\sigma)}}(m_2) \cap N_{\mathcal{F}_{D(\sigma)}}(m_3)\right) = \left(N_{\mathcal{F}_{D(\sigma)}}(m_1) \cup N_{\mathcal{F}_{D(\sigma)}}(M_2)\right)$$
$$\cap \left(N_{\mathcal{F}_{D(\sigma)}}(m_1) \cup N_{\mathcal{F}_{D(\sigma)}}(M_3)\right)$$

Similarly, we can prove

$$N_{\mathcal{F}_{D(\sigma)}}(m_1) \cap \left(N_{\mathcal{F}_{D(\sigma)}}(m_2) \cup N_{\mathcal{F}_{D(\sigma)}}(m_3)\right) = \left(N_{\mathcal{F}_{D(\sigma)}}(m_1) \cap N_{\mathcal{F}_{D(\sigma)}}(m_2)\right)$$
$$\cup \left(N_{\mathcal{F}_{D(\sigma)}}(m_1) \cap N_{\mathcal{F}_{D(\sigma)}}(m_3)\right).$$

**4. Involution law**

Let $N_{\mathcal{F}_{D(\sigma)}}(m)$ be an ENTSS. $N_{\mathcal{F}_{D(\sigma)}}(m)\prime$ be the complement of $N_{\mathcal{F}_{D(\sigma)}}(m)$.

$$N_{\mathcal{F}_{D(\sigma)}}(m) = \{<x, T_{\mathcal{F}_{D(\sigma)}}(m(x)), I_{\mathcal{F}_{D(\sigma)}}(m(x)), F_{\mathcal{F}_{D(\sigma)}}(m(x))> : x \in \Theta\}$$

$$N_{\mathcal{F}_{D(\sigma)}}(m)\prime = \{<x, F_{\mathcal{F}_{D(\sigma)}}(m(x)), 1 - I_{\mathcal{F}_{D(\sigma)}}(m(x)), T_{\mathcal{F}_{D(\sigma)}}(m(x))> : x \in \Theta\}$$

$$(N_{\mathcal{F}_{D(\sigma)}}(m)\prime)\prime = \{<x, T_{\mathcal{F}_{D(\sigma)}}(m(x)), I_{\mathcal{F}_{D(\sigma)}}(m(x)), F_{\mathcal{F}_{D(\sigma)}}(m(x))> : x \in \Theta\}$$

Hence,

$(N_{\mathcal{F}_{D(\sigma)}}(m)\prime)\prime = N_{\mathcal{F}_{D(\sigma)}}(m)$ has been proved.

**5. Idempotent law**

Let $N_{\mathcal{F}_{D(\sigma)}}(m)$ be an ENTSS. Since $\cdot$ & $*$ are commutative and associative, then

$$N_{\mathcal{F}_{D(\sigma)}}(m) = \{<x, T_{\mathcal{F}_{D(\sigma)}}(m(x)), I_{\mathcal{F}_{D(\sigma)}}(m(x)), F_{\mathcal{F}_{D(\sigma)}}(m(x))> : x \in \Theta\}$$

Now,

$$N_{\mathcal{F}_{D(\sigma)}}(m(x)) \cup N_{\mathcal{F}_{D(\sigma)}}(m(x)) \Rightarrow$$

$$T_{\mathcal{F}_{D(\sigma)}}(m(x)) \cdot T_{\mathcal{F}_{D(\sigma)}}(m(x)) = T_{\mathcal{F}_{D(\sigma)}}(m(x)) \, (\because x \cdot x = x)\,,$$

$$I_{\mathcal{F}_{D(\sigma)}}(m(x)) * I_{\mathcal{F}_{D(\sigma)}}(m(x)) = I_{\mathcal{F}_{D(\sigma)}}(m(x)) \, (\because x * x = x)\,,$$

$$F_{\mathcal{F}_{D(\sigma)}}(m(x)) * F_{\mathcal{F}_{D(\sigma)}}(m(x)) = F_{\mathcal{F}_{D(\sigma)}}(m(x)) \, (\because x * x = x)\,,$$

$$= \{\{<x, T_{\mathcal{F}_{D(\sigma)}}(m(x)), I_{\mathcal{F}_{D(\sigma)}}(m(x)), F_{\mathcal{F}_{D(\sigma)}}(m(x))> : x \in \Theta\}$$

$$= N_{\mathcal{F}_{D(\sigma)}}(m)$$

$$\therefore N_{\mathcal{F}_{D(\sigma)}}(m) \cup N_{\mathcal{F}_{D(\sigma)}}(m) = N_{\mathcal{F}_{D(\sigma)}}(m)$$

Similarly,

$$N_{\mathcal{F}_{D(\sigma)}}(m) \cap N_{\mathcal{F}_{D(\sigma)}}(m) = N_{\mathcal{F}_{D(\sigma)}}(m).$$

**6. Identity law**

Let $N_{\mathcal{F}_{D(\sigma)}}(m)$ be an ENTSS. Since $\cdot$ & $*$ are commutative and associative, then we have

$$N_{\mathcal{F}_{D(\sigma)}}(m) \cup \emptyset \Rightarrow$$

$$T_{\mathcal{F}_{D(\sigma)}}(m(x)) \cdot \emptyset = T_{\mathcal{F}_{D(\sigma)}}(m(x)) \, (a \cdot b = max(a, b))$$

$$I_{\mathcal{F}_{D(\sigma)}}(m(x)) * \emptyset = I_{\mathcal{F}_{D(\sigma)}}(m(x)) \, (a * b = min(a, b))$$

$$F_{\mathcal{F}_{D(\sigma)}}(m(x)) * \emptyset = F_{\mathcal{F}_{D(\sigma)}}(m(x)) \, (a * b = min(a, b))$$

$$\therefore N_{\mathcal{F}_{D(\sigma)}}(m) \cup \emptyset = N_{\mathcal{F}_{D(\sigma)}}(m)$$

Similarly, we prove
$N_{\mathcal{F}_{D(\sigma)}}(m) \cap \emptyset = \emptyset.$

**7. De Morgan's law**

Let $x \in \Theta$ and $N_{\mathcal{F}_{D(\sigma)}}(m_1)$ and $N_{\mathcal{F}_{D(\sigma)}}(m_2)$ be two ENTSSs. Since $\cdot$ & $*$ is commutative and associative then

LHS:

$$(N_{\mathcal{F}_{D(\sigma)}}(m_1) \cup N_{\mathcal{F}_{D(\sigma)}}(m_2))' \Rightarrow (T_{\mathcal{F}_{D(\sigma)}}(m_1(x)) \cdot T_{\mathcal{F}_{D(\sigma)}}(m_2(x)),$$
$$I_{\mathcal{F}_{D(\sigma)}}(m_1(x)) * I_{\mathcal{F}_{D(\sigma)}}(m_2(x)), F_{\mathcal{F}_{D(\sigma)}}(m_1(x)) * F_{\mathcal{F}_{D(\sigma)}}(m_2(x)))'$$
$$= [F_{\mathcal{F}_{D(\sigma)}}(m_1(x)) * F_{\mathcal{F}_{D(\sigma)}}(m_2(x)), 1 - (I_{\mathcal{F}_{D(\sigma)}}(m_1(x)) \cdot I_{\mathcal{F}_{D(\sigma)}}(m_2(x))),$$
$$T_{\mathcal{F}_{D(\sigma)}}(m_1(x)) \cdot T_{\mathcal{F}_{D(\sigma)}}(m_2(x))])$$

RHS:

$$N_{\mathcal{F}_{D(\sigma)}}(m_1)' = \{<x, F_{\mathcal{F}_{D(\sigma)}}(m_1(x)), 1 - I_{\mathcal{F}_{D(\sigma)}}(m_1(x)), T_{\mathcal{F}_{D(\sigma)}}(m_1(x))> : x \in \Theta\}$$

$$N_{\mathcal{F}_{D(\sigma)}}(m_2') = \{<x, F_{\mathcal{F}_{D(\sigma)}}(m_2(x)), 1 - I_{\mathcal{F}_{D(\sigma)}}(m_2(x)), T_{\mathcal{F}_{D(\sigma)}}(m_2(x))> : x \in \Theta\}$$

$$N_{\mathcal{F}_{D(\sigma)}}(m_1') \cap N_{\mathcal{F}_{D(\sigma)}}(m_2') = (F_{\mathcal{F}_{D(\sigma)}}(m_1(x)) * F_{\mathcal{F}_{D(\sigma)}}(m_2(x)), 1 - (I_{\mathcal{F}_{D(\sigma)}}(m_1(x)) \cdot I_{\mathcal{F}_{D(\sigma)}}(m_2(x))), T_{\mathcal{F}_{D(\sigma)}}(m_1(x)) \cdot T_{\mathcal{F}_{D(\sigma)}}(m_2(x)))$$

$\therefore\ LHS = RHS \Rightarrow (N_{\mathcal{F}_{D(\sigma)}}(m_1) \cup N_{\mathcal{F}_{D(\sigma)}}(m_2))' = N_{\mathcal{F}_{D(\sigma)}}(m_1)' \cap N_{\mathcal{F}_{D(\sigma)}}(m_2)'$

Hence proved.

Similarly, we can prove

$$(N_{\mathcal{F}_{D(\sigma)}}(m_1) \cap N_{\mathcal{F}_{D(\sigma)}}(m_2)' = N_{\mathcal{F}_{D(\sigma)}}(m_1)' \cup N_{\mathcal{F}_{D(\sigma)}}(m_2)'.$$

### 14.3.2 Pseudo-code algorithm

**Input:** The input consists of monthly data collected over 24 months for each profile (A, B, and C) at every station, with three measurements recorded per profile.

- Obtain values of groundwater parameters
- Put in data[month][profile][station]

**Output:** The outcome is the combined final BPA for all profiles.

**Algorithm 1** Pseudo-code of ENTSS in water quality assessments.

```
EVENT analysing water quality assessments DO
    Define Parameter Tree = Grouped parameters
    FOR EACH obtain values of Groundwater parameters DO
        Assign Neutrosophic value = (T, I, F) (each parameter for all profiles)
        T = degree_good (∀T ∈ [0,1])
        I = degree_uncertain (∀I ∈ [0,1])
        F = degree_bad (∀F ∈ [0,1])
Calculate Neutrosophic to BPA for each group of parameters using equation
(1.6)
    END FOR EACH
    Combine all BPA for all profiles
```

### 14.3.3 Step-by-step algorithmic procedures

An ENTSS is constructed in the following main steps:

- **Step 1:** Collecting the data
  Collect values of physico-chemical parameters on three separate profiles (A, B, and C). Arrange the data that have been collected in table form in order to have consistency and continue to analyse.

- **Step 2:** Build tree of physico-chemical parameters
  Detect and differentiate all the appropriate parameters (e.g., pH, TDS, BOD, and so on). Present these parameters in a tree structure in a hierarchical form to be able to analyse them logically and group them. This tree provides the basis of assigning neutrosophic values.

- **Step 3:** Assign neutrosophic values
  Provide neutrosophic values in every parameter in all the profiles:
  "Truth ($T$)" represents the goodness of the parameter.
  "Indeterminacy ($I$)" is the degree of uncertainty.
  "Falsity ($F$)" represents the degree of the badness of the parameter.

- **Step 4:** Calculate combined BPA based on evidence theory
  Apply DSET that combines evidence across profiles. Compute the total BPA to make a final decision of each parameter or subset. Apply conflict resolution in the case that conflicting evidence is encountered.

### 14.3.4 Comparison analysis

Table 14.1 clearly show that the tree soft sets (TSS) are a simple model of decision-making whose membership values lie in $[0, 1]$ but do not model

**TABLE 14.1**: Comparison of TSS, NTSS, DSET, and ENTSS.

| **Properties** | **TSS** | **NTSS** | **DSET** | **ENTSS (proposed)** |
|---|---|---|---|---|
| Membership values | [0,1] | Triple $T, I, F \in [0,1], 0 \leq \sum(T,I,F) \leq 3$ | Mass values $m(A) \in [0,1]$, $\sum m(A) = 1$ | Triple $T, I, F \in [0,1]$, $\sum(T,I,F) = 1$ and $\sum N(m(A)) = 1$ |
| Uncertainty handling | None | Yes, It is represented as Indeterminacy | BPA structure handles uncertainty via multi-subset assignments | Both indeterminacy and BPA framework handle uncertainty (dual-layer modelling) |
| Evidence combination | Not applicable | Not defined in original NTSS | Dempster–Shafer combination rule | Dempster–Shafer applied to triples (T,I,F) - fusion at each tree node |
| Main advantage | Simple and easy to compute | Captures truth, indeterminacy, falsity | Handles evidence with uncertainty | Integrates parameter hierarchy, neutrosophic uncertainty, and evidence theory for high richer decision-making |
| Limitation | Cannot model graded uncertainty | Indeterminacy only qualitative | No explicit parameter tree structure | Conflict handling needed for $K = 0$ case; requires more computation than NTSS |

graded uncertainty. Neutrosophic tree soft sets (NTSSs) are generalization of TSS, where an additional representation $(T, I, F)$ is provided to represent truth, indeterminacy, and falsity, respectively, and model basic uncertainty. DSET uses mass assignments $m(A)$ over subsets with $\sum m(A) = 1$. It permits representation of uncertainty in terms of multi-level probabilities over sets and provides a guiding set combination rule, but has no built-in length scale structure as parameter tree. The new ENTSSs unite the hierarchy of parameters of TSS/NTSS with the triple-valued uncertainty representation of NTSS and belief set representation of DSET. This two-layered consideration would allow richer decisions since it would combine indeterminacy and evidential uncertainty and it needs more computation and conflict management with high evidence disagreement $(K \approx 0)$.

## 14.4 Case Study

Groundwater samples were taken monthly with the help of 15 randomly chosen open wells and boreholes in and around Nagapattinam region in three different profiles (A, B, and C) over 24 months (August 2022–July 2024). It was in regard to the direction of flow of river water and effluent of shrimp farming that the profiles were chosen. At every profile, three samples were taken at an

**TABLE 14.2**: Water quality parameters and WHO limits.

| Parameters | WHO limit | Minimum value | Maximum value | Average value |
|---|---|---|---|---|
| pH | 6.5–8.5 | 7.02 | 8.51 | 7.94 |
| Ca (ppm) | 100 | 32 | 146 | 90.6 |
| Cl (ppm) | 250 | 175 | 2189 | 776 |
| Mg (ppm) | 200 | 16 | 187 | 58 |
| Bicarbonates (ppm) | — | — | — | — |
| K (ppm) | — | 4 | 240 | 79 |
| $SO_4$ (ppm) | 400 | 20 | 630 | 250 |
| Iron (ppm) | 0.3 | 3 | 50 | 10 |
| $HCO_3$ (ppm) | — | — | — | — |
| Na (ppm) | 6.5 | 92 | 1088 | 296 |

intermittent distance of approximately 1 km to three different stations, hence making approximately 75 $km^2$ as the area of study.

It was the physico-chemical parameters that included pH, electrical conductivity, total dissolved solids (TDS), total hardness (TH) and a number of other parameters such as bicarbonate ($HCO_3$), sulphate ($SO_4$), chloride (Cl), dissolved oxygen (DO), fluoride (F), nitrate ($NO_3$),carbonate ($CO_3$), phosphate ($PO_4$), sodium (Na), calcium (Ca), magnesium (Mg), and biochemical oxygen demand (BOD).

Accompanying the pace of social industrialization, it is gradually becoming the interest of people to improve the nature of environmental pollution, particularly the anti-pollution of water environment; the major enterprises and institutions that are interested in using online water quality detection systems are also on the rise. Most of the points that supply small amounts of water are widely spread in distant places and thus information cannot be easily gathered and it would need much investment. The problem of sparse water supply points and long distances can be well resolved by the user acquisition terminal equipment of the wireless network system and local transmission. The conductivity in water solution is not merely an overall indicator in quantification of water quality but also an indication of quantity of ionizable matter in water. As a result of this simple test, it finds wide application in earthquake monitoring as well as in the analysis of water quality. As science and technology advanced, different and new techniques have been applied in the monitoring and inspection of water quality.

### 14.4.1 Quality of sample

Particularly, 1 L of sample suffices most of the physio-chemical tests. In particular special determinations, however, greater quantities of sample are taken. Table 14.2 presents the water quality parameters measured in the study alongside the corresponding WHO limits. The groundwater samples were obtained over a two-year time period between August 2022 and July 2024. The selec-

**TABLE 14.3**: Selected profile samples.

| Profile name | Samples |
|---|---|
| A | Eruma saalai water |
| B | Pappakovil pipeline water |
| C | Pappakovil groundwater |

tion of the three profiles where the three samples were derived was done on the basis of flow of Pappakovil area in the Nagapattinam district. At every profile, the samples were taken at three distinct points at regular interval of approximate 1 km. There were total of three samples that were taken, and the stations are mentioned in Table 14.3.

The Nagapattinam district is located in the eastern coast of the state of Tamil Nadu, India. It lies south of the Cuddalore district and some portions of it lie even farther south of the Karaikal and Tiruvarur districts. The northern border of the district starts about 75 km to the south of the district headquarters of Cuddalore. It has Thanjavur and Tiruvarur districts to the west and the eastern and southern boundaries, respectively, to the Bay of Bengal. The geographical location of the Nagapattinam district is between $10°25'$ and $11°40'$ north latitude, and $76°49'$ and $80°01'$ east longitude. The Uppanar river is located in the heart of the area. The effluent coming directly from shrimp farming operations is classified as summer season because the night and day temperatures during the periods are around 25° C. The results obtained during the period (August 2022–July 2024) are shown in Table 14.4.

**TABLE 14.4**: Three different profiles' water quality parameters.

| Minerals | Profile A | Profile B | Profile C |
|---|---|---|---|
| Level of salt | 2.48 | 1.06 | 1.28 |
| $HCO_3$ | 10.3 | 8.7 | 0.3 |
| $CO_3$ | — | 0.2 | 0.5 |
| pH | 7.82 | 8.51 | 8.2 |
| Ca | 1.7 | 0.8 | 1.0 |
| Cl | 13.2 | 5.4 | 4.8 |
| $SO_4$ | 0.1 | 0.24 | 0.1 |
| K | 1.05 | 0.18 | 0.08 |
| Mg | 7.4 | 6.8 | 3.2 |
| Na | 15.9 | 4.74 | 7.41 |
| Residual $NaCO_3$ | 1.2 | — | 4.6 |
| Na retention rate | 7.44 | 2.61 | 5.16 |

Figure 14.1 is a hierarchical structure of the physico-chemical parameters in a tree diagram, where the parameters can be seen in their relationships and detailed level. Separation of these parameters into groups in Table 14.5 forms the basis of the construction of the hierarchical tree to allow orderly analysis and systematized evaluation in the ENTSS concept.

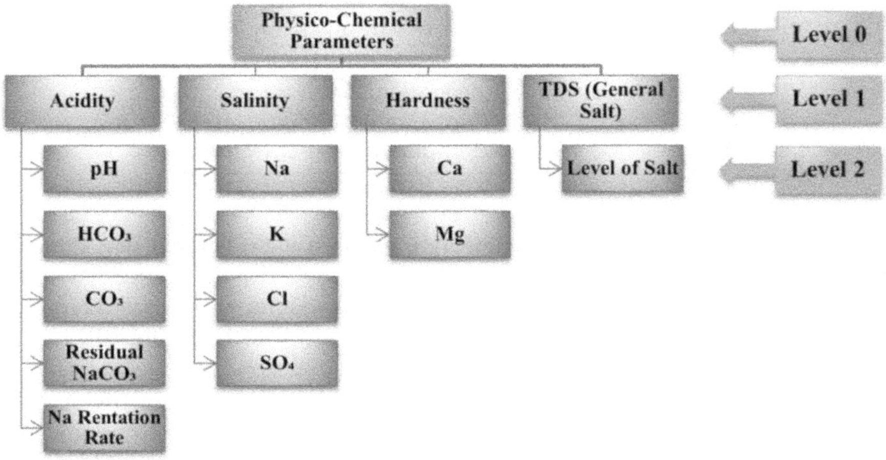


**FIGURE 14.1**: Physico-chemical parameters tree.

**TABLE 14.5**: Group of the parameters.

| **Group name** | **Parameters** |
|---|---|
| Acidity | pH, $HCO_3$, $CO_3$, Residual $NaCO_3$, Na Retention Rate |
| Salinity | Na, K, Cl, $SO_4$ |
| Hardness | Ca, Mg |
| TDS (general salt) | Level of Salt |

Table 14.6 derives the ideal threshold value of each physico-chemical parameter in line with the World Health Organization (WHO) drinking water quality guidelines. These threshold values would be used as reference points guiding the attainment of the neutrosophic membership degrees of each parameter.

## 14.4.2 Assigning neutrosophic values

To find the truth value,

$$T = 1 - \frac{(x - \{ideal\})}{range}$$

where range = difference between the standard bounds or a logical span.

For low uncertainty as it's still in range, assume the indeterminacy value $I = 0.05$.

To find the falsity,

$$F = 1 - T - I$$

**TABLE 14.6**: Water quality parameters with WHO standards and ideal values.

| Parameter | WHO guideline/ standard | Basis for assumption | Ideal value (ppm) |
|---|---|---|---|
| **pH** | Acceptable range: 6.5–8.5 | Midpoint of the range | 7.5 |
| **Ca** | Up to 100 | 80% of the permissible limit | 80 |
| **Mg** | Up to 200 | 80% of the permissible limit | 160 |
| **Cl** | Up to 250 | 80% of the permissible limit | 200 |
| **$SO_4$** | Up to 400 | 80% of the permissible limit | 320 |
| **Na** | Max 6.5 (low preferred) | Half of the upper limit used as ideal | 3.25 |
| **K** | No fixed limit | Average taken from sample data | 50 |
| **$HCO_3$** | No fixed limit | Mean from dataset assumed as ideal | 5 |
| **Total salts** | Not specified by WHO | Based on general freshwater composition | 1.5 |
| **Na retention** | No defined standard | Selected from dataset range | 4 |
| **$Na_2CO_3$** | No prescribed value | Based on dataset range | 2.0 |
| **$CO_3$** | Not clearly defined | Typical value from dataset | 0 |

Let's apply this for profile A having pH = 7.82

WHO range = 6.5–8.5; total span = 2.0; ideal = 7.5

Now normalize:

Normalized $T$-value $= 1 - \frac{(7.82-7.5)}{2.0} = 1 - \frac{0.32}{2.0} = 0.84$.

Therefore, $T = 0.84$

Assume $I = 0.05$

$F = 1 - T - I = 1 - 0.84 - 0.05 = 0.11$

1. pH = 7.82, ideal = 7.5, range = 2.0

$$T = 0.84, \quad I = 0.05, \quad F = 0.11$$

2. Ca = 1.7 ppm, ideal = 80, range = 100

$$T = \text{1-78.3/100=0.217}, \quad I = 0.05, \quad F = 0.733$$

3. Mg = 7.4 ppm, ideal = 160, range = 200

$$T = \text{1-152.6/200=0.237}, \quad I = 0.05, \quad F = 0.713$$

4. Cl = 13.2 ppm, Ideal = 200, Range = 250

$$T = \text{1-186.8/250=0.253}, \quad I = 0.05, \quad F = 0.697$$

5. $SO_4$ = 0.1 ppm, ideal = 320, range = 400

$$T = \text{1-319.9/400=0.200}, \quad I = 0.05, \quad F = 0.75$$

6. Na = 15.9 ppm, ideal = 3.25, range = 6.5 This exceeds even the maximum among all the three profiles. So, assign the truth value as 0.00 and the indeterminacy of $I = 0.10$.

$$T = 0.0, \quad I = 0.10, \quad F = 0.90$$

7. K = 1.05 ppm, ideal = 50, range = 100

$$T = \text{1-48.95/100=0.510}, \quad I = 0.05, \quad F = 0.44$$

8. $HCO_3$ = 10.3 ppm, Ideal = 5, Range = 20

$$T = \text{1-5.3/20=0.735}, \quad I = 0.05, \quad F = 0.215$$

9. Salt = 2.48, ideal = 1.5, range = 3

$$T = \text{1-0.98/3=0.673}, \quad I = 0.05, \quad F = 0.277$$

10. Na retention = 7.44, ideal = 4, range = 8

$$T = \text{1-3.44/8=0.57}, \quad I = 0.05, \quad F = 0.38$$

11. Residual $NaCO_3$ = 1.2, ideal = 2, range = 4

$$T = \text{1-0.8/4=0.80}, \quad I = 0.05, \quad F = 0.15$$

12. $CO_3$ = NaN, ideal = 0, range = 1
    The values are not defined in profile A; it clearly represents the high range of uncertainty, and the WHO recommends no standard limit. That is, the ideal = 0. Assign the truth value as 0.05 and the value of F = 0.05.

$$T = 0.05, \quad I = 0.9, \quad F = 0.05$$

Similarly, calculate Profile B's and C's parameters.

### 14.4.3 Defining water quality parameters as masses

The triplets $(T,I,F)$ are assigned to measured data, with truth, indeterminacy, and falsity representing the extent of consistency or inconsistency with the WHO guidelines. Subsequently, the basic probability assignments (BPAs) are given in terms of individual water quality parameters expressed as $m_1$–$m_{12}$, where $m_1 = pH$,
$m_2 = HCO_3$,
$m_3 = Na\ Retention,$
$m_4 = Residual\ NaCO_3$,
$m_5 = CO_3$,
$m_6 = Cl$,
$m_7 = SO_4$,
$m_8 = Na$,
$m_9 = K$,
$m_{10} = Ca$,
$m_{11} = Mg$
$m_{12} = Salt$.

Table 14.7 contains these parameter-specific BPAs, together with their neutrosophic values. The next stage would be the unification of identity BPAs of parameters with the same label in the hierarchical tree representation as presented in Tables 14.8–14.10, showing the BPA outcomes of level-1 nodes of these profiles (A, B, C), respectively. Lastly, the Dempster–Shafer combination rule is checked iteratively over the tree to generate accumulated BPA on every profile. Table 14.11 reports the final combined values of BPA values expressing a total assessment of water quality.

---

## 14.5 Results and Discussion

Figure 14.2 shows the graphical representation of final BPA values. The groundwater quality determination in Nagapattinam region conducted on three profiles (A, B, and C) during the period of 24 months indicated that there are high spatial and temporal differences. Significant influence of seasonality, especially in cases of monsoon and post-monsoon seasons, was found in several physico-chemical parameters. TDS, EC, chloride, and nitrate were also observed to have high levels during the months after monsoon, which is possibly caused by more surface run-off and infusion of agricultural and aquaculture effluents. Of the three profiles, profile A, which was upstream relative to river flow and shrimp farm discharge, consistently provided improved water quality compared to profiles B and C. Profile A demonstrated the highest belief mass in "Good" water quality (51%) and the lowest in "Bad" water quality (49%) when compared to profiles B and C.

**TABLE 14.7**: Neutrosophic values for three profiles.

| Mass | Profile A | | | Profile B | | | Profile C | | |
|---|---|---|---|---|---|---|---|---|---|
| | $T$ (Good) | $I$ (Uncertain) | $F$ (Bad) | $T$ (Good) | $I$ (Uncertain) | $F$ (Bad) | $T$ (Good) | $I$ (Uncertain) | $F$ (Bad) |
| $m_1$ | 0.84 | 0.05 | 0.11 | 0.75 | 0.05 | 0.21 | 0.85 | 0.05 | 0.10 |
| $m_2$ | 0.74 | 0.05 | 0.22 | 0.69 | 0.05 | 0.26 | 0.57 | 0.05 | 0.38 |
| $m_3$ | 0.77 | 0.05 | 0.18 | 0.45 | 0.05 | 0.50 | 0.49 | 0.05 | 0.46 |
| $m_4$ | 0.80 | 0.05 | 0.15 | 0.05 | 0.90 | 0.05 | 0.65 | 0.05 | 0.30 |
| $m_5$ | 0.05 | 0.90 | 0.05 | 0.80 | 0.05 | 0.15 | 0.50 | 0.05 | 0.45 |
| $m_6$ | 0.25 | 0.05 | 0.70 | 0.23 | 0.05 | 0.72 | 0.23 | 0.05 | 0.72 |
| $m_7$ | 0.20 | 0.05 | 0.75 | 0.18 | 0.05 | 0.77 | 0.19 | 0.05 | 0.77 |
| $m_8$ | 0.00 | 0.10 | 0.90 | 0.00 | 0.10 | 0.90 | 0.00 | 0.10 | 0.90 |
| $m_9$ | 0.51 | 0.05 | 0.44 | 0.45 | 0.05 | 0.50 | 0.42 | 0.05 | 0.53 |
| $m_{10}$ | 0.22 | 0.05 | 0.73 | 0.20 | 0.05 | 0.75 | 0.21 | 0.05 | 0.75 |
| $m_{11}$ | 0.24 | 0.05 | 0.71 | 0.21 | 0.05 | 0.74 | 0.28 | 0.05 | 0.67 |
| $m_{12}$ | 0.67 | 0.05 | 0.28 | 0.64 | 0.05 | 0.32 | 0.66 | 0.05 | 0.29 |

**TABLE 14.8**: Combined values of parameter groups of profile A.

| **Profile A** | **Acidity** ($m_1$–$m_5$) | **Salinity** ($m_6$–$m_9$) | **Hardness** ($m_{10}$–$m_{11}$) | **TDS** ($m_{12}$) |
|---|---|---|---|---|
| $T$ (Good) | 0.99601 | 0.01424 | 0.11038 | 0.67 |
| $I$ (Uncertain) | 0.00001 | 0.00004 | 0.00372 | 0.05 |
| $F$ (Bad) | 0.00398 | 0.98572 | 0.88589 | 0.28 |

**TABLE 14.9**: Combined values of parameter groups of profile B.

| **Profile B** | **Acidity** ($m_1$–$m_5$) | **Salinity** ($m_6$–$m_9$) | **Hardness** ($m_{10}$–$m_{11}$) | **TDS** ($m_{12}$) |
|---|---|---|---|---|
| $T$ (Good) | 0.96639 | 0.00902 | 0.09127 | 0.64 |
| $I$ (Uncertain) | 0.00002 | 0.00004 | 0.00361 | 0.05 |
| $F$ (Bad) | 0.03359 | 0.99095 | 0.90512 | 0.32 |

**TABLE 14.10**: Combined values of parameter groups of profile C.

| **Profile C** | **Acidity** ($m_1$–$m_5$) | **Salinity** ($m_6$–$m_9$) | **Hardness** ($m_{10}$–$m_{11}$) | **TDS** ($m_{12}$) |
|---|---|---|---|---|
| $T$ (Good) | 0.95272 | 0.00839 | 0.12484 | 0.66 |
| $I$ (Uncertain) | 0.00000 | 0.00003 | 0.00382 | 0.05 |
| $F$ (Bad) | 0.04727 | 0.99157 | 0.87134 | 0.29 |

**TABLE 14.11**: Final BPA values.

| **Parameter's BPA** | **Profile A** | **Profile B** | **Profile C** |
|---|---|---|---|
| $T$ (Good) | 0.50635 | 0.04896 | 0.04953 |
| $I$ (Uncertain) | 0.00000 | 0.00000 | 0.00000 |
| $F$ (Bad) | 0.49365 | 0.95104 | 0.95047 |

The ENTSS framework arranged these values in a tree hierarchical descent allowing evidence to be accumulated across related attributes and finally combining them into a profile-level information. This was done following parameter-level neutrosophic modelling. This tree model played a significant role in capturing the levels of dependence and relative distances between the indicators of water quality as the higher weights were established on the parameters that were strongly linked with the risks to the health of population, including $Na^+$ and $Cl^-$ concentrations.

The comparison of the water quality assessment was based on the ENTSS model integrating the principles of the neutrosophic logic, a soft set theory, and evidential reasoning to manage imprecision and uncertainty that accompany the lack of complete information in environmental data. The analysis was carried out between some water quality subsidiaries of profiles (A, B, and C), such as pH, Mg, Cl, $SO_4$, Na, K, $HCO_3$, $CO_3$, and other chemical indicators. All the parameters were initially converted as a neutrosophic triplet. The $T$,

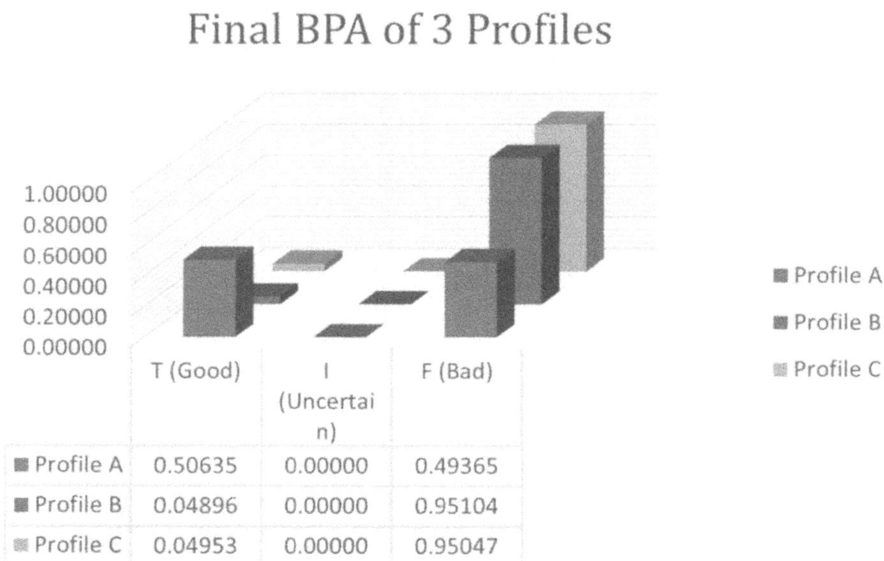


| | T (Good) | I (Uncertain) | F (Bad) |
|---|---|---|---|
| Profile A | 0.50635 | 0.00000 | 0.49365 |
| Profile B | 0.04896 | 0.00000 | 0.95104 |
| Profile C | 0.04953 | 0.00000 | 0.95047 |

**FIGURE 14.2**: Illustration of final BPA of all profiles.

*I*, *F* act as the membership of truth, indeterminacy, and falsity, respectively. These triplets were obtained by normalization of raw values with respect to the WHO standards/domain-specific threshold or ideal value as obtained in a contextual fashion.

### 14.5.1 Discussion of parameters in three profiles

Potential of hydrogen (pH) is the measurement of the acidity or alkalinity of water. The average pH level of the three profiles in the alternating months between August 2022 and June 2024 lies between 7.8 and 8.2. The pH of profile A is 7.8–8.1. In profile B, it lies between 8.0 and 8.2 and in profile C, it lies between 7.9 and 8.1. The mean pH in the rainy season and during summer is 8.0 and 8.1, respectively. The highest mean pH value, 8.2, was recorded in profile B during August 2022 (rainy season), June 2023 (summer), and October 2023 (rainy season) across all profiles (Table 14.4). The lowest mean value, 7.8, occurred in profile A in August 2022 and October 2023.

The high EC values recorded in all stations can be attributed to the high groundwater contamination by leaching action of rainfall and agricultural activities, which move surface contamination to groundwater.

The high value of TDS is recorded in the three profiles because of lack of high rate of contamination of groundwater by leaching action of rainfall and agricultural activities that transfer pollution on the surface to groundwater.

The TDS values of the three profiles are seasonally variable. All the stations have increased TDS contents during rainy seasons than during summer. It is more because of the leaching process by surface water and leaching of solid waste by rain water.

The overall hardness of B and C profiles is significantly greater than that of A. Overall, the high hardness can be largely attributed to pollution with river water that is rich in Ca, $SO_4$, Cl, and heavy metals.

The sodium and potassium levels in profile A are within the WHO range of 200 and 12 ppm, respectively, and can be used in the domestic applications. However, in profile B and C, they display greater values of WHO limits.

The overall hardness of the groundwater taken in the three profiles is above the allowable threshold. This could be attributed to high level of calcium and magnesium salts which were formed as a result of dissolution of rock minerals. The salt contents of the effluents of the agricultural, domestic, and shrimp farming activities may also add to the total hardness of the groundwater. This is particularly noted in profiles B and C which are more available to the above pollution. Cationic exchange with sodium ions can explain the calcium and magnesium concentrations in certain stations of A and B. This is also evidenced by the low sodium levels in such stations.

The concentration of nitrate, phosphate, and sulphate exceeded the acceptable limit during the period of analysis. As is typical, this high value is found more so in profiles B and C. The large contribution towards this high concentration could be the intensive agricultural activities whereby there is a large use of nitrate, phosphate, and sulphate fertilizers. They increase their levels in sewage and pollute groundwater. All this points to organic and inorganic groundwater pollution. Infiltration of riverine sewage also contributes to the concentration of sulphate and phosphate in particular. This could be as a result of fertilizers used in farming activities. This leads to increased concentration of phosphate and sulphate in agriculture and shrimp farming. These ions become concentrated when this sewage is not treated well and flow into the groundwater aquifer.

It is also observed when comparing the profiles that generally profile A is less affected than B & C. This could be caused by the fact that the river flows close to this region (profile A) and the running water serves to dilute these pollutants and lower their concentration in the groundwater. The low values could also be attributed to the upward slant of this profile as opposed to the steep slant of B & C. This will eliminate infiltration of wastes, hence curbing pollution.

### 14.5.2 Recommendation

Groundwater is never in sight and hence out of mind, and this remains a point of general ignorance. It is not common, and it has never been so, to monitor the groundwater levels or quality on a regular basis. It seems that the groundwater resources of Nagapattinam are largely influenced by the discharges of shrimp farming activities, agricultural processes, and the wrong disposal of domestic sewage. Considering the above observations, the following recommendations are made with regard to these aspects to enhance the quality of groundwater.

This valuable resource of groundwater should be preserved and conserved. There are a number of interventions that should be applied to regulate and reduce contamination of different wastages. These include appropriate treatment and disposal of effluents, good management of domestic and agricultural wastewater, decrease in the utilization of chemical and hazardous fertilizers, as well as keeping the environment clean and sanitary. It is also important to create awareness to the communities and inform them about personal hygiene and environment care. It is a basic need to ensure the accessibility of clean and drinkable water by every individual.

## 14.6 Conclusion

This chapter proposes the ENTSS approach to the effective analysis of multi-parameter water quality in uncertainty, which combines measures of truth, indeterminacy, and falsity with hierarchical evidence fusion. The results show that profile A is the most appropriate choice and that it has the highest belief in good quality and the lowest badness, whereas profiles B and C had greater risks and lesser suitability. The approach is not just a ranking of alternatives but also the identification of key parameters influencing the overall quality and providing a strong decision-support framework in uncertain and incomplete information settings in monitoring water quality. Besides this, since this method addresses the problems of uncertainty in the most explicit way, it can be particularly successful when used in the context of long-term groundwater monitoring in remote areas or under low resources conditions, where in-time data recording and transmission may be logistically difficult.

The groundwater of Nagapattinam area is sampled and analysed in this chapter to determine the various physio-chemical parameters and trace metals. The outcome of the application reveals that the majority of the physio-chemical parameters such as EC, Cl, $HCO_3$, COD, $NO_3$, $SO_4$, $PO_4$, Na, K, BOD, etc. are far beyond the permissible level. The trace metals examined are also marginally above the values recommended in the WHO guidelines. These findings suggest that groundwater in this area is significantly affected by various pollutants. Of these pollutants, effluent draining out of the shrimp culture and the agricultural wastes are the major contributors. All these are set to contaminate the groundwater resource of this region in their respective manners. The other cause of pollution is municipal waste disposal on land, which then impacts groundwater.

The last outcome reveals that there is an increased level of contamination across profiles B and C as compared to profile A. This can be attributed to intensive human activities in those areas, such as boat-building, agriculture, and other anthropogenic influences, which have also been validated by earlier research. This is also indicative of the fact that proper protection of water source is not observed in this area. It is also determined that nearly all the parameters exhibit seasonal variations with their values rising during monsoon due to the leaching of surface contaminants by rainwater. The values are high in summer, which might be attributed to evaporation. Therefore, it is time to become aware of the situation and clean up our own hygienic lifestyles. Protecting the potable water should be the responsibility of the people.

---

## Acronyms

| | |
|---|---|
| **BOD** | Biochemical Oxygen Demand |
| **BPA** | Basic Probability Assignment |
| **Ca** | Calcium |
| **Cl** | Chloride |
| **$CO_3$** | Carbonate |
| **DO** | Dissolved Oxygen |
| **DSET** | Dempster–Shafer Evidence Theory |
| **EC** | Electrical Conductivity |
| **F** | Fluoride |
| **FOD** | Frame of Discernment |
| **$HCO_3$** | Bicarbonate |
| **K** | Potassium |
| **ML** | Machine Learning |
| **Mg** | Magnesium |
| **Na** | Sodium |
| **$Na_2CO_3$** | Sodium Carbonate |
| **$NO_3$** | Nitrate |
| **pH** | Potential of Hydrogen |
| **ppm** | Parts per Million |
| **$PO_4$** | Phosphate |
| **$SO_4$** | Sulphate |
| **T, I, F** | Truth, Indeterminacy, Falsity |
| **TDS** | Total Dissolved Solids |
| **TH** | Total Hardness |
| **WHO** | World Health Organization |
| **XGBoost** | Extreme Gradient Boosting |

## Bibliography

[1] L. M. Bexfield, P. L. Toccalino, K. Belitz, W. T. Foreman, and E. T. Furlong. Hormones and pharmaceuticals in groundwater used as a source of drinking water across the united states. *Environmental Science & Technology*, 53(6):2950–2960, 2019. doi: 10.1021/acs.est.8b05592.

[2] S. M. Saghebian, M. A. G. Sefidkouhi, S. Darbandi et al. Ground water quality classification by decision tree method in ardebil region, iran. *Arabian Journal of Geosciences*, 7:4767–4777, 2014. doi: 10.1007/s12517-013-1042-y.

[3] N. F. C. Nordin et al. Groundwater quality forecasting modelling using ai. *Groundwater for Sustainable Development*, 14:100643, 2021.

[4] M. S. Fram and K. Belitz. Occurrence and concentrations of pharmaceutical compounds in groundwater used for public drinking-water supply in california. *Science of the Total Environment*, 409(18):3409–3417, 2011. doi: 10.1016/j.scitotenv.2011.05.053.

[5] H. Raheja, A. Goel, and M. Pal. Prediction of groundwater quality indices using machine learning algorithms. *Water Practice and Technology*, 17 (1):336–351, 2022. doi: 10.2166/wpt.2021.120.

[6] A. K. Das, N. Gupta, C. Granados, R. Das, and S. Das. Neutrosophic approach to water quality assessment: A case study of gomati river, the largest river in tripura, india. *Neutrosophic Systems with Applications*, 22(1), 2024. doi: 10.61356/j.nswa.2024.22386.

[7] K. Neeti and R. Singh. Groundwater quality assessment and health risks from fluoride in jamui, bihar. *Engineering, Technology & Applied Science Research*, 13(1):10204–10208, 2023. URL `https://etasr.com/index.php/ETASR/article/view/5576`.

[8] V. K. Singh, D. S. Bikundia, A. Sarswat, and D. Mohan. Groundwater quality assessment in the village of lutfullapur nawada, loni, district ghaziabad, uttar pradesh, india. *Environmental Monitoring and Assessment*, 184(7):4473–4488, 2012. doi: 10.1007/s10661-011-2279-0.

[9] A. K. Tiwari and A. K. Singh. Hydrogeochemical investigation and groundwater quality assessment of pratapgarh district, uttar pradesh. *Journal of the Geological Society of India*, 83:329–343, 2014. doi: 10.1007/s12594-014-0045-y.

[10] J.; Przysucha B.; Cel W. Kulisz, M.; Kujawska. Forecasting water quality index in groundwater using artificial neural network. *Energies*, 14:,

number = 18, pages = 5875, url = https://doi.org/10.3390/en14185875,, 2021.

[11] J. Jiao and V. Post. *Coastal hydrogeology.* Cambridge University Press, 2019.

[12] N. Subba Rao. Seasonal variation of groundwater quality in a part of guntur district, andhra pradesh, india. *Environmental Geology*, 49:413–429, 2006. doi: 10.1007/s00254-005-0089-9.

[13] B. F. Jones, A. Vengosh, E. Rosenthal, and Y. Yechieli. Geochemical investigations of seawater intrusion. In *Seawater Intrusion in Coastal Aquifers.* Springer, 1999.

[14] C. A. J. Appelo and D. Postma. *Geochemistry, Groundwater and Pollution.* A. A. Balkema, 2005.

[15] Ha H.D. Trinh N.T. et al. Nguyen, D.P. Application of artificial intelligence for forecasting surface quality index of irrigation systems in the red river delta, vietnam. *Environmental Systems Research*, 12(24), 2023. URL `https://doi.org/10.1186/s40068-023-00307-6`.

[16] Araya D. & Berg M. Podgorski, J. Geogenic manganese and iron in groundwater of southeast asia and bangladesh machine learning spatial prediction modeling and comparison with arsenic. *Science of the Total Environment*, 833:155131, 2022. URL `https://doi.org/10.1016/j.scitotenv.2022.155131`.

[17] Luo P. Wu P. Song C. & Chen X. Xu, C. Predicting river dissolved oxygen time series based on stand-alone models and hybrid wavelet-based models. *Journal of Environmental Management*, 2021.

[18] M. Gazzaz et al. Gis and machine learning technique for groundwater quality assessment in saudi arabia. *Water*, 2021.

[19] D.K. Roy et al. Automatic model selection based ml approach to predict seawater intrusion. *Environmental Earth Sciences*, 83:287, 2024.

[20] Ouyang T. Wang X. Yang H. Zhou B. Wu Z. & Shang M. han, K. Temporal prediction of algal parameters in three gorges reservoir based on highly time-resolved monitoring and long short-term memory network. *Journal of Hydrology*, 2022.

[21] B. Lombard et al. Rf and sgb modeling arsenic in us rivers. *Environmental Modelling & Software*, 2021.

[22] Y. Zeng, X. Zhang, W. Cao, J. Deng, and H. Zeng. An evidential reasoning method of comprehensive evaluation of water quality based on gaussian distribution. *Sustainability*, 16(20):9079, 2024. doi: 10.3390/su16209079.

[23] A. Neshat and B. Pradhan. Risk assessment of groundwater pollution with a new methodological framework: Application of dempster–shafer theory and gis. *Natural Hazards*, 78:1565–1585, 2015. doi: 10.1007/s11069-015-1788-5.

[24] A. M. Al-Abadi. The application of dempster–shafer theory of evidence for assessing groundwater vulnerability at galal badra basin, wasit governorate, east of iraq. *Applied Water Science*, 7:1725–1740, 2015. doi: 10.1007/s13201-015-0342-7.

[25] X. Li et al. *Water quality models using GRU, LSTM, RNN with Dempster–Shafer evidence theory. IWA Publishing*, 2019.

[26] T. Garai, H. Garg, and G. A. Biswas. Fraction ranking-based multi-criteria decision-making method for water resource management under bipolar neutrosophic fuzzy environment. *Artificial Intelligence Review*, 56:14865–14906, 2023. doi: 10.1007/s10462-023-10514-3.

[27] I. Deli, M. Ali, and F. Smarandache. Bipolar neutrosophic sets and their application based on multi-criteria decision making problems. *In 2015 International conference on advanced mechatronic systems (ICAMechS)*, 2015, August.

[28] F. Karaaslan. Neutrosophic soft sets with applications in decision making. *Infinite Study*, 2014.

[29] C. Parkash, S. Kaur, R. Dev, and M. Bali. Neutrosophic entropy-based fluoride contamination indices for community health risk assessment from groundwater of kangra county, north india. *Neutrosophic Sets and Systems*, 49(1), 2022. URL `https://digitalrepository.unm.edu/nss_journal/vol49/iss1/13`.

[30] A. Abdel-Monem and A. Abdelhafeez. Neutrosophic hybrid mcdm framework to evaluate the risks of excavation system. *Neutrosophic Sets and Systems*, 55:173–186, 2023. URL `https://fs.unm.edu/nss8/index.php/111/article/view/3185`.

[31] F. Ahmad, A. Y. Adhami, and F. Smarandache. Neutrosophic optimization model and computational algorithm for optimal shale gas water management under uncertainty. *Symmetry*, 11(4):544, 2019. doi: 10.3390/sym11040544.

[32] H. Wang, F. Smarandache, Y. Q. Zhang, and R. Sunderraman. Interval neutrosophic sets and logic: Theory and applications in computing. *Infinite Study*, 5, 2005.

[33] G. Shafer. *A Mathematical Theory of Evidence.* Princeton University Press, 1976.

[34] D. Molodtsov. Soft set theory—first results. *Computers & Mathematics with Applications*, 37(4–5):19–31, 1999. doi: 10.1016/S0898-1221(99)00056-5.

[35] F. Smarandache. *Neutrosophy: Neutrosophic Probability, Set, and Logic. Analytic Synthesis & Synthetic Analysis.* American Research Press, 1998.

[36] L. A. Zadeh. Fuzzy sets. *Information and Control*, 8(3):338–353, 1965. doi: 10.1016/S0019-9958(65)90241-X.

# Chapter 15

# *An Innovative Trapezoidal Neutrosophic CRADIS Algorithm Using Multiple-Attribute Group Decision-Making for Sharia Bank Branch Location Selection*

**Irvanizam Irvanizam, Edrian Hadinata, Razief Perucha Fauzie Afidh, Fahrul Radhi, and Denny Helzami**

## 15.1 Introduction

The selection of an appropriate location for establishing a new branch is a crucial strategic decision for any banking institution, including Sharia banks [1]. The location of a branch significantly affects customer accessibility, operational efficiency, market share, and long-term profitability. For Sharia-compliant financial institutions, branch office location selection (BOLS) is even more critical, as it must consider not only economic and demographic factors but also compliance with Islamic financial principles and community characteristics [2, 3]. Therefore, identifying the optimal location requires a comprehensive and systematic evaluation process capable of effectively balancing multiple, often conflicting, criteria.

Despite its importance, determining the most suitable location for a new Sharia bank branch (SBB) remains a challenging task for management committees [4]. The decision involves multiple quantitative and qualitative factors, including customer density, accessibility, market potential, competition, infrastructure, and socio-economic compatibility. Moreover, decision-makers (DMs) often face uncertainty and subjectivity when evaluating these criteria. The uncertainty arises from incomplete, imprecise, or inconsistent information, while subjectivity stems from individual judgments, preferences, and experiences among experts. These challenges complicate the decision-making process and may lead to suboptimal or biased outcomes if not properly

DOI: 10.1201/9781003683339-15

addressed. Additionally, variations in expert opinions and the dynamic nature of market conditions further intensify the complexity and uncertainty associated with the BOLS problem.

To address the issue of uncertainty in decision-making, researchers have widely adopted the fuzzy set theory introduced by Zadeh [5] in 1965. The fuzzy set approach allows for the representation of vague and imprecise information by assigning membership degrees to elements, providing a flexible framework for handling uncertainty in real-world problems. Over the decades, fuzzy theory has evolved beyond its original formulation, expanding its capability to model imprecision, vagueness, and incompleteness in human judgments [6]. However, conventional fuzzy sets remain limited in representing the indeterminacy component that frequently arises in group decision-making contexts.

To overcome these limitations, Smarandache [7] introduced the concept of neutrosophic sets (NSs) in 1999, extending fuzzy and intuitionistic fuzzy theories by incorporating three independent membership functions: truth, indeterminacy, and falsity. This triad provides a more comprehensive structure for modeling uncertainty and indeterminacy in complex decision-making environments. Building upon this foundation, researchers have developed various extensions of NSs to enhance their flexibility and applicability. One notable extension is the trapezoidal neutrosophic set (TraNS), first introduced by Ye [8], which employs trapezoidal membership functions to represent uncertain data more effectively. The trapezoidal form captures gradual transitions in membership values and offers improved interpretability, precision, and computational efficiency in modeling uncertain linguistic assessments. Consequently, the TraNS framework has gained increasing attention in multi-criteria and multi-attribute decision-making problems.

Several existing studies have successfully applied NSs and TraNS to various multiple attribute decision-making (MADM) problems. For instance, Dasan et al. [9] employed a sine-function-based approach to propose a novel distance measure for single-valued neutrosophic sets (SVNSs) in the context of student career evaluation. Rahman et al. [10] developed a new hybrid mathematical model that integrates the strengths of possibility NSs and hypersoft sets to more effectively address uncertainty, indeterminacy, and MADM challenges. Nivedita et al. [11] introduced an adaptive teaching learning-based optimization (TLBO) model based on TraNS that accurately handles uncertainty and imprecision in medical data, achieving highly reliable diabetes prediction. Their model demonstrated superior performance (99.89% accuracy) compared to existing fuzzy and neuro-fuzzy diagnostic systems while significantly reducing diagnostic errors. Tan et al. [12] designed a new MADM method that integrates the decision-making trial and evaluation laboratory (DEMATEL), fuzzy distance, and the linear assignment method (LAM) within a TraNS framework to effectively manage uncertainty, indeterminacy, and inconsistency in typhoon disaster-related decision data. Jana et al. [13] defined operational rules, score and accuracy functions, and proposed two novel aggregation operators for TraNS, accompanied by a detailed analysis of their properties. They further developed an MADM method based on these operators and validated it using a numerical example. Collectively, these studies demonstrate that NS- and TraNS-based modeling provides an effective means of addressing complex decision-making problems involving uncertain, vague, and inconsistent information.

Another major challenge in branch location selection is reducing subjectivity among decision-makers (DMs). Because the evaluation of location alternatives often involves multiple experts with different backgrounds, preferences, and levels of experience, it is essential to adopt a method that can objectively consolidate these diverse opinions. The multiple attribute group decision-making (MAGDM) framework provides an effective solution to this challenge. MAGDM integrates evaluations from multiple DMs to reach a consensus that reflects collective intelligence, thereby minimizing the influence of individual biases. This approach enhances the reliability, transparency, and fairness of the decision-making process by systematically aggregating multiple perspectives and weighting their contributions.

The MAGDM approach has been extensively developed and applied across various decision-making contexts. For example, Jamil et al. [14] introduced the induced bipolar neutrosophic (BN) Einstein aggregation operators by integrating Einstein operations into the NS environment for a robot selection problem. You et al. [15] proposed a comprehensive consensus-reaching model for SVNS-based MAGDM, which incorporates the generalized Shapley neutrosophic Choquet integral for aggregating correlated criteria, a projection-based consensus measure for evaluating agreement levels, and a personalized feedback mechanism that accounts for experts' psychological responses to group pressure. Irvanizam et al. [16] enhanced the additive ratio assessment (ARAS) method using SVNSs to address MAGDM issues in school evaluation. Wei et al. [17] extended the VIKOR method to solve MAGDM problems by employing two-tuple linguistic neutrosophic numbers (2TLNNs) and their interval-valued variants. Pramanik and Mallick [18] applied prospect theory to develop a TODIM strategy within a TraNS framework for evaluating investment companies under MAGDM settings. These studies collectively demonstrate that integrating neutrosophic theory with MAGDM offers a robust paradigm for making balanced and rational decisions under uncertain and subjective conditions.

In recent years, numerous researchers have explored the integration of advanced decision-making algorithms to enhance ranking accuracy and interpretability. Among these developments, the compromise ranking of alternatives from distance to ideal solution (CRADIS) method has emerged as a promising alternative to traditional approaches such as TOPSIS, VIKOR, and EDAS [19, 20]. Introduced by Puška et al. [21], the CRADIS method provides a more balanced compromise solution by simultaneously considering the distances of alternatives from both ideal and anti-ideal solutions. Unlike some existing techniques that may suffer from ranking reversals or high sensitivity to normalization, CRADIS ensures greater ranking stability, robustness, and discriminatory power among alternatives. Moreover, it maintains computational simplicity and interpretability while offering an equitable balance between benefit and cost criteria [22].

A growing body of literature has demonstrated the applicability of CRADIS in diverse decision-making scenarios. For example, Ashraf et al. [19] extended the CRADIS method to address a radioactive waste reduction problem under a disc spherical fuzzy (D-SF) environment with unknown weight information. Yuan et al. [23] employed picture fuzzy sets (PFSs) to propose a new distance measure based on the Jensen–Shannon divergence and integrated it with the CRADIS method for MADM applications. Yin et al. [24] constructed an innovative MADM framework that combines the CRADIS method with circular Pythagorean fuzzy sets (CPyFSs) to assess university students' career choices. Seikh and Mukherjee [25] developed an interactive MADM framework that incorporates Fermatean fuzzy Sets and the triangular divergence distance measure within the CRADIS methodology to support decision-making in online platform selection. However, most existing CRADIS-based models have been developed under deterministic, fuzzy, or intuitionistic fuzzy environments, with limited attention to NS or TraNS frameworks. As a result, uncertainty and indeterminacy in expert evaluations remain insufficiently addressed in previous CRADIS studies.

To address this gap, this chapter proposes an innovative CRADIS algorithm based on TraNS within a MAGDM framework. The proposed approach integrates the strengths of TraNS—capable of effectively representing uncertainty, vagueness, and indeterminacy—with the robust ranking capabilities of the CRADIS method. By combining these two paradigms, the model aims to produce more accurate, consistent, and interpretable ranking results for SBB location selection, while simultaneously reducing subjective bias and enhancing group consensus among DMs.

The main contributions of this study are threefold.

1. It introduces a novel trapezoidal neutrosophic CRADIS algorithm for addressing the SBB location selection problem, capable of efficiently managing uncertainty and indeterminacy in expert judgments.

2. It incorporates the proposed algorithm into a MAGDM framework, offering a structured mechanism to reduce subjectivity and harmonize the diverse opinions of multiple experts.

3. It provides a comprehensive comparative analysis to demonstrate the effectiveness and superiority of the proposed approach over existing methods, thereby contributing to the advancement of decision-support systems in Islamic banking management.

The remainder of this chapter is organized as follows. Section 15.2 reviews the foundational concepts of single-valued neutrosophic sets (SVNSs), trapezoidal neutrosophic sets (TraNSs), trapezoidal neutrosophic numbers (TraNNs), and the associated mathematical operations. Section 15.3 presents the methodological framework and the formulation of the trapezoidal neutrosophic CRADIS algorithm. Section 15.4 illustrates the application of the proposed model to the SBB location selection problem through a detailed case study. Section 15.5 provides the results, comparative analyses, and discussion. Finally, Section 15.6 concludes the chapter by summarizing key findings, managerial implications, and potential directions for future research.

---

## 15.2 Preliminaries

This section provides a concise overview of the fundamental notions of SVNSs, TraNS, TraNNs, and some mathematics operations for TraNNs. [26] Let $N$ denote a universe of discourse. An SVNS $A$ on $N$ is defined as

$$A = \{\langle n, T_A(n), I_A(n), F_A(n)\rangle \mid n \in N\}, \tag{15.1}$$

where $T_A(n)$, $I_A(n)$, and $F_A(n)$ represent the truth-membership, indeterminacy-membership, and falsity-membership functions of the element $n$ with respect to set $A$, respectively. Each of these functions satisfies the condition $0 \leq T_A(n) + \ I_A(n) + \ F_A(n) \leq 3$ and $T_A(n),\ I_A(n),\ F_A(n) \in [0,1]$ [8]. Let $\tilde{T}$ be a universe of discourse. A TraNS $\tilde{A}$ on $\tilde{T}$ is depicted as

$$\tilde{A} = \{(t, T_{\tilde{A}}(t), I_{\tilde{A}}(t), F_{\tilde{A}}(t)) \mid t \in \tilde{T}\}, \tag{15.2}$$

where

$$T_{\tilde{A}}(t) = \left\langle \overset{a}{T}_{\tilde{A}}(t), \overset{b}{T}_{\tilde{A}}(t), \overset{c}{T}_{\tilde{A}}(t), \overset{d}{T}_{\tilde{A}}(t) \right\rangle,$$

$$I_{\tilde{A}}(t) = \left\langle \overset{a}{I}_{\tilde{A}}(t), \overset{b}{I}_{\tilde{A}}(t), \overset{c}{I}_{\tilde{A}}(t), \overset{d}{I}_{\tilde{A}}(t) \right\rangle,$$

and $F_{\tilde{A}}(t) = \left\langle \overset{a}{F}_{\tilde{A}}(t), \overset{b}{F}_{\tilde{A}}(t), \overset{c}{F}_{\tilde{A}}(t), \overset{d}{F}_{\tilde{A}}(t) \right\rangle$ are three trapezoidal fuzzy numbers (TraFNs) where $\tilde{T} \to [0,1]$ and represent the truth-membership, indeterminacy-membership, and falsity-membership functions of the element $t$ with respect to the set $\tilde{T}$, respectively. All of these functions satisfy the condition $0 \leq \overset{d}{T}_{\tilde{A}}(t) + \overset{d}{I}_{\tilde{A}}(t) + \overset{d}{F}_{\tilde{A}}(t) \leq 3$, $t \in \tilde{T}$.

For simplicity, the three TraFNs are represented as $T_{\tilde{A}}(t) = \left\langle \overset{a}{T}_{\tilde{A}}, \overset{b}{T}_{\tilde{A}}, \overset{c}{T}_{\tilde{A}}, \overset{d}{T}_{\tilde{A}} \right\rangle$, $I_{\tilde{A}}(t) = \left\langle \overset{a}{I}_{\tilde{A}}, \overset{b}{I}_{\tilde{A}}, \overset{c}{I}_{\tilde{A}}, \overset{d}{I}_{\tilde{A}} \right\rangle$, and $F_{\tilde{A}}(t) = \left\langle \overset{a}{F}_{\tilde{A}}, \overset{b}{F}_{\tilde{A}}, \overset{c}{F}_{\tilde{A}}, \overset{d}{F}_{\tilde{A}} \right\rangle$. Consequently, a TraNN can be expressed as $\tilde{n} = \left(\left\langle \overset{a}{T_{\tilde{n}}}, \overset{b}{T_{\tilde{n}}}, \overset{c}{T_{\tilde{n}}}, \overset{d}{T_{\tilde{n}}} \right\rangle, \left\langle \overset{a}{I_{\tilde{n}}}, \overset{b}{I_{\tilde{n}}}, \overset{c}{I_{\tilde{n}}}, \overset{d}{I_{\tilde{n}}} \right\rangle, \left\langle \overset{a}{F_{\tilde{n}}}, \overset{b}{F_{\tilde{n}}}, \overset{c}{F_{\tilde{n}}}, \overset{d}{F_{\tilde{n}}} \right\rangle\right)$, which serves as a fundamental element of the TRaNS [8].

Let $\tilde{x}_1 = \left(\left\langle \overset{a}{T_{\tilde{x}_1}}, \overset{b}{T_{\tilde{x}_1}}, \overset{c}{T_{\tilde{x}_1}}, \overset{d}{T_{\tilde{x}_1}} \right\rangle, \left\langle \overset{a}{I_{\tilde{x}_1}}, \overset{b}{I_{\tilde{x}_1}}, \overset{c}{I_{\tilde{x}_1}}, \overset{d}{I_{\tilde{x}_1}} \right\rangle, \left\langle \overset{a}{F_{\tilde{x}_1}}, \overset{b}{F_{\tilde{x}_1}}, \overset{c}{F_{\tilde{x}_1}}, \overset{d}{F_{\tilde{x}_1}} \right\rangle\right)$ and $\tilde{x}_2 = \left(\left\langle \overset{a}{T_{\tilde{x}_2}}, \overset{b}{T_{\tilde{x}_2}}, \overset{c}{T_{\tilde{x}_2}}, \overset{d}{T_{\tilde{x}_2}} \right\rangle, \left\langle \overset{a}{I_{\tilde{x}_2}}, \overset{b}{I_{\tilde{x}_2}}, \overset{c}{I_{\tilde{x}_2}}, \overset{d}{I_{\tilde{x}_2}} \right\rangle, \left\langle \overset{a}{F_{\tilde{x}_2}}, \overset{b}{F_{\tilde{x}_2}}, \overset{c}{F_{\tilde{x}_2}}, \overset{d}{F_{\tilde{x}_2}} \right\rangle\right)$ be two TraNNs. Then the expressions of the operations below hold true.

1. Addition

$$\begin{aligned}\tilde{x}_1 \oplus \tilde{x}_2 = \Bigg(&\left\langle \overset{a}{T_{\tilde{x}_1}} + \overset{a}{T_{\tilde{x}_2}} - \overset{a}{T_{\tilde{x}_1}}\overset{a}{T_{\tilde{x}_2}}, \overset{b}{T_{\tilde{x}_1}} + \overset{b}{T_{\tilde{x}_2}} - \overset{b}{T_{\tilde{x}_1}}\overset{b}{T_{\tilde{x}_2}}, \overset{c}{T_{\tilde{x}_1}} + \overset{c}{T_{\tilde{x}_2}} - \overset{c}{T_{\tilde{x}_1}}\overset{c}{T_{\tilde{x}_2}}, \right. \\ &\left. \overset{d}{T_{\tilde{x}_1}} + \overset{d}{T_{\tilde{x}_2}} - \overset{d}{T_{\tilde{x}_1}}\overset{d}{T_{\tilde{x}_2}} \right\rangle, \left\langle \overset{a}{I_{\tilde{x}_1}}\overset{a}{I_{\tilde{x}_2}}, \overset{b}{I_{\tilde{x}_1}}\overset{b}{I_{\tilde{x}_2}}, \overset{c}{I_{\tilde{x}_1}}\overset{c}{I_{\tilde{x}_2}}, \overset{d}{I_{\tilde{x}_1}}\overset{d}{I_{\tilde{x}_2}} \right\rangle, \\ &\left\langle \overset{a}{F_{\tilde{x}_1}}\overset{a}{F_{\tilde{x}_2}}, \overset{b}{F_{\tilde{x}_1}}\overset{b}{F_{\tilde{x}_2}}, \overset{c}{F_{\tilde{x}_1}}\overset{c}{F_{\tilde{x}_2}}, \overset{d}{F_{\tilde{x}_1}}\overset{d}{F_{\tilde{x}_2}} \right\rangle\Bigg)\end{aligned} \tag{15.3}$$

2. Multiplication

$$\begin{aligned}\tilde{x}_1 \otimes \tilde{x}_2 = \Bigg(&\left\langle \overset{a}{T_{\tilde{x}_1}}\overset{a}{T_{\tilde{x}_2}}, \overset{b}{T_{\tilde{x}_1}}\overset{b}{T_{\tilde{x}_2}}, \overset{c}{T_{\tilde{x}_1}}\overset{c}{T_{\tilde{x}_2}}, \overset{d}{T_{\tilde{x}_1}}\overset{d}{T_{\tilde{x}_2}} \right\rangle, \left\langle \overset{a}{I_{\tilde{x}_1}} + \overset{a}{I_{\tilde{x}_2}} - \overset{a}{I_{\tilde{x}_1}}\overset{a}{I_{\tilde{x}_2}}, \right. \\ &\left. \overset{b}{I_{\tilde{x}_1}} + \overset{b}{I_{\tilde{x}_2}} - \overset{b}{I_{\tilde{x}_1}}\overset{b}{I_{\tilde{x}_2}}, \overset{c}{I_{\tilde{x}_1}} + \overset{c}{I_{\tilde{x}_2}} - \overset{c}{I_{\tilde{x}_1}}\overset{c}{I_{\tilde{x}_2}}, \overset{d}{I_{\tilde{x}_1}} + \overset{d}{I_{\tilde{x}_2}} - \overset{d}{I_{\tilde{x}_1}}\overset{d}{I_{\tilde{x}_2}} \right\rangle, \\ &\left\langle \overset{a}{F_{\tilde{x}_1}} + \overset{a}{F_{\tilde{x}_2}} - \overset{a}{F_{\tilde{x}_1}}\overset{a}{F_{\tilde{x}_2}}, \overset{b}{F_{\tilde{x}_1}} + \overset{b}{F_{\tilde{x}_2}} - \overset{b}{F_{\tilde{x}_1}}\overset{b}{F_{\tilde{x}_2}}, \overset{c}{F_{\tilde{x}_1}} + \overset{c}{F_{\tilde{x}_2}} - \overset{c}{F_{\tilde{x}_1}}\overset{c}{F_{\tilde{x}_2}}, \right. \\ &\left. \overset{d}{F_{\tilde{x}_1}} + \overset{d}{F_{\tilde{x}_2}} - \overset{d}{F_{\tilde{x}_1}}\overset{d}{F_{\tilde{x}_2}} \right\rangle\Bigg)\end{aligned} \tag{15.4}$$

3. Multiple by a scalar $\hbar > 0$

$$\begin{aligned}\hbar\tilde{x}_1 = \Bigg(&\left\langle 1 - \left(1 - \overset{a}{T_{\tilde{x}_1}}\right)^{\hbar}, 1 - \left(1 - \overset{b}{T_{\tilde{x}_1}}\right)^{\hbar}, 1 - \left(1 - \overset{c}{T_{\tilde{x}_1}}\right)^{\hbar}, 1 - \left(1 - \overset{d}{T_{\tilde{x}_1}}\right)^{\hbar} \right\rangle, \\ &\left\langle \left(\overset{a}{I_{\tilde{x}_1}}\right)^{\hbar}, \left(\overset{b}{I_{\tilde{x}_1}}\right)^{\hbar}, \left(\overset{c}{I_{\tilde{x}_1}}\right)^{\hbar}, \left(\overset{d}{I_{\tilde{x}_1}}\right)^{\hbar} \right\rangle, \left\langle \left(\overset{a}{F_{\tilde{x}_1}}\right)^{\hbar}, \left(\overset{b}{F_{\tilde{x}_1}}\right)^{\hbar}, \left(\overset{c}{F_{\tilde{x}_1}}\right)^{\hbar}, \left(\overset{d}{F_{\tilde{x}_1}}\right)^{\hbar} \right\rangle\Bigg)\end{aligned} \tag{15.5}$$

4. Power by a scalar $\hbar > 0$

$$(\tilde{x}_1)^{\hbar} = \left(\left\langle \left(\overset{a}{T}_{\tilde{x}_1}\right)^{\hbar}, \left(\overset{b}{T}_{\tilde{x}_1}\right)^{\hbar}, \left(\overset{c}{T}_{\tilde{x}_1}\right)^{\hbar}, \left(\overset{d}{T}_{\tilde{x}_1}\right)^{\hbar} \right\rangle,\right.$$
$$\left\langle 1-\left(1-\overset{a}{I}_{\tilde{x}_1}\right)^{\hbar}, 1-\left(1-\overset{b}{I}_{\tilde{x}_1}\right)^{\hbar}, 1-\left(1-\overset{c}{I}_{\tilde{x}_1}\right)^{\hbar}, 1-\left(1-\overset{d}{I}_{\tilde{x}_1}\right)^{\hbar} \right\rangle,$$
$$\left.\left\langle 1-\left(1-\overset{a}{F}_{\tilde{x}_1}\right)^{\hbar}, 1-\left(1-\overset{b}{F}_{\tilde{x}_1}\right)^{\hbar}, 1-\left(1-\overset{c}{F}_{\tilde{x}_1}\right)^{\hbar}, 1-\left(1-\overset{d}{F}_{\tilde{x}_1}\right)^{\hbar} \right\rangle\right) \tag{15.6}$$

[27]

**Definition 15.1:** Let $\tilde{x}_1 = \left(\left\langle \overset{a}{T}_{\tilde{x}_1}, \overset{b}{T}_{\tilde{x}_1}, \overset{c}{T}_{\tilde{x}_1}, \overset{d}{T}_{\tilde{x}_1} \right\rangle, \left\langle \overset{a}{I}_{\tilde{x}_1}, \overset{b}{I}_{\tilde{x}_1}, \overset{c}{I}_{\tilde{x}_1}, \overset{d}{I}_{\tilde{x}_1} \right\rangle, \left\langle \overset{a}{F}_{\tilde{x}_1}, \overset{b}{F}_{\tilde{x}_1}, \overset{c}{F}_{\tilde{x}_1}, \overset{d}{F}_{\tilde{x}_1} \right\rangle\right)$ and $\tilde{x}_2 = \left(\left\langle \overset{a}{T}_{\tilde{x}_2}, \overset{b}{T}_{\tilde{x}_2}, \overset{c}{T}_{\tilde{x}_2}, \overset{d}{T}_{\tilde{x}_2} \right\rangle, \left\langle \overset{a}{I}_{\tilde{x}_2}, \overset{b}{I}_{\tilde{x}_2}, \overset{c}{I}_{\tilde{x}_2}, \overset{d}{I}_{\tilde{x}_2} \right\rangle, \left\langle \overset{a}{F}_{\tilde{x}_2}, \overset{b}{F}_{\tilde{x}_2}, \overset{c}{F}_{\tilde{x}_2}, \overset{d}{F}_{\tilde{x}_2} \right\rangle\right)$ be two TraNNs. Then the distance between $\tilde{x}_1$ and $\tilde{x}_1$ is defined as

$$Ds(\tilde{x}_1, \tilde{x}_1) = \frac{1}{9}\sqrt{\begin{array}{l}\left[\hat{T}_{\tilde{x}_1} + \hat{I}_{\tilde{x}_1} + \hat{F}_{\tilde{x}_1} - \hat{T}_{\tilde{x}_2} - \hat{I}_{\tilde{x}_2} - \hat{F}_{\tilde{x}_2}\right. \\ \left. - \left(\triangle^T_{\tilde{x}_1} - \triangle^T_{\tilde{x}_2}\right) - \left(\triangle^I_{\tilde{x}_1} - \triangle^I_{\tilde{x}_2}\right) - \left(\triangle^F_{\tilde{x}_1} - \triangle^F_{\tilde{x}_2}\right)\right]^2 \\ + \left[\left(\triangle^T_{\tilde{x}_1} - \triangle^T_{\tilde{x}_2}\right) + \left(\triangle^I_{\tilde{x}_1} - \triangle^I_{\tilde{x}_2}\right) + \left(\triangle^F_{\tilde{x}_1} - \triangle^F_{\tilde{x}_2}\right)\right]^2\end{array}} \tag{15.7}$$

where $\hat{T}_{\tilde{x}_1} = \sum_{k\in\{a,b,c,d\}} \overset{k}{T}_{\tilde{x}_1}$, $\hat{I}_{\tilde{x}_1} = \sum_{k\in\{a,b,c,d\}} \overset{k}{I}_{\tilde{x}_1}$, $\hat{F}_{\tilde{x}_1} = \sum_{k\in\{a,b,c,d\}} \overset{k}{F}_{\tilde{x}_1}$, $\hat{T}_{\tilde{x}_2} = \sum_{k\in\{a,b,c,d\}} \overset{k}{T}_{\tilde{x}_2}$, $\hat{I}_{\tilde{x}_2} = \sum_{k\in\{a,b,c,d\}} \overset{k}{I}_{\tilde{x}_2}$, $\hat{F}_{\tilde{x}_2} = \sum_{k\in\{a,b,c,d\}} \overset{k}{F}_{\tilde{x}_2}$,

$\triangle^T_{\tilde{x}_1} = \dfrac{\overset{d}{T}_{\tilde{x}_1}\overset{c}{T}_{\tilde{x}_1} - \overset{a}{T}_{\tilde{x}_1}\overset{b}{T}_{\tilde{x}_1}}{\left(\overset{d}{T}_{\tilde{x}_1} + \overset{c}{T}_{\tilde{x}_1}\right) - \left(\overset{a}{T}_{\tilde{x}_1} + \overset{b}{T}_{\tilde{x}_1}\right)}$, $\triangle^I_{\tilde{x}_1} = \dfrac{\overset{d}{I}_{\tilde{x}_1}\overset{c}{I}_{\tilde{x}_1} - \overset{a}{I}_{\tilde{x}_1}\overset{b}{I}_{\tilde{x}_1}}{\left(\overset{d}{I}_{\tilde{x}_1} + \overset{c}{I}_{\tilde{x}_1}\right) - \left(\overset{a}{I}_{\tilde{x}_1} + \overset{b}{I}_{\tilde{x}_1}\right)}$,

$\triangle^F_{\tilde{x}_1} = \dfrac{\overset{d}{F}_{\tilde{x}_1}\overset{c}{F}_{\tilde{x}_1} - \overset{a}{F}_{\tilde{x}_1}\overset{b}{F}_{\tilde{x}_1}}{\left(\overset{d}{F}_{\tilde{x}_1} + \overset{c}{F}_{\tilde{x}_1}\right) - \left(\overset{a}{F}_{\tilde{x}_1} + \overset{b}{F}_{\tilde{x}_1}\right)}$, $\triangle^T_{\tilde{x}_2} = \dfrac{\overset{d}{T}_{\tilde{x}_2}\overset{c}{T}_{\tilde{x}_2} - \overset{a}{T}_{\tilde{x}_2}\overset{b}{T}_{\tilde{x}_2}}{\left(\overset{d}{T}_{\tilde{x}_2} + \overset{c}{T}_{\tilde{x}_2}\right) - \left(\overset{a}{T}_{\tilde{x}_2} + \overset{b}{T}_{\tilde{x}_2}\right)}$,

$\triangle^I_{\tilde{x}_2} = \dfrac{\overset{d}{I}_{\tilde{x}_2}\overset{c}{I}_{\tilde{x}_2} - \overset{a}{I}_{\tilde{x}_2}\overset{b}{I}_{\tilde{x}_2}}{\left(\overset{d}{I}_{\tilde{x}_2} + \overset{c}{I}_{\tilde{x}_2}\right) - \left(\overset{a}{I}_{\tilde{x}_2} + \overset{b}{I}_{\tilde{x}_2}\right)}$, and $\triangle^F_{\tilde{x}_2} = \dfrac{\overset{d}{F}_{\tilde{x}_2}\overset{c}{F}_{\tilde{x}_2} - \overset{a}{F}_{\tilde{x}_2}\overset{b}{F}_{\tilde{x}_2}}{\left(\overset{d}{F}_{\tilde{x}_2} + \overset{c}{F}_{\tilde{x}_2}\right) - \left(\overset{a}{F}_{\tilde{x}_2} + \overset{b}{F}_{\tilde{x}_2}\right)}$.

## 15.3 The Proposed TraNN-CRADIS Algorithm for Multi-Attribute Group Decision-Making

This section introduces a methodological framework for estimating expert and attribute weights and for ranking the alternatives in a TraNN environment with incomplete weight information. For a MAGDM problem, let $L = (L_1, L_2, ..., L_p)$ denote the set of alternatives and $G = (G_1, G_2, ..., G_q)$ represent the set of evaluation attributes. Suppose $\Gamma = [\ell_{ij}^k]$ is the initial decision-making matrix, where each element $\ell_{ij}^k = \left(\left\langle \overset{a}{T}{}_{ij}^k, \overset{b}{T}{}_{ij}^k, \overset{c}{T}{}_{ij}^k, \overset{d}{T}{}_{ij}^k \right\rangle, \left\langle \overset{a}{I}{}_{ij}^k, \overset{b}{I}{}_{ij}^k, \overset{c}{I}{}_{ij}^k, \overset{d}{I}{}_{ij}^k \right\rangle, \left\langle \overset{a}{F}{}_{ij}^k, \overset{b}{F}{}_{ij}^k, \overset{c}{F}{}_{ij}^k, \overset{d}{F}{}_{ij}^k \right\rangle\right)$ corresponds to the TraNN of alternative $L_i$ with respect to attribute $C_j$. Specifically, $\left\langle \overset{a}{T}{}_{ij}^k, \overset{b}{T}{}_{ij}^k, \overset{c}{T}{}_{ij}^k, \overset{d}{T}{}_{ij}^k \right\rangle$, $\left\langle \overset{a}{I}{}_{ij}^k, \overset{b}{I}{}_{ij}^k, \overset{c}{I}{}_{ij}^k, \overset{d}{I}{}_{ij}^k \right\rangle$, and $\left\langle \overset{a}{F}{}_{ij}^k, \overset{b}{F}{}_{ij}^k, \overset{c}{F}{}_{ij}^k, \overset{d}{F}{}_{ij}^k \right\rangle$ represent the trapezoidal truth-, indeterminacy-, and falsity-membership functions, respectively, with all parameters taking values in the interval $[0, 1]$ for $i = 1, 2, ..., p$, $j = 1, 2, ..., q$, and $k = 1, 2, ..., r$. Accordingly, each DM constructs a corresponding decision-making matrix, as illustrated in Expression 15.8.

$$\Gamma_k = \left(\ell_{ij}^k\right)_{p \times q} = \begin{matrix} \\ L_1 \\ L_2 \\ \vdots \\ L_p \end{matrix} \begin{matrix} G_1 & G_2 & \cdots & G_q \\ \end{matrix} \begin{pmatrix} \ell_{11}^k & \ell_{12}^k & \cdots & \ell_{1q}^k \\ \ell_{21}^k & \ell_{22}^k & \cdots & \ell_{2q}^k \\ \vdots & \vdots & \ddots & \vdots \\ \ell_{p1}^k & \ell_{p2}^k & \cdots & \ell_{pq}^k \end{pmatrix}, \forall k = 1, 2, ..., r. \tag{15.8}$$

Subsequently, the proposed CRADIS approach is formulated under the condition where both DM and attribute weights are entirely unknown. The computational procedure of the proposed method is outlined as follows and presented in Figure 15.1.

**Step 1:** Let $\lambda = (\lambda_1, \lambda_2, ..., \lambda_r)$ be a set of DM weights and $\eta = (\eta_1, \eta_2, ..., \eta_q)$ be a set of attribute weights, calculated by using the ordinary priority approach (OPA) proposed by Ataei et al. (2020) [28, 29].

**Step 2:** Form an aggregated decision-making matrix $A = \left(\widehat{a}_{ij}\right)_{p \times q}$, in which each element $\widehat{a}_{ij} = \left(\left\langle \overset{a}{T}_{ij}, \overset{b}{T}_{ij}, \overset{c}{T}_{ij}, \overset{d}{T}_{ij}, \right\rangle, \left\langle \overset{a}{I}_{ij}, \overset{b}{I}_{ij}, \overset{c}{I}_{ij}, \overset{d}{I}_{ij} \right\rangle, \left\langle \overset{a}{F}_{ij}, \overset{b}{F}_{ij}, \overset{c}{F}_{ij}, \overset{d}{F}_{ij} \right\rangle\right)$ is a TraNN derived through the aggregation of all individual decision-making matrices as defined in Expression 15.9.

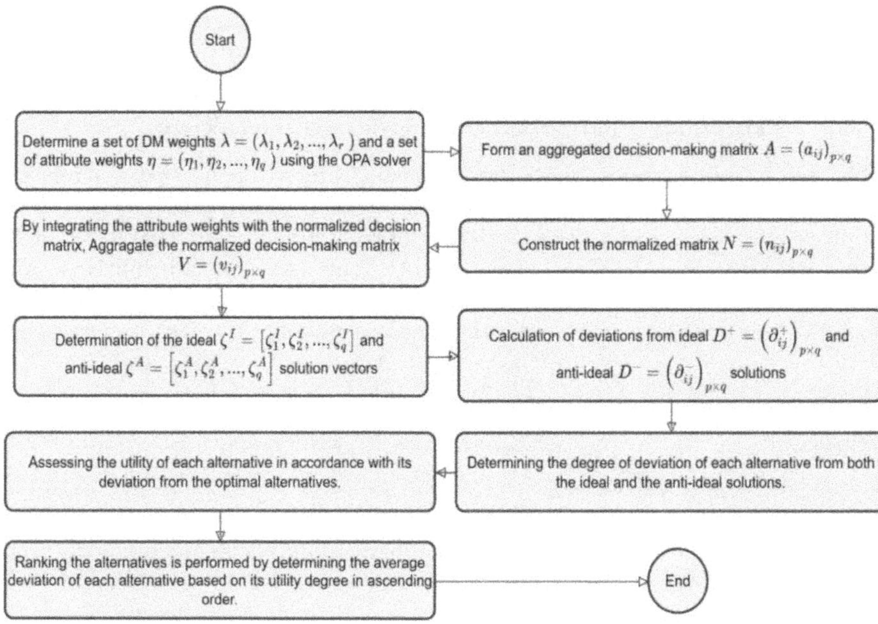


**FIGURE 15.1**: Flowchart of the proposed algorithm.

$$
\begin{aligned}
\widehat{a}_{ij} = \Bigg( \Bigg\langle & 1 - \prod_{k=1}^{r} \left(1 - \overset{a}{T}{}_{ij}^{k}\right)^{\lambda_k}, 1 - \prod_{k=1}^{r} \left(1 - \overset{b}{T}{}_{ij}^{k}\right)^{\lambda_k}, 1 - \prod_{k=1}^{r} \left(1 - \overset{c}{T}{}_{ij}^{k}\right)^{\lambda_k}, \\
& 1 - \prod_{k=1}^{r} \left(1 - \overset{d}{T}{}_{ij}^{k}\right)^{\lambda_k} \Bigg\rangle, \Bigg\langle \prod_{k=1}^{r} \left(\overset{a}{I}{}_{ij}^{k}\right)^{\lambda_k}, \prod_{k=1}^{r} \left(\overset{b}{I}{}_{ij}^{k}\right)^{\lambda_k}, \prod_{k=1}^{r} \left(\overset{c}{I}{}_{ij}^{k}\right)^{\lambda_k}, \\
& \prod_{k=1}^{r} \left(\overset{d}{I}{}_{ij}^{k}\right)^{\lambda_k} \Bigg\rangle, \Bigg\langle \prod_{k=1}^{r} \left(\overset{a}{F}{}_{ij}^{k}\right)^{\lambda_k}, \prod_{k=1}^{r} \left(\overset{b}{F}{}_{ij}^{k}\right)^{\lambda_k}, \prod_{k=1}^{r} \left(\overset{c}{F}{}_{ij}^{k}\right)^{\lambda_k}, \prod_{k=1}^{r} \left(\overset{d}{F}{}_{ij}^{k}\right)^{\lambda_k} \Bigg\rangle \Bigg).
\end{aligned}
\tag{15.9}
$$

**Step 3:** Construct the normalized matrix $N = \left(\widehat{n}_{ij}\right)_{p\times q}$, in which each element $\widehat{n}_{ij} = \Bigg( \Bigg\langle \overset{a}{T'}_{ij}, \overset{b}{T'}_{ij}, \overset{c}{T'}_{ij}, \overset{d}{T'}_{ij}, \Bigg\rangle, \Bigg\langle \overset{a}{I'}_{ij}, \overset{b}{I'}_{ij}, \overset{c}{I'}_{ij}, \overset{d}{I'}_{ij} \Bigg\rangle, \Bigg\langle \overset{a}{F'}_{ij}, \overset{b}{F'}_{ij}, \overset{c}{F'}_{ij}, \overset{d}{F'}_{ij} \Bigg\rangle \Bigg)$ represents a TraNN derived from the normalization of the aggregated decision-making matrix according to Expression 15.10, where $\Psi_B$ and $\Psi_C$ denote benefit and cost criteria, respectively.

$$\widehat{n}_{ij} = \begin{cases} \left( \left\langle \overset{a}{T}_{ij}, \overset{b}{T}_{ij}, \overset{c}{T}_{ij}, \overset{d}{T}_{ij}, \right\rangle, \left\langle \overset{a}{I}_{ij}, \overset{b}{I}_{ij}, \overset{c}{I}_{ij}, \overset{d}{I}_{ij} \right\rangle, \left\langle \overset{a}{F}_{ij}, \overset{b}{F}_{ij}, \overset{c}{F}_{ij}, \overset{d}{F}_{ij} \right\rangle \right), & G_j \in \Psi_B \\ \left( \left\langle \overset{a}{F}_{ij}, \overset{b}{F}_{ij}, \overset{c}{F}_{ij}, \overset{d}{F}_{ij} \right\rangle, \left\langle \overset{a}{I}_{ij}, \overset{b}{I}_{ij}, \overset{c}{I}_{ij}, \overset{d}{I}_{ij} \right\rangle, \left\langle \overset{a}{T}_{ij}, \overset{b}{T}_{ij}, \overset{c}{T}_{ij}, \overset{d}{T}_{ij}, \right\rangle \right), & G_j \in \Psi_C \end{cases} \tag{15.10}$$

**Step 4:** The next step is aggravation of the normalized decision-making matrices. In this step, the aggregated normalized decision-matrix $V = (\widehat{v}_{ij})_{p\times q}$ is constructed by integrating the attribute weights with the normalized decision matrix, where each resulting value $\widehat{v}_{ij} = \left( \left\langle \overset{a}{\widehat{T}}_{ij}, \overset{b}{\widehat{T}}_{ij}, \overset{c}{\widehat{T}}_{ij}, \overset{d}{\widehat{T}}_{ij} \right\rangle, \left\langle \overset{a}{\widehat{I}}_{ij}, \overset{b}{\widehat{I}}_{ij}, \overset{c}{\widehat{I}}_{ij}, \overset{d}{\widehat{I}}_{ij} \right\rangle, \left\langle \overset{a}{\widehat{F}}_{ij}, \overset{b}{\widehat{F}}_{ij}, \overset{c}{\widehat{F}}_{ij}, \overset{d}{\widehat{F}}_{ij} \right\rangle \right)$ is represented as a TraNN and can be calculated by using Equation 15.11.

$$\begin{aligned} \widehat{v}_{ij} = \Bigg( \Bigg\langle & 1 - \left(1 - \overset{a}{T'}_{ij},\right)^{\eta_j}, 1 - \left(1 - \overset{b}{T'}_{ij},\right)^{\eta_j}, 1 - \left(1 - \overset{c}{T'}_{ij},\right)^{\eta_j}, \\ & 1 - \left(1 - \overset{d}{T'}_{ij},\right)^{\eta_j} \Bigg\rangle, \left\langle \left(\overset{a}{I'}_{ij}\right)^{\eta_j}, \left(\overset{b}{I'}_{ij}\right)^{\eta_j}, \left(\overset{c}{I'}_{ij}\right)^{\eta_j}, \left(\overset{d}{I'}_{ij}\right)^{\eta_j} \right\rangle, \\ & \left\langle \left(\overset{a}{F'}_{ij}\right)^{\eta_j}, \left(\overset{b}{F'}_{ij}\right)^{\eta_j}, \left(\overset{c}{F'}_{ij}\right)^{\eta_j}, \left(\overset{d}{F'}_{ij}\right)^{\eta_j} \right\rangle \Bigg) \end{aligned} \tag{15.11}$$

**Step 5:** This step involves determination of the ideal $\zeta^I = [\zeta_1^I, \zeta_2^I, ..., \zeta_q^I]$ and anti-ideal $\zeta^A = [\zeta_1^A, \zeta_2^A, ..., \zeta_q^A]$ solution vectors, in which each element of

$$\begin{aligned} \zeta_j^I = \Bigg( & \left\langle \overset{a}{T}_{ij}^{\zeta_j^I}, \overset{b}{T}_{ij}^{\zeta_j^I}, \overset{c}{T}_{ij}^{\zeta_j^I}, \overset{d}{T}_{ij}^{\zeta_j^I} \right\rangle, \left\langle \overset{a}{I}_{ij}^{\zeta_j^I}, \overset{b}{I}_{ij}^{\zeta_j^I}, \overset{c}{I}_{ij}^{\zeta_j^I}, \overset{d}{I}_{ij}^{\zeta_j^I} \right\rangle, \\ & \left\langle \overset{a}{F}_{ij}^{\zeta_j^I}, \overset{b}{F}_{ij}^{\zeta_j^I}, \overset{c}{F}_{ij}^{\zeta_j^I}, \overset{d}{F}_{ij}^{\zeta_j^I} \right\rangle \Bigg) \end{aligned}$$

and

$$\begin{aligned} \zeta_j^A = \Bigg( & \left\langle \overset{a}{T}_{ij}^{\zeta_j^A}, \overset{b}{T}_{ij}^{\zeta_j^A}, \overset{c}{T}_{ij}^{\zeta_j^A}, \overset{d}{T}_{ij}^{\zeta_j^A} \right\rangle, \left\langle \overset{a}{I}_{ij}^{\zeta_j^A}, \overset{b}{I}_{ij}^{\zeta_j^A}, \overset{c}{I}_{ij}^{\zeta_j^A}, \overset{d}{I}_{ij}^{\zeta_j^A} \right\rangle, \\ & \left\langle \overset{a}{F}_{ij}^{\zeta_j^A}, \overset{b}{F}_{ij}^{\zeta_j^A}, \overset{c}{F}_{ij}^{\zeta_j^A}, \overset{d}{F}_{ij}^{\zeta_j^A} \right\rangle \Bigg) \end{aligned}$$

for $j = 1, 2, ..., q$, is a TraNN. To identify the ideal solution, the maximum values of $\widehat{v}_{ij}$ for each attribute $j$ are extracted from the aggregated normalized decision matrix $V$. Conversely, the anti-ideal solution is determined by identifying the minimum values of $\widehat{v}_{ij}$ for each attribute $j$ within the same matrix. The ideal and anti-ideal values can be derived by using Equations 15.12 and 15.13, respectively.

$$\zeta_j^I = \left( \left\langle \max_{1\leq i\leq p}\left\{\overset{a}{\widehat{T}_{ij}}\right\}, \max_{1\leq i\leq p}\left\{\overset{b}{\widehat{T}_{ij}}\right\}, \max_{1\leq i\leq p}\left\{\overset{c}{\widehat{T}_{ij}}\right\}, \max_{1\leq i\leq p}\left\{\overset{d}{\widehat{T}_{ij}}\right\} \right\rangle, \right.$$
$$\left\langle \max_{1\leq i\leq p}\left\{\overset{a}{\widehat{I}_{ij}}\right\}, \max_{1\leq i\leq p}\left\{\overset{b}{\widehat{I}_{ij}}\right\}, \max_{1\leq i\leq p}\left\{\overset{c}{\widehat{I}_{ij}}\right\}, \max_{1\leq i\leq p}\left\{\overset{d}{\widehat{I}_{ij}}\right\} \right\rangle,$$
$$\left. \left\langle \min_{1\leq i\leq p}\left\{\overset{a}{\widehat{F}_{ij}}\right\}, \min_{1\leq i\leq p}\left\{\overset{b}{\widehat{F}_{ij}}\right\}, \min_{1\leq i\leq p}\left\{\overset{c}{\widehat{F}_{ij}}\right\}, \min_{1\leq i\leq p}\left\{\overset{d}{\widehat{F}_{ij}}\right\} \right\rangle \right) \tag{15.12}$$

$$\zeta_j^A = \left( \left\langle \min_{1\leq i\leq p}\left\{\overset{a}{\widehat{T}_{ij}}\right\}, \min_{1\leq i\leq p}\left\{\overset{b}{\widehat{T}_{ij}}\right\}, \min_{1\leq i\leq p}\left\{\overset{c}{\widehat{T}_{ij}}\right\}, \min_{1\leq i\leq p}\left\{\overset{d}{\widehat{T}_{ij}}\right\} \right\rangle, \right.$$
$$\left\langle \min_{1\leq i\leq p}\left\{\overset{a}{\widehat{I}_{ij}}\right\}, \min_{1\leq i\leq p}\left\{\overset{b}{\widehat{I}_{ij}}\right\}, \min_{1\leq i\leq p}\left\{\overset{c}{\widehat{I}_{ij}}\right\}, \min_{1\leq i\leq p}\left\{\overset{d}{\widehat{I}_{ij}}\right\} \right\rangle,$$
$$\left. \left\langle \max_{1\leq i\leq p}\left\{\overset{a}{\widehat{F}_{ij}}\right\}, \max_{1\leq i\leq p}\left\{\overset{b}{\widehat{F}_{ij}}\right\}, \max_{1\leq i\leq p}\left\{\overset{c}{\widehat{F}_{ij}}\right\}, \max_{1\leq i\leq p}\left\{\overset{d}{\widehat{F}_{ij}}\right\} \right\rangle \right) \tag{15.13}$$

**Step 6:** This step involves calculation of deviations from ideal $D^+ = \left(\partial_{ij}^+\right)_{p\times q}$ and anti-ideal $D^- = \left(\partial_{ij}^-\right)_{p\times q}$ solutions by using distance measurement between two TrNNs. The deviations of ideal and anti-ideal elements can be measured by using Equations 15.14 and 15.15, respectively.

$$\partial_{ij}^+ = \frac{1}{9}\sqrt{\begin{array}{l}\left[\hat{T}_{ij}^I + \hat{I}_{ij}^I + \hat{F}_{ij}^I - \hat{T}_{ij}^L - \hat{I}_{ij}^L - \hat{F}_{ij}^L \right. \\ \left. - \left(\Delta_{ij}^{TI} - \Delta_{ij}^{TL}\right) - \left(\Delta_{ij}^{II} - \Delta_{ij}^{IL}\right) - \left(\Delta_{ij}^{FI} - \Delta_{ij}^{FL}\right)\right]^2 \\ + \left[\left(\Delta_{ij}^{TI} - \Delta_{ij}^{TL}\right) + \left(\Delta_{ij}^{II} - \Delta_{ij}^{IL}\right) + \left(\Delta_{ij}^{FI} - \Delta_{ij}^{FL}\right)\right]^2\end{array}} \tag{15.14}$$

where $\hat{T}_{ij}^I = \sum_{k\in\{a,b,c,d\}} \overset{k}{T}{}_{ij}^{\zeta_j^I}$, $\hat{I}_{ij}^I = \sum_{k\in\{a,b,c,d\}} \overset{k}{I}{}_{ij}^{\zeta_j^I}$, $\hat{F}_{ij}^I = \sum_{k\in\{a,b,c,d\}} \overset{k}{F}{}_{ij}^{\zeta_j^I}$,

$\hat{T}_{ij}^L = \sum_{k\in\{a,b,c,d\}} \overset{k}{\widehat{T}_{ij}}$, $\hat{I}_{ij}^L = \sum_{k\in\{a,b,c,d\}} \overset{k}{\widehat{I}_{ij}}$, $\hat{F}_{ij}^L = \sum_{k\in\{a,b,c,d\}} \overset{k}{\widehat{F}_{ij}}$,

$$\triangle_{ij}^{TI}=\frac{\overset{d}{T}_{ij}^{\varsigma_j^I}\,\overset{c}{T}_{ij}^{\varsigma_j^I}-\overset{a}{T}_{ij}^{\varsigma_j^I}\,\overset{b}{T}_{ij}^{\varsigma_j^I}}{\left(\overset{d}{T}_{ij}^{\varsigma_j^I}+\overset{c}{T}_{ij}^{\varsigma_j^I}\right)-\left(\overset{a}{T}_{ij}^{\varsigma_j^I}+\overset{b}{T}_{ij}^{\varsigma_j^I}\right)},\ \triangle_{ij}^{TL}=\frac{\overset{d}{\widehat{T}}_{ij}\,\overset{c}{\widehat{T}}_{ij}-\overset{a}{\widehat{T}}_{ij}\,\overset{b}{\widehat{T}}_{ij}}{\left(\overset{d}{\widehat{T}}_{ij}+\overset{c}{\widehat{T}}_{ij}\right)-\left(\overset{a}{\widehat{T}}_{ij}+\overset{b}{\widehat{T}}_{ij}\right)},$$

$$\triangle_{ij}^{II}=\frac{\overset{d}{I}_{ij}^{\varsigma_j^I}\,\overset{c}{I}_{ij}^{\varsigma_j^I}-\overset{a}{I}_{ij}^{\varsigma_j^I}\,\overset{b}{I}_{ij}^{\varsigma_j^I}}{\left(\overset{d}{I}_{ij}^{\varsigma_j^I}+\overset{c}{I}_{ij}^{\varsigma_j^I}\right)-\left(\overset{a}{I}_{ij}^{\varsigma_j^I}+\overset{b}{I}_{ij}^{\varsigma_j^I}\right)},\ \triangle_{ij}^{IL}=\frac{\overset{d}{\widehat{I}}_{ij}\,\overset{c}{\widehat{I}}_{ij}-\overset{a}{\widehat{I}}_{ij}\,\overset{b}{\widehat{I}}_{ij}}{\left(\overset{d}{\widehat{I}}_{ij}+\overset{c}{\widehat{I}}_{ij}\right)-\left(\overset{a}{\widehat{I}}_{ij}+\overset{b}{\widehat{I}}_{ij}\right)},$$

$$\triangle_{ij}^{FI}=\frac{\overset{d}{F}_{ij}^{\varsigma_j^I}\,\overset{c}{F}_{ij}^{\varsigma_j^I}-\overset{a}{F}_{ij}^{\varsigma_j^I}\,\overset{b}{F}_{ij}^{\varsigma_j^I}}{\left(\overset{d}{F}_{ij}^{\varsigma_j^I}+\overset{c}{F}_{ij}^{\varsigma_j^I}\right)-\left(\overset{a}{F}_{ij}^{\varsigma_j^I}+\overset{b}{F}_{ij}^{\varsigma_j^I}\right)},\ \text{and}\ \triangle_{ij}^{FL}=\frac{\overset{d}{\widehat{F}}_{ij}\,\overset{c}{\widehat{F}}_{ij}-\overset{a}{\widehat{F}}_{ij}\,\overset{b}{\widehat{F}}_{ij}}{\left(\overset{d}{\widehat{F}}_{ij}+\overset{c}{\widehat{F}}_{ij}\right)-\left(\overset{a}{\widehat{F}}_{ij}+\overset{b}{\widehat{F}}_{ij}\right)}.$$

$$\partial_{ij}^{-}=\frac{1}{9}\sqrt{\begin{array}{l}\left[\hat{T}_{ij}^{A}+\hat{I}_{ij}^{A}+\hat{F}_{ij}^{A}-\hat{T}_{ij}^{L}-\hat{I}_{ij}^{L}-\hat{F}_{ij}^{L}\right.\\ \left.-\left(\triangle_{ij}^{TA}-\triangle_{ij}^{TL}\right)-\left(\triangle_{ij}^{IA}-\triangle_{ij}^{IL}\right)-\left(\triangle_{ij}^{FA}-\triangle_{ij}^{FL}\right)\right]^{2}\\ +\left[\left(\triangle_{ij}^{TA}-\triangle_{ij}^{TL}\right)+\left(\triangle_{ij}^{IA}-\triangle_{ij}^{IL}\right)+\left(\triangle_{ij}^{FA}-\triangle_{ij}^{FL}\right)\right]^{2}\end{array}} \tag{15.15}$$

where $\hat{T}_{ij}^{A}=\sum_{k\in\{a,b,c,d\}}\overset{k}{T}_{ij}^{\varsigma_j^A}$, $\hat{I}_{ij}^{A}=\sum_{k\in\{a,b,c,d\}}\overset{k}{I}_{ij}^{\varsigma_j^A}$, $\hat{F}_{ij}^{A}=\sum_{k\in\{a,b,c,d\}}\overset{k}{F}_{ij}^{\varsigma_j^A}$, $\hat{T}_{ij}^{L}=\sum_{k\in\{a,b,c,d\}}\overset{k}{\widehat{T}}_{ij}$, $\hat{I}_{ij}^{L}=\sum_{k\in\{a,b,c,d\}}\overset{k}{\widehat{I}}_{ij}$, $\hat{F}_{ij}^{L}=\sum_{k\in\{a,b,c,d\}}\overset{k}{\widehat{F}}_{ij}$,

$$\triangle_{ij}^{TA}=\frac{\overset{d}{T}_{ij}^{\varsigma_j^A}\,\overset{c}{T}_{ij}^{\varsigma_j^A}-\overset{a}{T}_{ij}^{\varsigma_j^A}\,\overset{b}{T}_{ij}^{\varsigma_j^A}}{\left(\overset{d}{T}_{ij}^{\varsigma_j^A}+\overset{c}{T}_{ij}^{\varsigma_j^A}\right)-\left(\overset{a}{T}_{ij}^{\varsigma_j^A}+\overset{b}{T}_{ij}^{\varsigma_j^A}\right)},\ \triangle_{ij}^{TL}=\frac{\overset{d}{\widehat{T}}_{ij}\,\overset{c}{\widehat{T}}_{ij}-\overset{a}{\widehat{T}}_{ij}\,\overset{b}{\widehat{T}}_{ij}}{\left(\overset{d}{\widehat{T}}_{ij}+\overset{c}{\widehat{T}}_{ij}\right)-\left(\overset{a}{\widehat{T}}_{ij}+\overset{b}{\widehat{T}}_{ij}\right)},$$

$$\triangle_{ij}^{IA}=\frac{\overset{d}{I}_{ij}^{\varsigma_j^A}\,\overset{c}{I}_{ij}^{\varsigma_j^A}-\overset{a}{I}_{ij}^{\varsigma_j^A}\,\overset{b}{I}_{ij}^{\varsigma_j^A}}{\left(\overset{d}{I}_{ij}^{\varsigma_j^A}+\overset{c}{I}_{ij}^{\varsigma_j^A}\right)-\left(\overset{a}{I}_{ij}^{\varsigma_j^A}+\overset{b}{I}_{ij}^{\varsigma_j^A}\right)},\ \triangle_{ij}^{IL}=\frac{\overset{d}{\widehat{I}}_{ij}\,\overset{c}{\widehat{I}}_{ij}-\overset{a}{\widehat{I}}_{ij}\,\overset{b}{\widehat{I}}_{ij}}{\left(\overset{d}{\widehat{I}}_{ij}+\overset{c}{\widehat{I}}_{ij}\right)-\left(\overset{a}{\widehat{I}}_{ij}+\overset{b}{\widehat{I}}_{ij}\right)},$$

$$\triangle_{ij}^{FA}=\frac{\overset{d}{F}_{ij}^{\varsigma_j^A}\,\overset{c}{F}_{ij}^{\varsigma_j^A}-\overset{a}{F}_{ij}^{\varsigma_j^A}\,\overset{b}{F}_{ij}^{\varsigma_j^A}}{\left(\overset{d}{F}_{ij}^{\varsigma_j^A}+\overset{c}{F}_{ij}^{\varsigma_j^A}\right)-\left(\overset{a}{F}_{ij}^{\varsigma_j^A}+\overset{b}{F}_{ij}^{\varsigma_j^A}\right)},\ \text{and}\ \triangle_{ij}^{FL}=\frac{\overset{d}{\widehat{F}}_{ij}\,\overset{c}{\widehat{F}}_{ij}-\overset{a}{\widehat{F}}_{ij}\,\overset{b}{\widehat{F}}_{ij}}{\left(\overset{d}{\widehat{F}}_{ij}+\overset{c}{\widehat{F}}_{ij}\right)-\left(\overset{a}{\widehat{F}}_{ij}+\overset{b}{\widehat{F}}_{ij}\right)}.$$

**Step 7:** This step involves determining the degree of deviation of each alternative from both the ideal and the anti-ideal solutions. Equations 15.16 and 15.17 are employed to compute the deviation degrees of each alternative from the ideal and anti-ideal solutions, respectively.

$$S_i^+ = \sum_{j=1}^{q} \partial_{ij}^+, \forall i = 1, 2, ..., p \tag{15.16}$$

$$S_i^- = \sum_{j=1}^{q} \partial_{ij}^-, \forall i = 1, 2, ..., p \tag{15.17}$$

**Step 8:** This step involves assessing the utility of each alternative in accordance with its deviation from the optimal alternatives. The optimal alternatives, denoted as $S_0^+$ and$S_0^-$, are identified as the ones having the shortest distance to the ideal solution and the longest distance to the anti-ideal solution, respectively. Equations 15.18) and 15.19 are utilized to obtain the utility degrees of each alternative from the ideal and anti-ideal solutions, respectively.

$$U_i^+ = \frac{S_0^+}{S_i^+}, \forall i = 1, 2, ..., p \tag{15.18}$$

$$U_i^- = \frac{S_i^-}{S_0^-}, \forall i = 1, 2, ..., p \tag{15.19}$$

**Step 9:** This step involves ranking the alternatives is performed by determining the average deviation of each alternative based on its utility degree in ascending order. Equation 15.20 is employed to compute the average deviation of each alternative.

$$\bar{U}_i = \frac{\left(U_i^+ + U_i^-\right)}{2}, \forall i = 1, 2, ..., p \tag{15.20}$$

---

## 15.4 Numerical Example

To illustrate the applicability of the proposed method, a MAGDM problem is considered. Suppose the President Director of a bank intends to establish a new SBB office. Six potential locations, denoted as $L_1$, $L_2$, $L_3$, $L_4$, $L_5$, and $L_6$, are evaluated. Each location is assessed based on five criteria relevant to branch establishment, namely: accessibility and transportation $(G_1)$, market potential $(G_2)$, infrastructure and utilities $(G_3)$, cost factors $(G_4)$, and safety and environmental conditions $(G_5)$. The President Director appoints three members of the Board of Directors, denoted as $E_k$ $(k = 1, 2, 3)$, to serve as DMs for the evaluation. The DMs agree that criteria $G_1$, $G_2$, $G_3$, and $G_5$ represent benefit attributes, whereas $G_4$ is classified as a cost attribute.

The President Director evaluates the performance ratings of the alternatives $L_i$ $(i = 1, 2, \ldots, 6)$ with respect to the attributes $G_j$ $(j = 1, 2, \ldots, 5)$. Accordingly, the linguistic decision matrices $D = (\tilde{a}_{ij})$ provided by each DM are presented in Tables 15.1–15.3.

**TABLE 15.1**: The linguistic term decision matrix assinged by DM $E_1$.

| **Locations** | $G_1$ | $G_2$ | $G_3$ | $G_4$ | $G_5$ |
|---|---|---|---|---|---|
| $L_1$ | EGP | VPP | VGP | EPP | GP |
| $L_2$ | VPP | EPP | GP | PP | EGP |
| $L_3$ | VGP | PP | EPP | FP | FP |
| $L_4$ | GP | PP | FP | GP | VPP |
| $L_5$ | EGP | VGP | FP | GP | VGP |
| $L_6$ | PP | PP | VGP | EPP | GP |

**TABLE 15.2**: The linguistic term decision matrix assinged by DM $E_2$.

| **Locations** | $G_1$ | $G_2$ | $G_3$ | $G_4$ | $G_5$ |
|---|---|---|---|---|---|
| $L_1$ | VPP | FP | GP | EPP | EGP |
| $L_2$ | VPP | VGP | EGP | GP | FP |
| $L_3$ | FP | EGP | FP | VGP | VPP |
| $L_4$ | GP | EPP | GP | GP | PP |
| $L_5$ | VPP | EGP | PP | FP | VGP |
| $L_6$ | GP | VGP | EGP | VGP | EGP |

**TABLE 15.3**: The linguistic term decision matrix assinged by DM $E_3$.

| **Locations** | $G_1$ | $G_2$ | $G_3$ | $G_4$ | $G_5$ |
|---|---|---|---|---|---|
| $L_1$ | EPP | EGP | FP | VGP | VPP |
| $L_2$ | PP | GP | GP | EPP | VPP |
| $L_3$ | PP | FP | GP | VPP | VGP |
| $L_4$ | PP | PP | GP | VPP | VGP |
| $L_5$ | EGP | GP | EGP | VGP | GP |
| $L_6$ | EPP | GP | EPP | EGP | GP |

Subsequently, all linguistic decision matrices are converted into trapezoidal neutrosophic decision matrices using the linguistic term transformation scale—such as extremely poor priority (EPP), very poor priority (VPP), poor priority (PP), fair priority (FP), good priority (GP), very good priority (VGP), and extremely good priority (EGP)—as defined in Table 15.4. Accordingly, the linguistic decision matrices in Tables 15.1– 15.3 are transformed into the corresponding TraNN decision matrices Γ1, Γ2, and $\Gamma_3$, which are presented in Tables 15.5–15.7, respectively.

Based on the proposed method, the most suitable alternative is then identified by following the steps outlined below:

**Step 1:** The OPA solver developed by Mahmoudi et al. [29] is applied using the input data shown in Table 15.8, yielding the weights of the DMs and criteria as summarized in Table 15.9.

**TABLE 15.4**: The linguistic term values for TraNNs.

| Symbols | TraNNs |
|---|---|
| EPP | (<0.05, 0.10, 0.20, 0.30>, <0.65, 0.70, 0.75, 0.80>, <0.55, 0.60, 0.65, 0.70>) |
| VPP | (<0.10, 0.18, 0.28, 0.38>, <0.60, 0.66, 0.72, 0.78>, <0.50, 0.55, 0.60, 0.65>) |
| PP | (<0.18, 0.25, 0.35, 0.45>, <0.55, 0.62, 0.68, 0.74>, <0.44, 0.50, 0.55, 0.60>) |
| FP | (<0.30, 0.38, 0.48, 0.58>, <0.46, 0.52, 0.58, 0.64>, <0.36, 0.40, 0.45, 0.50>) |
| GP | (<0.42, 0.50, 0.60, 0.68>, <0.38, 0.44, 0.50, 0.56>, <0.28, 0.32, 0.36, 0.40>) |
| VGP | (<0.55, 0.62, 0.72, 0.78>, <0.28, 0.34, 0.40, 0.46>, <0.18, 0.22, 0.26, 0.30>) |
| EGP | (<0.68, 0.75, 0.88, 0.95>, <0.12, 0.18, 0.22, 0.28>, <0.08, 0.10, 0.12, 0.15>) |

**TABLE 15.5**: The TraNN decision matrix assinged by DM $E_1$.

| $L_i$ | $G_1$ | $G_2$ | $G_3$ | $G_4$ | $G_5$ |
|---|---|---|---|---|---|
| $L_1$ | (<0.68, 0.75, 0.88, 0.95>, <0.12, 0.18, 0.22, 0.28>, <0.08, 0.1, 0.12, 0.15>) | (<0.1, 0.18, 0.28, 0.38>, <0.6, 0.66, 0.72, 0.78>, <0.5, 0.55, 0.6, 0.65>) | (<0.55, 0.62, 0.72, 0.78>, <0.28, 0.34, 0.4, 0.46>, <0.18, 0.22, 0.26, 0.3>) | (<0.05, 0.1, 0.2, 0.3>, <0.65, 0.7, 0.75, 0.8>, <0.55, 0.6, 0.65, 0.7>) | (<0.42, 0.5, 0.6, 0.68>, <0.38, 0.44, 0.5, 0.56>, <0.28, 0.32, 0.36, 0.4>) |
| $L_2$ | (<0.1, 0.18, 0.28, 0.38>, <0.6, 0.66, 0.72, 0.78>, <0.5, 0.55, 0.6, 0.65>) | (<0.05, 0.1, 0.2, 0.3>, <0.65, 0.7, 0.75, 0.8>, <0.55, 0.6, 0.65, 0.7>) | (<0.42, 0.5, 0.6, 0.68>, <0.38, 0.44, 0.5, 0.56>, <0.28, 0.32, 0.36, 0.4>) | (<0.18, 0.25, 0.35, 0.45>, <0.55, 0.62, 0.68, 0.74>, <0.44, 0.5, 0.55, 0.6>) | (<0.68, 0.75, 0.88, 0.95>, <0.12, 0.18, 0.22, 0.28>, <0.08, 0.1, 0.12, 0.15>) |
| $L_3$ | (<0.55, 0.62, 0.72, 0.78>, <0.28, 0.34, 0.4, 0.46>, <0.18, 0.22, 0.26, 0.3>) | (<0.18, 0.25, 0.35, 0.45>, <0.55, 0.62, 0.68, 0.74>, <0.44, 0.5, 0.55, 0.6>) | (<0.05, 0.1, 0.2, 0.3>, <0.65, 0.7, 0.75, 0.8>, <0.55, 0.6, 0.65, 0.7>) | (<0.3, 0.38, 0.48, 0.58>, <0.46, 0.52, 0.58, 0.64>, <0.36, 0.4, 0.45, 0.5>) | (<0.3, 0.38, 0.48, 0.58>, <0.46, 0.52, 0.58, 0.64>, <0.36, 0.4, 0.45, 0.5>) |
| $L_4$ | (<0.42, 0.5, 0.6, 0.68>, <0.38, 0.44, 0.5, 0.56>, <0.28, 0.32, 0.36, 0.4>) | (<0.18, 0.25, 0.35, 0.45>, <0.55, 0.62, 0.68, 0.74>, <0.44, 0.5, 0.55, 0.6>) | (<0.3, 0.38, 0.48, 0.58>, <0.46, 0.52, 0.58, 0.64>, <0.36, 0.4, 0.45, 0.5>) | (<0.42, 0.5, 0.6, 0.68>, <0.38, 0.44, 0.5, 0.56>, <0.28, 0.32, 0.36, 0.4>) | (<0.1, 0.18, 0.28, 0.38>, <0.6, 0.66, 0.72, 0.78>, <0.5, 0.55, 0.6, 0.65>) |
| $L_5$ | (<0.68, 0.75, 0.88, 0.95>, <0.12, 0.18, 0.22, 0.28>, <0.08, 0.1, 0.12, 0.15>) | (<0.55, 0.62, 0.72, 0.78>, <0.28, 0.34, 0.4, 0.46>, <0.18, 0.22, 0.26, 0.3>) | (<0.3, 0.38, 0.48, 0.58>, <0.46, 0.52, 0.58, 0.64>, <0.36, 0.4, 0.45, 0.5>) | (<0.42, 0.5, 0.6, 0.68>, <0.38, 0.44, 0.5, 0.56>, <0.28, 0.32, 0.36, 0.4>) | (<0.55, 0.62, 0.72, 0.78>, <0.28, 0.34, 0.4, 0.46>, <0.18, 0.22, 0.26, 0.3>) |
| $L_6$ | (<0.18, 0.25, 0.35, 0.45>, <0.55, 0.62, 0.68, 0.74>, <0.44, 0.5, 0.55, 0.6>) | (<0.18, 0.25, 0.35, 0.45>, <0.55, 0.62, 0.68, 0.74>, <0.44, 0.5, 0.55, 0.6>) | (<0.55, 0.62, 0.72, 0.78>, <0.28, 0.34, 0.4, 0.46>, <0.18, 0.22, 0.26, 0.3>) | (<0.05, 0.1, 0.2, 0.3>, <0.65, 0.7, 0.75, 0.8>, <0.55, 0.6, 0.65, 0.7>) | (<0.42, 0.5, 0.6, 0.68>, <0.38, 0.44, 0.5, 0.56>, <0.28, 0.32, 0.36, 0.4>) |

**TABLE 15.6**: The TraNN decision matrix assinged by DM $E_2$.

| $L_i$ | $G_1$ | $G_2$ | $G_3$ | $G_4$ | $G_5$ |
|---|---|---|---|---|---|
| $L_1$ | (<0.1, 0.18, 0.28, 0.38>, <0.6, 0.66, 0.72, 0.78>, <0.5, 0.55, 0.6, 0.65>) | (<0.3, 0.38, 0.48, 0.58>, <0.46, 0.52, 0.58, 0.64>, <0.36, 0.4, 0.45, 0.5>) | (<0.42, 0.5, 0.6, 0.68>, <0.38, 0.44, 0.5, 0.56>, <0.28, 0.32, 0.36, 0.4>) | (<0.05, 0.1, 0.2, 0.3>, <0.65, 0.7, 0.75, 0.8>, <0.55, 0.6, 0.65, 0.7>) | (<0.68, 0.75, 0.88, 0.95>, <0.12, 0.18, 0.22, 0.28>, <0.08, 0.1, 0.12, 0.15>) |
| $L_2$ | (<0.1, 0.18, 0.28, 0.38>, <0.6, 0.66, 0.72, 0.78>, <0.5, 0.55, 0.6, 0.65>) | (<0.55, 0.62, 0.72, 0.78>, <0.28, 0.34, 0.4, 0.46>, <0.18, 0.22, 0.26, 0.3>) | (<0.68, 0.75, 0.88, 0.95>, <0.12, 0.18, 0.22, 0.28>, <0.08, 0.1, 0.12, 0.15>) | (<0.42, 0.5, 0.6, 0.68>, <0.38, 0.44, 0.5, 0.56>, <0.28, 0.32, 0.36, 0.4>) | (<0.3, 0.38, 0.48, 0.58>, <0.46, 0.52, 0.58, 0.64>, <0.36, 0.4, 0.45, 0.5>) |
| $L_3$ | (<0.3, 0.38, 0.48, 0.58>, <0.46, 0.52, 0.58, 0.64>, <0.36, 0.4, 0.45, 0.5>) | (<0.68, 0.75, 0.88, 0.95>, <0.12, 0.18, 0.22, 0.28>, <0.08, 0.1, 0.12, 0.15>) | (<0.3, 0.38, 0.48, 0.58>, <0.46, 0.52, 0.58, 0.64>, <0.36, 0.4, 0.45, 0.5>) | (<0.55, 0.62, 0.72, 0.78>, <0.28, 0.34, 0.4, 0.46>, <0.18, 0.22, 0.26, 0.3>) | (<0.1, 0.18, 0.28, 0.38>, <0.6, 0.66, 0.72, 0.78>, <0.5, 0.55, 0.6, 0.65>) |
| $L_4$ | (<0.42, 0.5, 0.6, 0.68>, <0.38, 0.44, 0.5, 0.56>, <0.28, 0.32, 0.36, 0.4>) | (<0.05, 0.1, 0.2, 0.3>, <0.65, 0.7, 0.75, 0.8>, <0.55, 0.6, 0.65, 0.7>) | (<0.42, 0.5, 0.6, 0.68>, <0.38, 0.44, 0.5, 0.56>, <0.28, 0.32, 0.36, 0.4>) | (<0.42, 0.5, 0.6, 0.68>, <0.38, 0.44, 0.5, 0.56>, <0.28, 0.32, 0.36, 0.4>) | (<0.18, 0.25, 0.35, 0.45>, <0.55, 0.62, 0.68, 0.74>, <0.44, 0.5, 0.55, 0.6>) |
| $L_5$ | (<0.1, 0.18, 0.28, 0.38>, <0.6, 0.66, 0.72, 0.78>, <0.5, 0.55, 0.6, 0.65>) | (<0.68, 0.75, 0.88, 0.95>, <0.12, 0.18, 0.22, 0.28>, <0.08, 0.1, 0.12, 0.15>) | (<0.18, 0.25, 0.35, 0.45>, <0.55, 0.62, 0.68, 0.74>, <0.44, 0.5, 0.55, 0.6>) | (<0.3, 0.38, 0.48, 0.58>, <0.46, 0.52, 0.58, 0.64>, <0.36, 0.4, 0.45, 0.5>) | (<0.55, 0.62, 0.72, 0.78>, <0.28, 0.34, 0.4, 0.46>, <0.18, 0.22, 0.26, 0.3>) |
| $L_6$ | (<0.42, 0.5, 0.6, 0.68>, <0.38, 0.44, 0.5, 0.56>, <0.28, 0.32, 0.36, 0.4>) | (<0.55, 0.62, 0.72, 0.78>, <0.28, 0.34, 0.4, 0.46>, <0.18, 0.22, 0.26, 0.3>) | (<0.68, 0.75, 0.88, 0.95>, <0.12, 0.18, 0.22, 0.28>, <0.08, 0.1, 0.12, 0.15>) | (<0.55, 0.62, 0.72, 0.78>, <0.28, 0.34, 0.4, 0.46>, <0.18, 0.22, 0.26, 0.3>) | (<0.68, 0.75, 0.88, 0.95>, <0.12, 0.18, 0.22, 0.28>, <0.08, 0.1, 0.12, 0.15>) |

**TABLE 15.7**: The TraNN decision matrix assinged by DM $E_3$.

| $L_i$ | $G_1$ | $G_2$ | $G_3$ | $G_4$ | $G_5$ |
|---|---|---|---|---|---|
| $L_1$ | (<0.05, 0.1, 0.2, 0.3>, <0.65, 0.7, 0.75, 0.8>, <0.55, 0.6, 0.65, 0.7>) | (<0.68, 0.75, 0.88, 0.95>, <0.12, 0.18, 0.22, 0.28>, <0.08, 0.1, 0.12, 0.15>) | (<0.3, 0.38, 0.48, 0.58>, <0.46, 0.52, 0.58, 0.64>, <0.36, 0.4, 0.45, 0.5>) | (<0.55, 0.62, 0.72, 0.78>, <0.28, 0.34, 0.4, 0.46>, <0.18, 0.22, 0.26, 0.3>) | (<0.1, 0.18, 0.28, 0.38>, <0.6, 0.66, 0.72, 0.78>, <0.5, 0.55, 0.6, 0.65>) |
| $L_2$ | (<0.18, 0.25, 0.35, 0.45>, <0.55, 0.62, 0.68, 0.74>, <0.44, 0.5, 0.55, 0.6>) | (<0.42, 0.5, 0.6, 0.68>, <0.38, 0.44, 0.5, 0.56>, <0.28, 0.32, 0.36, 0.4>) | (<0.42, 0.5, 0.6, 0.68>, <0.38, 0.44, 0.5, 0.56>, <0.28, 0.32, 0.36, 0.4>) | (<0.05, 0.1, 0.2, 0.3>, <0.65, 0.7, 0.75, 0.8>, <0.55, 0.6, 0.65, 0.7>) | (<0.1, 0.18, 0.28, 0.38>, <0.6, 0.66, 0.72, 0.78>, <0.5, 0.55, 0.6, 0.65>) |
| $L_3$ | (<0.18, 0.25, 0.35, 0.45>, <0.55, 0.62, 0.68, 0.74>, <0.44, 0.5, 0.55, 0.6>) | (<0.3, 0.38, 0.48, 0.58>, <0.46, 0.52, 0.58, 0.64>, <0.36, 0.4, 0.45, 0.5>) | (<0.42, 0.5, 0.6, 0.68>, <0.38, 0.44, 0.5, 0.56>, <0.28, 0.32, 0.36, 0.4>) | (<0.1, 0.18, 0.28, 0.38>, <0.6, 0.66, 0.72, 0.78>, <0.5, 0.55, 0.6, 0.65>) | (<0.55, 0.62, 0.72, 0.78>, <0.28, 0.34, 0.4, 0.46>, <0.18, 0.22, 0.26, 0.3>) |
| $L_4$ | (<0.18, 0.25, 0.35, 0.45>, <0.55, 0.62, 0.68, 0.74>, <0.44, 0.5, 0.55, 0.6>) | (<0.18, 0.25, 0.35, 0.45>, <0.55, 0.62, 0.68, 0.74>, <0.44, 0.5, 0.55, 0.6>) | (<0.42, 0.5, 0.6, 0.68>, <0.38, 0.44, 0.5, 0.56>, <0.28, 0.32, 0.36, 0.4>) | (<0.1, 0.18, 0.28, 0.38>, <0.6, 0.66, 0.72, 0.78>, <0.5, 0.55, 0.6, 0.65>) | (<0.55, 0.62, 0.72, 0.78>, <0.28, 0.34, 0.4, 0.46>, <0.18, 0.22, 0.26, 0.3>) |
| $L_5$ | (<0.68, 0.75, 0.88, 0.95>, <0.12, 0.18, 0.22, 0.28>, <0.08, 0.1, 0.12, 0.15>) | (<0.42, 0.5, 0.6, 0.68>, <0.38, 0.44, 0.5, 0.56>, <0.28, 0.32, 0.36, 0.4>) | (<0.68, 0.75, 0.88, 0.95>, <0.12, 0.18, 0.22, 0.28>, <0.08, 0.1, 0.12, 0.15>) | (<0.55, 0.62, 0.72, 0.78>, <0.28, 0.34, 0.4, 0.46>, <0.18, 0.22, 0.26, 0.3>) | (<0.42, 0.5, 0.6, 0.68>, <0.38, 0.44, 0.5, 0.56>, <0.28, 0.32, 0.36, 0.4>) |
| $L_6$ | (<0.05, 0.1, 0.2, 0.3>, <0.65, 0.7, 0.75, 0.8>, <0.55, 0.6, 0.65, 0.7>) | (<0.42, 0.5, 0.6, 0.68>, <0.38, 0.44, 0.5, 0.56>, <0.28, 0.32, 0.36, 0.4>) | (<0.05, 0.1, 0.2, 0.3>, <0.65, 0.7, 0.75, 0.8>, <0.55, 0.6, 0.65, 0.7>) | (<0.68, 0.75, 0.88, 0.95>, <0.12, 0.18, 0.22, 0.28>, <0.08, 0.1, 0.12, 0.15>) | (<0.42, 0.5, 0.6, 0.68>, <0.38, 0.44, 0.5, 0.56>, <0.28, 0.32, 0.36, 0.4>) |

**TABLE 15.8**: Assessment of the priority order of DMs based on their experiences and criteria from the DMs' perspective

| **Decision-makers** | Rank | Criteria rank | | | | |
|---|---|---|---|---|---|---|
| | | $G_1$ | $G_2$ | $G_3$ | $G_4$ | $G_5$ |
| $E_1$ | 1 | 2 | 2 | 3 | 1 | 3 |
| $E_2$ | 1 | 1 | 2 | 3 | 2 | 3 |
| $E_3$ | 2 | 1 | 2 | 2 | 3 | 2 |

**TABLE 15.9**: Weights calculated by the OPA solver.

| DMs' weight | | | Criteria weight | | | | |
|---|---|---|---|---|---|---|---|
| $E_1$ | $E_2$ | $E_3$ | $G_1$ | $G_2$ | $G_3$ | $G_4$ | $G_5$ |
| 0.391 | 0.349 | 0.26 | 0.266 | 0.214 | 0.189 | 0.2 | 0.131 |

**Step 2:** By applying the TraNNWAA operator, the aggregated decision-making matrix is obtained, as presented in Table 15.10.

**Step 3:** The normalized matrix $N$ is established considering that criteria $G_1$, $G_2$, $G_3$, and $G_5$ are benefit criteria, while $G_4$ is a cost attribute. The resulting normalized matrix is presented in Table 15.11.

**Step 4:** The process of integrating the attribute weights ($\eta_1 = 0.266$, $\eta_2 = 0.214$, $\eta_3 = 0.189$, $\eta_4 = 0.2$, and $\eta_5 = 0.131$) with the normalized decision matrix yields the aggravated values of the normalized decision-making matrices, as illustrated in Table 15.12.

**Step 5:** The ideal ($\zeta^I$) and anti-ideal ($\zeta^A$) solution vectors are established, and their respective elements are presented in Table 15.13.

**Step 6:** For each alternative, the deviation values from the ideal and anti-ideal solutions are computed, and the results are presented in Table 15.14.

**Step 7:** Subsequently, the degree of deviation for each alternative from both the ideal and anti-ideal solutions is calculated, and the results are shown in Tables 15.15.

**Step 8:** Based on the results obtained in Step 7, the utility of each alternative is evaluated according to its deviation from the optimal alternatives. The optimal alternatives, denoted as $S_0^+$ and$S_0^-$, are identified as the ones having the shortest distance to the ideal solution and the longest distance to the anti-ideal solution, respectively. After obtaining the optimal alternatives, $S_0^+ = 0.0375$ and $S_0^- = 0.1451$, we calculated the resulting values, as presented in Tables 15.16.

**Step 9:** After calculating the average deviation of each alternative based on

**TABLE 15.10**: The aggregated TraNN decision matrix.

| $L_i$ | $G_1$ | $G_2$ | $G_3$ | $G_4$ | $G_5$ |
|---|---|---|---|---|---|
| $L_1$ | (<0.39, 0.47, 0.63, 0.76>, <0.33, 0.4, 0.46, 0.53>, <0.25, 0.29, 0.33, 0.37>) | (<0.37, 0.45, 0.6, 0.72>, <0.36, 0.43, 0.49, 0.56>, <0.28, 0.32, 0.36, 0.41>) | (<0.45, 0.53, 0.63, 0.7>, <0.35, 0.42, 0.48, 0.54>, <0.25, 0.29, 0.34, 0.38>) | (<0.22, 0.28, 0.39, 0.48>, <0.52, 0.58, 0.64, 0.69>, <0.41, 0.46, 0.51, 0.56>) | (<0.47, 0.55, 0.69, 0.8>, <0.29, 0.36, 0.41, 0.48>, <0.21, 0.25, 0.28, 0.32>) |
| $L_2$ | (<0.12, 0.2, 0.3, 0.4>, <0.59, 0.65, 0.71, 0.77>, <0.48, 0.54, 0.59, 0.64>) | (<0.36, 0.43, 0.54, 0.62>, <0.42, 0.48, 0.54, 0.6>, <0.31, 0.36, 0.4, 0.45>) | (<0.53, 0.61, 0.74, 0.83>, <0.25, 0.32, 0.38, 0.44>, <0.18, 0.21, 0.25, 0.28>) | (<0.25, 0.32, 0.42, 0.52>, <0.5, 0.57, 0.63, 0.69>, <0.4, 0.45, 0.5, 0.54>) | (<0.45, 0.53, 0.68, 0.8>, <0.29, 0.37, 0.42, 0.49>, <0.22, 0.25, 0.29, 0.33>) |
| $L_3$ | (<0.39, 0.46, 0.57, 0.65>, <0.4, 0.46, 0.52, 0.58>, <0.29, 0.34, 0.38, 0.43>) | (<0.43, 0.51, 0.66, 0.78>, <0.31, 0.38, 0.44, 0.51>, <0.23, 0.27, 0.31, 0.35>) | (<0.25, 0.32, 0.43, 0.52>, <0.5, 0.56, 0.62, 0.67>, <0.4, 0.44, 0.49, 0.54>) | (<0.36, 0.44, 0.54, 0.63>, <0.41, 0.48, 0.54, 0.6>, <0.31, 0.35, 0.4, 0.45>) | (<0.32, 0.4, 0.5, 0.59>, <0.44, 0.51, 0.57, 0.63>, <0.34, 0.38, 0.43, 0.48>) |
| $L_4$ | (<0.37, 0.44, 0.55, 0.63>, <0.42, 0.48, 0.54, 0.6>, <0.31, 0.36, 0.4, 0.44>) | (<0.14, 0.2, 0.3, 0.4>, <0.58, 0.65, 0.7, 0.76>, <0.48, 0.53, 0.58, 0.63>) | (<0.38, 0.46, 0.56, 0.64>, <0.41, 0.47, 0.53, 0.59>, <0.31, 0.35, 0.39, 0.44>) | (<0.35, 0.43, 0.53, 0.62>, <0.43, 0.49, 0.55, 0.61>, <0.33, 0.37, 0.41, 0.45>) | (<0.27, 0.35, 0.46, 0.55>, <0.48, 0.54, 0.61, 0.67>, <0.37, 0.42, 0.47, 0.52>) |
| $L_5$ | (<0.54, 0.62, 0.78, 0.88>, <0.21, 0.28, 0.33, 0.4>, <0.15, 0.18, 0.21, 0.25>) | (<0.57, 0.65, 0.77, 0.86>, <0.23, 0.29, 0.34, 0.41>, <0.15, 0.18, 0.22, 0.25>) | (<0.4, 0.48, 0.62, 0.73>, <0.35, 0.42, 0.48, 0.54>, <0.26, 0.3, 0.34, 0.39>) | (<0.42, 0.5, 0.6, 0.68>, <0.38, 0.44, 0.5, 0.56>, <0.27, 0.31, 0.36, 0.4>) | (<0.52, 0.59, 0.69, 0.76>, <0.3, 0.36, 0.42, 0.48>, <0.2, 0.24, 0.28, 0.32>) |
| $L_6$ | (<0.25, 0.32, 0.42, 0.52>, <0.5, 0.57, 0.63, 0.69>, <0.4, 0.45, 0.5, 0.54>) | (<0.39, 0.47, 0.57, 0.65>, <0.39, 0.46, 0.52, 0.58>, <0.29, 0.33, 0.38, 0.42>) | (<0.51, 0.59, 0.73, 0.82>, <0.26, 0.33, 0.38, 0.45>, <0.18, 0.22, 0.25, 0.29>) | (<0.45, 0.52, 0.66, 0.76>, <0.31, 0.38, 0.44, 0.5>, <0.23, 0.27, 0.3, 0.35>) | (<0.53, 0.61, 0.74, 0.83>, <0.25, 0.32, 0.38, 0.44>, <0.18, 0.21, 0.25, 0.28>) |

**TABLE 15.11**: The normalized aggregated TraNN decision matrix.

| $L_i$ | $G_1$ | $G_2$ | $G_3$ | $G_4$ | $G_5$ |
|---|---|---|---|---|---|
| $L_1$ | (<0.39, 0.47, 0.63, 0.76>, <0.33, 0.4, 0.46, 0.53>, <0.25, 0.29, 0.33, 0.37>) | (<0.37, 0.45, 0.6, 0.72>, <0.36, 0.43, 0.49, 0.56>, <0.28, 0.32, 0.36, 0.41>) | (<0.45, 0.53, 0.63, 0.7>, <0.35, 0.42, 0.48, 0.54>, <0.25, 0.29, 0.34, 0.38>) | (<0.41, 0.46, 0.51, 0.56>, <0.52, 0.58, 0.64, 0.69>, <0.22, 0.28, 0.39, 0.48>) | (<0.47, 0.55, 0.69, 0.8>, <0.29, 0.36, 0.41, 0.48>, <0.21, 0.25, 0.28, 0.32>) |
| $L_2$ | (<0.12, 0.2, 0.3, 0.4>, <0.59, 0.65, 0.71, 0.77>, <0.48, 0.54, 0.59, 0.64>) | (<0.36, 0.43, 0.54, 0.62>, <0.42, 0.48, 0.54, 0.6>, <0.31, 0.36, 0.4, 0.45>) | (<0.53, 0.61, 0.74, 0.83>, <0.25, 0.32, 0.38, 0.44>, <0.18, 0.21, 0.25, 0.28>) | (<0.4, 0.45, 0.5, 0.54>, <0.5, 0.57, 0.63, 0.69>, <0.25, 0.32, 0.42, 0.52>) | (<0.45, 0.53, 0.68, 0.8>, <0.29, 0.37, 0.42, 0.49>, <0.22, 0.25, 0.29, 0.33>) |
| $L_3$ | (<0.39, 0.46, 0.57, 0.65>, <0.4, 0.46, 0.52, 0.58>, <0.29, 0.34, 0.38, 0.43>) | (<0.43, 0.51, 0.66, 0.78>, <0.31, 0.38, 0.44, 0.51>, <0.23, 0.27, 0.31, 0.35>) | (<0.25, 0.32, 0.43, 0.52>, <0.5, 0.56, 0.62, 0.67>, <0.4, 0.44, 0.49, 0.54>) | (<0.31, 0.35, 0.4, 0.45>, <0.41, 0.48, 0.54, 0.6>, <0.36, 0.44, 0.54, 0.63>) | (<0.32, 0.4, 0.5, 0.59>, <0.44, 0.51, 0.57, 0.63>, <0.34, 0.38, 0.43, 0.48>) |
| $L_4$ | (<0.37, 0.44, 0.55, 0.63>, <0.42, 0.48, 0.54, 0.6>, <0.31, 0.36, 0.4, 0.44>) | (<0.14, 0.2, 0.3, 0.4>, <0.58, 0.65, 0.7, 0.76>, <0.48, 0.53, 0.58, 0.63>) | (<0.38, 0.46, 0.56, 0.64>, <0.41, 0.47, 0.53, 0.59>, <0.31, 0.35, 0.39, 0.44>) | (<0.33, 0.37, 0.41, 0.45>, <0.43, 0.49, 0.55, 0.61>, <0.35, 0.43, 0.53, 0.62>) | (<0.27, 0.35, 0.46, 0.55>, <0.48, 0.54, 0.61, 0.67>, <0.37, 0.42, 0.47, 0.52>) |
| $L_5$ | (<0.54, 0.62, 0.78, 0.88>, <0.21, 0.28, 0.33, 0.4>, <0.15, 0.18, 0.21, 0.25>) | (<0.57, 0.65, 0.77, 0.86>, <0.23, 0.29, 0.34, 0.41>, <0.15, 0.18, 0.22, 0.25>) | (<0.4, 0.48, 0.62, 0.73>, <0.35, 0.42, 0.48, 0.54>, <0.26, 0.3, 0.34, 0.39>) | (<0.27, 0.31, 0.36, 0.4>, <0.38, 0.44, 0.5, 0.56>, <0.42, 0.5, 0.6, 0.68>) | (<0.52, 0.59, 0.69, 0.76>, <0.3, 0.36, 0.42, 0.48>, <0.2, 0.24, 0.28, 0.32>) |
| $L_6$ | (<0.25, 0.32, 0.42, 0.52>, <0.5, 0.57, 0.63, 0.69>, <0.4, 0.45, 0.5, 0.54>) | (<0.39, 0.47, 0.57, 0.65>, <0.39, 0.46, 0.52, 0.58>, <0.29, 0.33, 0.38, 0.42>) | (<0.51, 0.59, 0.73, 0.82>, <0.26, 0.33, 0.38, 0.45>, <0.18, 0.22, 0.25, 0.29>) | (<0.23, 0.27, 0.3, 0.35>, <0.31, 0.38, 0.44, 0.5>, <0.45, 0.52, 0.66, 0.76>) | (<0.53, 0.61, 0.74, 0.83>, <0.25, 0.32, 0.38, 0.44>, <0.18, 0.21, 0.25, 0.28>) |

**TABLE 15.12**: The aggravated decision matrix.

| $L_i$ | $G_1$ | $G_2$ | $G_3$ | $G_4$ | $G_5$ |
|---|---|---|---|---|---|
| $L_1$ | (<0.12, 0.16, 0.23, 0.32>,<br><0.74, 0.79, 0.81, 0.84>,<br><0.69, 0.72, 0.74, 0.77>) | (<0.09, 0.12, 0.18, 0.24>,<br><0.8, 0.84, 0.86, 0.88>,<br><0.76, 0.78, 0.8, 0.82>) | (<0.11, 0.13, 0.17, 0.21>,<br><0.82, 0.85, 0.87, 0.89>,<br><0.77, 0.79, 0.81, 0.83>) | (<0.1, 0.12, 0.13, 0.15>,<br><0.88, 0.9, 0.91, 0.93>,<br><0.74, 0.78, 0.83, 0.86>) | (<0.08, 0.1, 0.14, 0.19>,<br><0.85, 0.87, 0.89, 0.91>,<br><0.82, 0.83, 0.85, 0.86>) |
| $L_2$ | (<0.03, 0.06, 0.09, 0.13>,<br><0.87, 0.89, 0.91, 0.93>,<br><0.82, 0.85, 0.87, 0.89>) | (<0.09, 0.11, 0.15, 0.19>,<br><0.83, 0.86, 0.88, 0.9>,<br><0.78, 0.8, 0.82, 0.84>) | (<0.13, 0.16, 0.22, 0.29>,<br><0.77, 0.81, 0.83, 0.86>,<br><0.72, 0.75, 0.77, 0.79>) | (<0.1, 0.11, 0.13, 0.14>,<br><0.87, 0.89, 0.91, 0.93>,<br><0.75, 0.79, 0.84, 0.88>) | (<0.08, 0.09, 0.14, 0.19>,<br><0.85, 0.88, 0.89, 0.91>,<br><0.82, 0.84, 0.85, 0.87>) |
| $L_3$ | (<0.12, 0.15, 0.2, 0.24>,<br><0.78, 0.81, 0.84, 0.87>,<br><0.72, 0.75, 0.77, 0.8>) | (<0.11, 0.14, 0.21, 0.28>,<br><0.78, 0.82, 0.84, 0.86>,<br><0.73, 0.76, 0.78, 0.8>) | (<0.05, 0.07, 0.1, 0.13>,<br><0.88, 0.9, 0.91, 0.93>,<br><0.84, 0.86, 0.87, 0.89>) | (<0.07, 0.08, 0.1, 0.11>,<br><0.84, 0.86, 0.88, 0.9>,<br><0.81, 0.85, 0.89, 0.91>) | (<0.05, 0.06, 0.09, 0.11>,<br><0.9, 0.91, 0.93, 0.94>,<br><0.87, 0.88, 0.9, 0.91>) |
| $L_4$ | (<0.11, 0.14, 0.19, 0.23>,<br><0.79, 0.82, 0.85, 0.87>,<br><0.74, 0.76, 0.78, 0.81>) | (<0.03, 0.05, 0.07, 0.1>,<br><0.89, 0.91, 0.93, 0.94>,<br><0.85, 0.87, 0.89, 0.91>) | (<0.09, 0.11, 0.14, 0.18>,<br><0.84, 0.87, 0.89, 0.91>,<br><0.8, 0.82, 0.84, 0.85>) | (<0.08, 0.09, 0.1, 0.11>,<br><0.84, 0.87, 0.89, 0.91>,<br><0.81, 0.85, 0.88, 0.91>) | (<0.04, 0.05, 0.08, 0.1>,<br><0.91, 0.92, 0.94, 0.95>,<br><0.88, 0.89, 0.91, 0.92>) |
| $L_5$ | (<0.19, 0.23, 0.33, 0.43>,<br><0.66, 0.71, 0.75, 0.78>,<br><0.61, 0.63, 0.66, 0.69>) | (<0.17, 0.2, 0.27, 0.34>,<br><0.73, 0.77, 0.8, 0.83>,<br><0.67, 0.7, 0.72, 0.75>) | (<0.09, 0.12, 0.17, 0.22>,<br><0.82, 0.85, 0.87, 0.89>,<br><0.78, 0.8, 0.82, 0.84>) | (<0.06, 0.07, 0.08, 0.1>,<br><0.82, 0.85, 0.87, 0.89>,<br><0.84, 0.87, 0.9, 0.93>) | (<0.09, 0.11, 0.14, 0.17>,<br><0.86, 0.88, 0.89, 0.91>,<br><0.81, 0.83, 0.85, 0.86>) |
| $L_6$ | (<0.07, 0.1, 0.14, 0.18>,<br><0.83, 0.86, 0.88, 0.9>,<br><0.78, 0.81, 0.83, 0.85>) | (<0.1, 0.13, 0.17, 0.2>,<br><0.82, 0.85, 0.87, 0.89>,<br><0.77, 0.79, 0.81, 0.83>) | (<0.13, 0.15, 0.22, 0.28>,<br><0.77, 0.81, 0.83, 0.86>,<br><0.72, 0.75, 0.77, 0.79>) | (<0.05, 0.06, 0.07, 0.08>,<br><0.79, 0.83, 0.85, 0.87>,<br><0.85, 0.88, 0.92, 0.95>) | (<0.09, 0.12, 0.16, 0.21>,<br><0.84, 0.86, 0.88, 0.9>,<br><0.8, 0.82, 0.83, 0.85>) |

**TABLE 15.13**: The ideal and anti-ideal solution vectors.

| — | $G_1$ | $G_2$ | $G_3$ | $G_4$ | $G_5$ |
|---|---|---|---|---|---|
| —$\zeta_j^I$ | (<0.19, 0.23, 0.33, 0.43>, <0.87, 0.89, 0.91, 0.93>, <0.61, 0.63, 0.66, 0.69>) | (<0.17, 0.2, 0.27, 0.34>, <0.89, 0.91, 0.93, 0.94>, <0.67, 0.7, 0.72, 0.75>) | (<0.13, 0.16, 0.22, 0.29>, <0.88, 0.9, 0.91, 0.93>, <0.72, 0.75, 0.77, 0.79>) | (<0.1, 0.12, 0.13, 0.15>, <0.88, 0.9, 0.91, 0.93>, <0.74, 0.78, 0.83, 0.86>) | (<0.09, 0.12, 0.16, 0.21>, <0.91, 0.92, 0.94, 0.95>, <0.8, 0.82, 0.83, 0.85>) |
| —$\zeta_j^A$ | (<0.03, 0.06, 0.09, 0.13>, <0.66, 0.71, 0.75, 0.78>, <0.82, 0.85, 0.87, 0.89>) | (<0.03, 0.05, 0.07, 0.1>, <0.73, 0.77, 0.8, 0.83>, <0.85, 0.87, 0.89, 0.91>) | (<0.05, 0.07, 0.1, 0.13>, <0.77, 0.81, 0.83, 0.86>, <0.84, 0.86, 0.87, 0.89>) | (<0.05, 0.06, 0.07, 0.08>, <0.79, 0.83, 0.85, 0.87>, <0.85, 0.88, 0.92, 0.95>) | (<0.04, 0.05, 0.08, 0.1>, <0.84, 0.86, 0.88, 0.9>, <0.88, 0.89, 0.91, 0.92>) |

**TABLE 15.14**: The deviation values from the ideal and anti-ideal solutions.

| $L_i$ | $D^+ = \left(\partial_{ij}^+\right)_{p\times q}$ | | | | | $D^- = \left(\partial_{ij}^-\right)_{p\times q}$ | | | | |
|---|---|---|---|---|---|---|---|---|---|---|
| | $G_1$ | $G_2$ | $G_3$ | $G_4$ | $G_5$ | $G_1$ | $G_2$ | $G_3$ | $G_4$ | $G_5$ |
| $L_1$ | 0.0384 | 0.0265 | 0.0173 | 0 | 0.0174 | 0.0264 | 0.025 | 0.0149 | 0.0115 | 0.005 |
| $L_2$ | 0.0033 | 0.0202 | 0.0307 | 0.0021 | 0.0168 | 0.0615 | 0.0312 | 0.0016 | 0.0136 | 0.0056 |
| $L_3$ | 0.0276 | 0.0336 | 0.0016 | 0.0014 | 0.0038 | 0.0371 | 0.0179 | 0.0307 | 0.01 | 0.0186 |
| $L_4$ | 0.0235 | 0.0025 | 0.0102 | 0.0001 | 0.0013 | 0.0413 | 0.049 | 0.0221 | 0.0115 | 0.0212 |
| $L_5$ | 0.0615 | 0.049 | 0.0174 | 0.0038 | 0.0169 | 0.0033 | 0.0025 | 0.0148 | 0.0077 | 0.0056 |
| $L_6$ | 0.0128 | 0.0226 | 0.0309 | 0.0115 | 0.0212 | 0.052 | 0.0289 | 0.0013 | 0 | 0.0013 |

**TABLE 15.15**: The degree of deviation of each alternative from both the ideal ($S_i^+$) and the anti-ideal ($S_i^-$) solutions.

| | $L_1$ | $L_2$ | $L_3$ | $L_4$ | $L_5$ | $L_6$ |
|---|---|---|---|---|---|---|
| $S_i^+$ | 0.0996 | 0.0731 | 0.068 | 0.0375 | 0.1486 | 0.0989 |
| $S_i^-$ | 0.0827 | 0.1135 | 0.1143 | 0.1451 | 0.0339 | 0.0835 |

**TABLE 15.16**: The utility functions of each alternative from both the ideal ($U_i^+$) and the anti-ideal ($U_i^-$) solutions.

| | $L_1$ | $L_2$ | $L_3$ | $L_4$ | $L_5$ | $L_6$ |
|---|---|---|---|---|---|---|
| $U_i^+$ | 0.3768 | 0.5134 | 0.5518 | 1 | 0.2528 | 0.3796 |
| $U_i^-$ | 0.5701 | 0.7824 | 0.7881 | 1 | 0.2336 | 0.5756 |

**TABLE 15.17**: The average deviation values of each alternative and their rank results.

| | $L_1$ | $L_2$ | $L_3$ | $L_4$ | $L_5$ | $L_6$ |
|---|---|---|---|---|---|---|
| $\bar{U}_i$ | 0.4735 | 0.6479 | 0.67 | 1 | 0.2432 | 0.4776 |
| Rank | 2 | 4 | 5 | 6 | 1 | 3 |

its utility degree, the alternatives are ranked in ascending order of their utility values. The final ranking, $L_5 < L_1 < L_6 < L_2 < L_3 < L_4$, indicates that alternative $L_5$ is the most suitable location for establishing a new SBB office. The average deviation values of each alternative and their rank results are provided in Table 15.17.

## 15.5 Comparative Analysis

In this section, a comparative evaluation is carried out to assess the performance of the proposed TraNN-CRADIS method relative to other TraNN aggregation operators and an alternative TraNN-based decision-making tech-

**TABLE 15.18**: The comparative analysis result based on their final scores.

| **Approaches** | | **Locations** | | | | | | **The best location** |
|---|---|---|---|---|---|---|---|---|
| | | $L_1$ | $L_2$ | $L_3$ | $L_4$ | $L_5$ | $L_6$ | |
| TraNNWAA | $Sc(L_i)$ | 0.1754 | 0.1457 | 0.1519 | 0.1309 | 0.2106 | 0.1549 | $L_5$ |
| | Result | | $L_5 < L_1 < L_6 < L_3 < L_2 < L_4$ | | | | | |
| TraNNWGA | $Sc(L_i)$ | 0.8593 | 0.8412 | 0.8562 | 0.8447 | 0.8871 | 0.8558 | $L_5$ |
| | Result | | $L_5 < L_1 < L_3 < L_6 < L_4 < L_2$ | | | | | |
| TraNN-TOPSIS | $R(L_i)$ | 0.4459 | 0.5741 | 0.5508 | 0.6677 | 0.2705 | 0.4699 | $L_5$ |
| | Result | | $L_5 < L_1 < L_6 < L_3 < L_2 < L_4$ | | | | | |
| The proposed TraNN-CRADIS | $\bar{U}_i$ | 0.4735 | 0.6479 | 0.67 | 1 | 0.2432 | 0.4776 | $L_5$ |
| | Result | | $L_5 < L_1 < L_6 < L_2 < L_3 < L_4$ | | | | | |

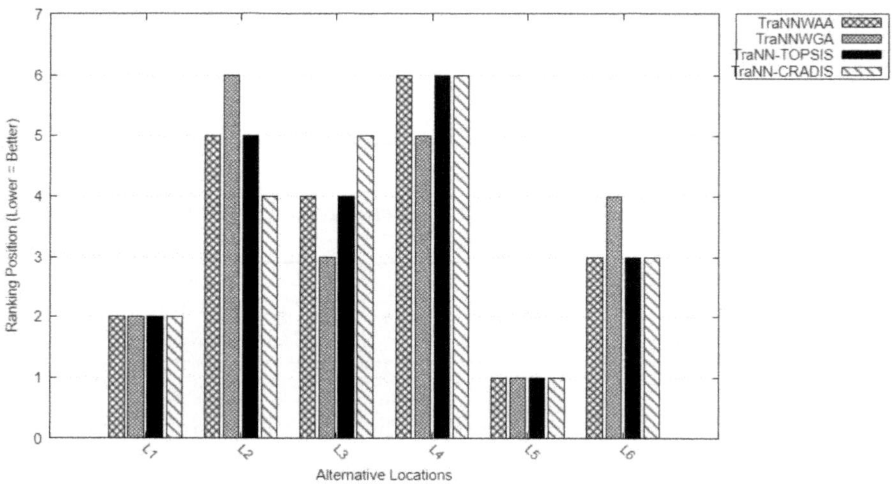


**FIGURE 15.2**: Comparative ranking of six potential Sharia Bank branch locations.

nique. This comparison aims to highlight the applicability, efficiency, and benefits of the TraNN-CRADIS approach in determining the most appropriate site for an SBB. The comparative analysis involves the TraNNWAA and TraNNWGA operators proposed by Ye [8], as well as the TraNN-TOPSIS method introduced by Giri et al. [30]. The analysis considers six candidate branch locations and five evaluation attributes—accessibility and transportation ($G_1$), market potential ($G_2$), infrastructure and utilities ($G_3$), cost factors ($G_4$), and safety and environmental conditions ($G_5$). The outcomes of the comparison are summarized in Table 15.18, while Figures 15.2 and 15.3 visualize the score functions and ranking results, respectively.

Table 15.18 presents the comparative results derived from each method. As illustrated, all approaches consistently identify $L_5$ as the most preferable alternative, suggesting that this location exhibits the best overall performance when evaluated against the defined criteria. This consistency across multiple

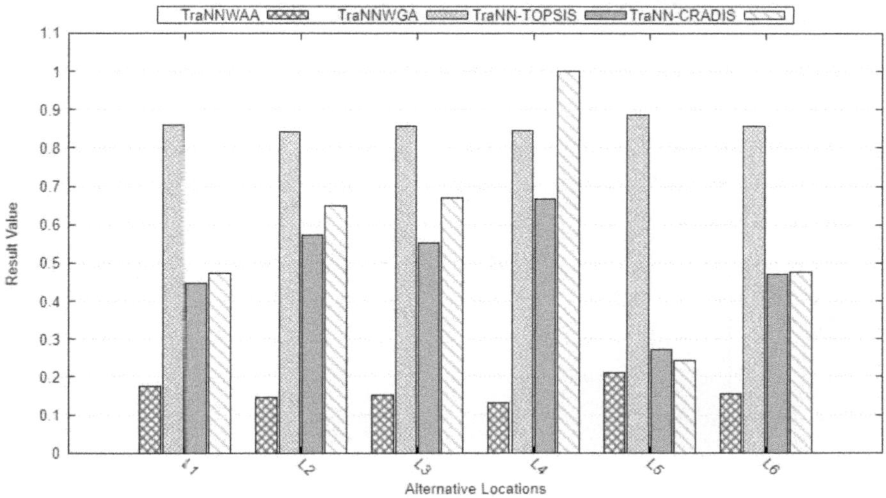


**FIGURE 15.3**: Comparison of result values for each method.

models reinforces the validity and credibility of the proposed TraNN-CRADIS method in addressing MCGDM problems. Furthermore, despite minor differences in the intermediate rankings among the remaining alternatives, the general ranking tendencies remain relatively stable across all methods.

A closer examination reveals that methods based on weighted averaging, such as TraNNWAA and TraNNWGA, produce relatively similar rank patterns due to their comparable aggregation structures. In contrast, TraNN-TOPSIS and TraNN-CRADIS, which employ distance-based evaluation mechanisms, generate slightly different rank distributions. These variations can be attributed to differences in normalization procedures, weighting integration, and the way each method measures proximity to ideal and anti-ideal solutions. Notably, the proposed TraNN-CRADIS method offers a more discriminative ranking pattern, capturing finer variations among the alternatives while maintaining consistency in identifying the best location.

To quantitatively validate the consistency between the proposed TraNN-CRADIS method and other approaches, the spearman rank correlation coefficient ($r_s$) was computed for each pairwise comparison of ranking results. The Spearman correlation coefficient measures the degree of association between two sets of rankings, with values ranging from $-1$ (perfect inverse correlation) to $+1$ (perfect positive correlation). Higher correlation values indicate greater similarity between the ranking outcomes of different methods. The pairwise correlation results among methods is provided in Figure 15.4.

The pairwise correlation results show strong positive correlations between the proposed TraNN-CRADIS method and the three comparative methods, with all correlation coefficients exceeding 0.710. Specifically, the correlation

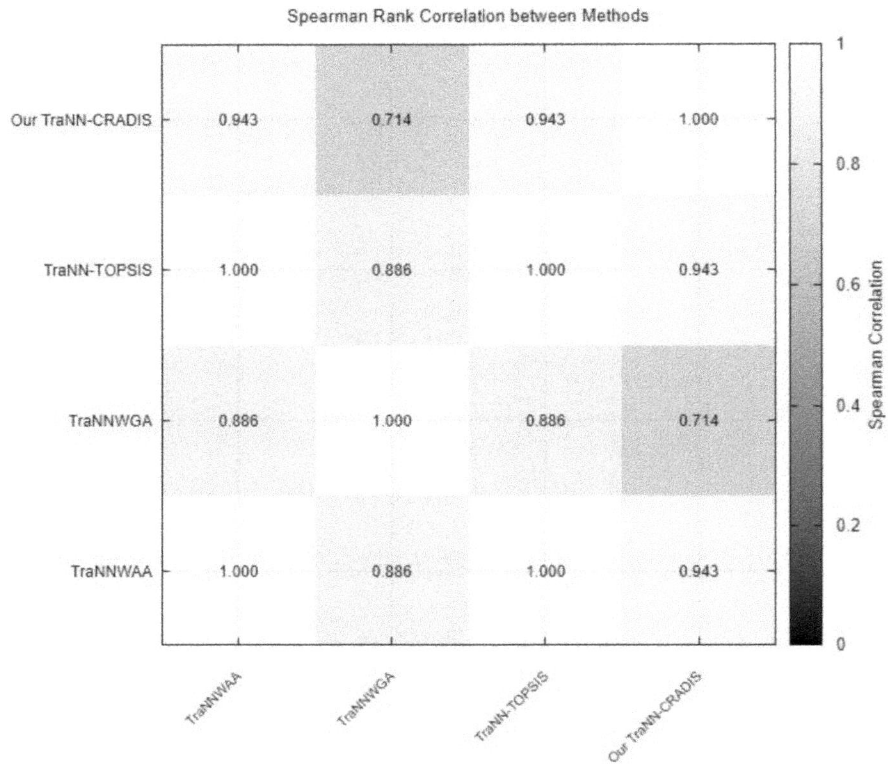


**FIGURE 15.4**: Confusion matrix for Spearman correlation.

between TraNN-CRADIS and TraNNWAA is the highest ($r_s = 0.943$), followed by TraNN-TOPSIS ($r_s = 0.943$), and TraNNWGA ($r_s = 0.714$). These results demonstrate that the proposed approach yields ranking outcomes that are highly consistent with established TraNN-based models, thereby confirming its stability and methodological soundness.

As a result, the comparative analysis confirms that the proposed TraNN-CRADIS method not only aligns closely with other TraNN-based decision-making approaches in identifying the best alternative ($L_5$) but also provides a more refined and discriminative ranking of all alternatives. The high Spearman correlation coefficients further substantiate the reliability and robustness of the proposed method, making it a viable and effective tool for practical decision-making problems, especially SBB location selection.

## 15.6 Conclusions and Future Studies

This chapter presents an enhanced CRADIS methodology as a robust and flexible tool for evaluating and prioritizing feasible alternatives in multi-criteria decision-making contexts. While the conventional CRADIS approach has demonstrated effectiveness in handling general decision-making problems, it faces notable limitations in risk assessment and ranking tasks under conditions of ambiguity and uncertainty, as it relies primarily on crisp numerical data.

To overcome these shortcomings, we have proposed a novel CRADIS framework based on the TraNN environment. The proposed TraNN-based CRADIS model extends the conventional CRADIS method to effectively evaluate the selection of SBB office locations under uncertain and imprecise conditions. Within this framework, an expert committee consisting of three DMs expresses uncertain judgments and preferences regarding six potential SBB office locations, which are represented using TraNNs.

Furthermore, the OPA method is employed to determine the relative weights of both the decision-makers and the evaluation criteria. The enhanced CRADIS model integrates TraNN information to capture the relationships among decision variables while maintaining flexibility in accounting for the degree of deviation of each location from both the ideal and anti-ideal solutions.

The numerical case study has verified the validity, practical applicability, and effectiveness of the proposed framework in evaluating and ranking SBB locations. Moreover, comparative analyses with existing decision-making approaches have substantiated the robustness, stability, and superior performance of the proposed TraNN-CRADIS method.

Although the proposed model demonstrates encouraging results, several challenges remain to be addressed. Future research could extend its application to a broader range of office location evaluation scenarios characterized by diverse forms of uncertainty and vagueness, including complex spherical fuzzy sets [31], hesitant fuzzy sets [32], fuzzy set-pair analysis [33], and bipolar neutrosophic sets [34]. Furthermore, subsequent studies may focus on developing an enhanced CRADIS framework that integrates consensus-building mechanisms and supports large-scale group decision-making within social network-based structures and complex linguistic environments.

## Bibliography


[1] Irvanizam Irvanizam and Denny Helzami. Bipolar neutrosophic ARAS method based on multiple attribute group decision-making to evaluate the service quality of sharia bank branch offices. In *Studies in Fuzziness and Soft Computing*, Studies in fuzziness and soft computing, pages 319–348. Springer Nature Switzerland, Cham, 2025.

[2] Simona Franzoni and Asma Ait Allali. Principles of islamic finance and principles of corporate social responsibility: What convergence? *Sustainability*, 10(3):637, February 2018.

[3] Irvanizam Irvanizam, Mahyuddin K M Nasution, Tulus Tulus, and Erna B Nababan. A hybrid decision support framework using MEREC-RAFSI with spherical fuzzy numbers for selecting banking financial aid recipients. *IEEE Access*, 13:33101–33123, 2025.

[4] Abdullah Hamzah Al-Hadrami, Sutan Emir Hidayat, and Mohamed Isa Al-Sharbti. The important selection criteria in choosing Islamic Banks: A survey in bahrain. *Al-Iqtishad J. Islam. Econ.*, 9(2):165–184, May 2017.

[5] L A Zadeh. Fuzzy sets. *Inf. Contr.*, 8(3):338–353, June 1965.

[6] Muhammad Kamran, Nadeem Salamat, Chiranjibe Jana, and Qin Xin. Decision-making technique with neutrosophic z-rough set approach for sustainable industry evaluation using sine trigonometric operators. *Appl. Soft Comput.*, 169(112539):112539, January 2025.

[7] Florentin Smarandache. *A Unifying Field in Logics: Neutrosophic Logic, Neutrosophic Set, Neutrosophic Probability and Statistics (fourth edition).* 2001, doi: 10.48550/arXiv.math/0101228.

[8] Jun Ye. Trapezoidal neutrosophic set and its application to multiple attribute decision-making. *Neural Comput. Appl.*, 26(5):1157–1166, July 2015.

[9] M Arockia Dasan, E Bementa, Muhammad Aslam, and V F Little Flower. Multi-attribute decision-making problem in career determination using single-valued neutrosophic distance measure. *Complex Intell. Syst.*, 10 (4):5411–5425, August 2024.

[10] Atiqe Ur Rahman, Muhammad Saeed, and Hamiden Abd El-Wahed Khalifa. Multi-attribute decision-making based on aggregations and similarity measures of neutrosophic hypersoft sets with possibility setting. *J. Exp. Theor. Artif. Intell.*, 36(2):1–26, July 2022.

[11] Seema Agrawal, Tarun Kumar, Kapil Kumar, M K Sharma, and Vishnu Narayan Mishra. Trapezoidal neutrosophic teaching learning-based optimization in enhancing accuracy of diabetes prognosis. *Results Control Optim.*, 17(100464):100464, December 2024.

[12] Ruipu Tan and Wende Zhang. Multiple attribute decision making method based on DEMATEL and fuzzy distance of trapezoidal fuzzy neutrosophic numbers and its application in typhoon disaster evaluation. *J. Intell. Fuzzy Syst.*, 39(3):3413–3439, October 2020.

[13] Chiranjibe Jana, Madhumangal Pal, Faruk Karaaslan, and Jian-Qiang Wang. Trapezoidal neutrosophic aggregation operators and its application in multiple attribute decision making process. *Sci. Iran.*, November 2018.

[14] Muhammad Jamil, Farkhanda Afzal, Ayesha Maqbool, Saleem Abdullah, Ali Akgül, and Abdul Bariq. Multiple attribute group decision making approach for selection of robot under induced bipolar neutrosophic aggregation operators. *Complex Intell. Syst.*, 10(2):2765–2779, April 2024.

[15] Xinli You, Fujun Hou, and Zhenkai Lou. Consensus building in multi-criteria group decision-making with single-valued neutrosophic sets. *Cognit. Comput.*, 13(6):1496–1514, November 2021.

[16] Irvanizam Irvanizam, Novi Zahara, and Marzuki Marzuki. Single-valued neutrosophic ARAS based on multi-criteria group decision-making for a school selection. In *Advances in Systems Analysis, Software Engineering, and High Performance Computing*, pages 307–332. IGI Global, Hershey, Pennsylvania, September 2024.

[17] Guiwu Wei, Jie Wang, Jianping Lu, Jiang Wu, Cun Wei, Fuad E Alsaadi, and Tasawar Hayat. VIKOR method for multiple criteria group decision making under 2-tuple linguistic neutrosophic environment. *Econ. Res.-Ekon. Istraž.*, 33(1):3185–3208, January 2020.

[18] Surapati Pramanik and Rama Mallick. TODIM strategy for multi-attribute group decision making in trapezoidal neutrosophic number environment. *Complex Intell. Syst.*, 5(4):379–389, December 2019.

[19] Shahzaib Ashraf, Wania Iqbal, Muhammad Shazib Hameed, Vladimir Simic, and Nebojsa Bacanin. An enhanced CRADIS decision model for optimizing radioactive waste reduction through transmutations based on disc spherical fuzzy information. *Appl. Soft Comput.*, 167(112289): 112289, December 2024.

[20] Weizhong Wang, Yi Wang, Shuyi Fan, Xiao Han, Qun Wu, and Dragan Pamucar. A complex spherical fuzzy CRADIS method based Fine-Kinney framework for occupational risk evaluation in natural gas pipeline construction. *J. Pet. Sci. Eng.*, 220(111246):111246, January 2023.

[21] Adis Puška, Željko Stević, and Dragan Pamučar. Evaluation and selection of healthcare waste incinerators using extended sustainability criteria and multi-criteria analysis methods. *Environ. Dev. Sustain.*, 24(9):11195–11225, 2022.

[22] Adis Puška, Darko Božanić, Zoran Mastilo, and Dragan Pamučar. Extension of MEREC-CRADIS methods with double normalization-case study selection of electric cars. *Soft Comput.*, 27(11):7097–7113, June 2023.

[23] Jiaqi Yuan, Zichun Chen, and Miaofeng Wu. A novel distance measure and CRADIS method in picture fuzzy environment. *Int. J. Comput. Intell. Syst.*, 16(1):186, November 2023.

[24] Shi Yin, Raiha Imran, Kifayat Ullah, Zeeshan Ali, and Izatmand Haleemzai. Responsible AI in student management: Preventing misdecision in career choice of university students under inaccurate guidance. *Sci. Rep.*, 15(1):38177, October 2025.

[25] Mijanur Rahaman Seikh and Arnab Mukherjee. Fermatean fuzzy CRADIS approach based on triangular divergence for selecting online shopping platform. *T*rans. Fuzzy Sets Syst., 6(2):108, 2024.

[26] Haibin Wang, Florentin Smarandache, Yanqing Zhang, and Rajshekhar Sunderraman. Single valued neutrosophic sets. *I*nfinite Study, 2010.

[27] Broumi Said, Malayalan Lathamaheswari, Ruipu Tan, Deivanayagampillai Nagarajan, Talea Mohamed, Florentin Smarandache, and Assia Bakali. A new distance measure for trapezoidal fuzzy neutrosophic numbers based on the centroids. *Neutrosophic Sets and Systems*, 35:478–502, 2020.

[28] Younes Ataei, Amin Mahmoudi, Mohammad Reza Feylizadeh, and Deng-Feng Li. Ordinal priority approach (OPA) in multiple attribute decision-making. *Appl. Soft Comput.*, 86(105893):105893, January 2020.

[29] Amin Mahmoudi, Mahsa Sadeghi, Xiaopeng Deng, and Pengcheng Pan. OPA solver: A web-based software for ordinal priority approach in multiple criteria decision analysis using JavaScript. *SoftwareX*, 24(101546):101546, December 2023.

[30] Molla Mahatab Uddin Giri, Bibhas C. and Pranab Biswas. Topsis method for madm based on interval trapezoidal neutrosophic number. *Neutrosophic Sets and Systems*, 22(1):151–167, 2018.

[31] D. Ramot, R. Milo, M. Friedman, and A. Kandel. Complex fuzzy sets. *IEEE Transactions on Fuzzy Systems*, 10(2):171–186, 2002. doi: 10.1109/91.995119.

[32] Huchang Liao and Zeshui Xu. Hesitant fuzzy set and its extensions. In *Hesitant Fuzzy Decision Making Methodologies and Applications*, pages 1–36. Springer Singapore, Singapore, 2017.

[33] Irvanizam Irvanizam, Intan Syahrini, Razief Perucha Fauzie Afidh, Muhammad Reki Andika, and Hizir Sofyan. Applying fuzzy multiple-attribute decision making based on set-pair analysis with triangular fuzzy number for decent homes distribution problem. In *2018 6th International Conference on Cyber and IT Service Management (CITSM)*, pages 1–7, 2018. doi: 10.1109/CITSM.2018.8674290.

[34] Irfan Deli, Mumtaz Ali, and Florentin Smarandache. Bipolar neutrosophic sets and their application based on multi-criteria decision making problems. In *2015 International Conference on Advanced Mechatronic Systems (ICAMechS)*, pages 249–254, 2015. doi: 10.1109/ICAMechS.2015.7287068.

# Chapter 16

## A Novel Approach for the Shortest Path Computation in Networks Using Quadripartitioned Single-Valued Neutrosophic Refined Numbers

**Arokia Pratheesha S. V. and R. Radha**

### 16.1 Introduction

In real-world problems, information is often inconsistent, incomplete, or indeterminate, making it difficult to represent with precise numbers. To address this, Zadeh [35] introduced fuzzy sets, later extended by Atanassov [1] into intuitionistic fuzzy sets (IFSs) by incorporating both membership and non-membership values. Building further, Smarandache [26, 28] developed the concept of neutrosophic sets, which generalize classical, fuzzy, and IFS through three components: truth ($T$), indeterminacy ($I$), and falsity ($F$). For practical use, Wang and his team proposed single-valued neutrosophic sets (SVNSs).

In parallel, Yager [32] introduced the idea of bags (multisets), allowing repeated elements, later formalized by Blizard and Calude. Shinoj et al. [4, 16, 29] combined fuzzy multisets with IFS to form intuitionistic fuzzy multisets (IFMSs). However, to better capture higher-order uncertainties, Chatterjee et al. [24] proposed the single-valued neutrosophic multiset (SVNMSs) [30], later refined through Smarandache $n$-valued neutrosophic logic, which decomposes $T$, $I$, and $F$ into multiple subcomponents. Deli et al. studied neutrosophic refined sets, while Ye et al. developed distance and similarity measures for SVNMSs with applications in fields like medical diagnosis.

Separately, Belnap's [6] four-valued logic (true, false, neither, both) inspired Smarandache to extend neutrosophic logic by splitting indeterminacy into contradiction ($C$) and unknown ($U$). Building on this, Chatterjee et al. introduced the quadripartitioned single-valued neutrosophic set (QSVNS) with four components: $T$, $C$, $U$, and $F$, providing a more precise tool for modeling complex uncertainty. The shortest path problem (SPP) is a classic combinato-

DOI: 10.1201/9781003683339-16 

rial optimization problem with wide applications in science and engineering, such as road networks, transportation, and logistics. It focuses on determining a path from a source node to a destination node with the minimum total weight, where each edge carries a cost, distance, or time value. Traditional SPP assumes that all parameters are known with certainty. However, in real-world networks, uncertainty often exists in edge values like distance or travel time.

To address this, researchers have extended SPP models using fuzzy sets, IFSs, and vague sets [19–23]. Among solution techniques, Dijkstra's algorithm [18] remains one of the most widely used methods for computing shortest paths. More recently, attention has shifted toward neutrosophic graph theory [7–10, 14, 15, 27], which better handles indeterminacy and incomplete information. Building on this, Broumi et al. [11–13] proposed algorithms for shortest path problems in networks where edge weights are expressed through neutrosophic numbers—including single-valued, bipolar, and interval-valued forms—thereby offering a more flexible framework for decision-making under uncertainty.

The structure of this chapter is organized as follows: Section 16.2 outlines the fundamental principles of $\mathcal{QSVNS}$ and $\mathcal{QSVNR}$. Section 16.3 describes the proposed algorithm for determining the optimal path and corresponding minimum distance in a $\mathcal{QSVNR}$ graph. Section 16.4 provides a numerical example to illustrate the procedure for identifying the minimum path and distance between the source and destination nodes. Finally, Section 16.5 summarizes the findings and outlines potential directions for future research.

## 16.2 Preliminaries

**Definition 16.1:** [24]: Let $\mathcal{A}^*_{\mathcal{QNS}}$ be a universe discourse. A $\mathcal{QSVNS}$ $\breve{\mathcal{H}}_{\mathcal{QNS}}$ on $\mathcal{A}^*_{\mathcal{QNS}}$ is defined by a truth-membership function $\breve{\mathcal{T}}_{\breve{\mathcal{H}}}(\mathfrak{a}_1)$, a contradiction membership function $\breve{\mathcal{D}}_{\breve{\mathcal{H}}}(\mathfrak{a}_1)$, an unknown membership function $\breve{\mathcal{Y}}_{\breve{\mathcal{H}}}(\mathfrak{a}_1)$, and a falsity membership function $\breve{\mathcal{F}}_{\breve{\mathcal{H}}}(\mathfrak{a}_1)$ such that for each $\mathfrak{a}_1 \in \mathcal{A}^*_{\mathcal{QNS}}$, $\breve{\mathcal{T}}_{\mathcal{H}}$,$\breve{\mathcal{D}}_{\mathcal{H}}$,$\breve{\mathcal{Y}}_{\mathcal{H}}$,$\breve{\mathcal{F}}_{\mathcal{H}} \in [0,1]$ and $0 \le \breve{\mathcal{T}}_{\mathcal{H}}(\mathfrak{a}_1)+\breve{\mathcal{T}}_{\mathcal{H}}(\mathfrak{a}_1)+\breve{\mathcal{T}}_{\mathcal{H}}(\mathfrak{a}_1)+\breve{\mathcal{T}}_{\mathcal{H}}(\mathfrak{a}_1) \le 4$.

**Definition 16.2:** [2]: Let $\mathcal{W}^*_{\mathcal{QNR}}$ be a universe. A $\mathcal{QSVNRS}$ $\breve{\mathcal{H}}_{\mathcal{QNR}}$ on $\mathcal{W}^*_{\mathcal{QNR}}$ is defined by

$$\breve{\mathcal{H}}_{\mathcal{QNR}} = \langle \breve{\mathcal{T}}^{\mathrm{j}}_{\breve{\mathcal{H}}_{\mathcal{QNR}}}, \breve{\mathcal{D}}^{\mathrm{j}}_{\breve{\mathcal{H}}_{\mathcal{QNR}}}, \breve{\mathcal{Y}}^{\mathrm{j}}_{\breve{\mathcal{H}}_{\mathcal{QNR}}}, \breve{\mathcal{F}}^{\mathrm{j}}_{\breve{\mathcal{H}}_{\mathcal{QNR}}} : \mathrm{j} = 1, 2 \ldots \mathfrak{p} \rangle \text{ and } \breve{\mathcal{L}}_{\mathcal{QNR}}$$
$$= \langle \breve{\mathcal{T}}^{\mathrm{j}}_{\breve{\mathcal{L}}_{\mathcal{QNR}}}, \breve{\mathcal{D}}^{\mathrm{j}}_{\breve{\mathcal{L}}_{\mathcal{QNR}}}, \breve{\mathcal{Y}}^{\mathrm{j}}_{\breve{\mathcal{L}}_{\mathcal{QNR}}}, \breve{\mathcal{F}}^{\mathrm{j}}_{\breve{\mathcal{L}}_{\mathcal{QNR}}} : \mathrm{j} = 1, 2 \ldots \mathfrak{p} \rangle$$

where $\breve{\mathcal{T}}^{\mathrm{j}}_{\breve{\mathcal{H}}_{\mathcal{QNR}}}$,$\breve{\mathcal{D}}^{\mathrm{j}}_{\breve{\mathcal{H}}_{\mathcal{QNR}}}$,$\breve{\mathcal{Y}}^{\mathrm{j}}_{\breve{\mathcal{H}}_{\mathcal{QNR}}}$,$\breve{\mathcal{F}}^{\mathrm{j}}_{\breve{\mathcal{H}}_{\mathcal{QNR}}} \rightarrow [0,1]$ such that

$0 \leq \breve{\mathcal{T}}^{\mathrm{j}}_{\breve{\mathcal{H}}_{\mathcal{QNR}}}(\mathfrak{a}_1) + \breve{\mathcal{D}}^{\mathrm{j}}_{\breve{\mathcal{H}}_{\mathcal{QNR}}}(\mathfrak{a}_1) + \breve{\mathcal{Y}}^{\mathrm{j}}_{\breve{\mathcal{H}}_{\mathcal{QNR}}}(\mathfrak{a}_1) + \breve{\mathcal{F}}^{\mathrm{j}}_{\breve{\mathcal{H}}_{\mathcal{QNR}}}(\mathfrak{a}_1) \leq 4 (\mathrm{j} = 1, 2, ...\mathfrak{p})$ for any $\mathfrak{a}_1 \in \mathcal{W}^*_{\mathcal{QNR}}$.

Here $\breve{\mathcal{T}}^{\mathrm{j}}_{\breve{\mathcal{H}}_{\mathcal{QNR}}}, \breve{\mathcal{D}}^{\mathrm{j}}_{\breve{\mathcal{H}}_{\mathcal{QNR}}}, \breve{\mathcal{Y}}^{\mathrm{j}}_{\breve{\mathcal{H}}_{\mathcal{QNR}}}, \breve{\mathcal{F}}^{\mathrm{j}}_{\breve{\mathcal{H}}_{\mathcal{QNR}}}$ is the truth, a contradiction, an unknown, and a falsity membership sequence of the element $\mathfrak{a}_1$, respectively. And $\mathfrak{p}$ is often referred to as $\mathcal{QSVNRS}(\breve{\mathcal{H}}_{\mathcal{QNR}})$ dimension.

**Definition 16.3:** [3]: Let $\breve{\mathcal{H}}_{\mathcal{QNR}} = \langle \breve{\mathcal{T}}^{\mathrm{j}}_{\breve{\mathcal{H}}_{\mathcal{QNR}}}, \breve{\mathcal{D}}^{\mathrm{j}}_{\breve{\mathcal{H}}_{\mathcal{QNR}}}, \breve{\mathcal{Y}}^{\mathrm{j}}_{\breve{\mathcal{H}}_{\mathcal{QNR}}}, \breve{\mathcal{F}}^{\mathrm{j}}_{\breve{\mathcal{H}}_{\mathcal{QNR}}} : \mathrm{j} = 1,2...\mathfrak{p} \rangle$ be a $\mathcal{QSVNRN}$s, and the score $(\mathcal{S}_{\breve{\mathcal{H}}_{\mathcal{QNR}}})$ and accuracy function $(\mathcal{P}_{\breve{\mathcal{H}}_{\mathcal{QNR}}})$ can be defined as follows:

$$\mathcal{S}_{\breve{\mathcal{H}}_{\mathcal{QNR}}} = \frac{1}{\ddot{\rho}} \sum_{\mathrm{j}=1}^{\mathfrak{p}} \frac{1}{4} (3 + \breve{\mathcal{T}}^{\mathrm{j}}_{\breve{\mathcal{H}}_{\mathcal{QNR}}} - \breve{\mathcal{D}}^{\mathrm{j}}_{\breve{\mathcal{H}}_{\mathcal{QNR}}} - \breve{\mathcal{Y}}^{\mathrm{j}}_{\breve{\mathcal{H}}_{\mathcal{QNR}}} - \breve{\mathcal{F}}^{\mathrm{j}}_{\breve{\mathcal{H}}_{\mathcal{QNR}}}), \mathcal{S}_{\breve{\mathcal{H}}_{\mathcal{QNR}}} \in [0, 1], \tag{16.1}$$

$$\mathcal{P}_{\breve{\mathcal{H}}_{\mathcal{QNR}}} = \frac{1}{\ddot{\rho}} \sum_{\mathrm{j}=1}^{\mathfrak{p}} (\breve{\mathcal{T}}^{\mathrm{j}}_{\breve{\mathcal{H}}_{\mathcal{QNR}}} - \breve{\mathcal{F}}^{\mathrm{j}}_{\breve{\mathcal{H}}_{\mathcal{QNR}}}), \mathcal{H}_{\breve{\mathcal{H}}_{\mathcal{QNR}}} \in [-1, 1] \tag{16.2}$$

**Definition 16.4:** [3]: Let $\breve{\mathcal{K}}_{\mathcal{QNR}} = \langle \breve{\mathcal{T}}^{\mathrm{j}}_{\breve{\mathcal{K}}_{\mathcal{QNR}}}, \breve{\mathcal{D}}^{\mathrm{j}}_{\breve{\mathcal{K}}_{\mathcal{QNR}}}, \breve{\mathcal{Y}}^{\mathrm{j}}_{\breve{\mathcal{K}}_{\mathcal{QNR}}}, \breve{\mathcal{F}}^{\mathrm{j}}_{\breve{\mathcal{K}}_{\mathcal{QNR}}} : \mathrm{j} = 1,2...\mathfrak{p} \rangle$ and $\breve{\mathcal{L}}_{\mathcal{QNR}} = \langle \breve{\mathcal{T}}^{\mathrm{j}}_{\breve{\mathcal{L}}_{\mathcal{QNR}}}, \breve{\mathcal{D}}^{\mathrm{j}}_{\breve{\mathcal{L}}_{\mathcal{QNR}}}, \breve{\mathcal{Y}}^{\mathrm{j}}_{\breve{\mathcal{L}}_{\mathcal{QNR}}}, \breve{\mathcal{F}}^{\mathrm{j}}_{\breve{\mathcal{L}}_{\mathcal{QNR}}} : \mathrm{j} = 1,2...\mathfrak{p} \rangle$ be two $\mathcal{QSVNRN}s$. Then the operations for $\mathcal{QSVNRN}$s can be defined as:

(i) $\breve{\mathcal{K}}_1 \oplus \breve{\mathcal{K}}_2 = (\breve{\mathcal{T}}^{\mathrm{j}}_1 + \breve{\mathcal{T}}^{\mathrm{j}}_2 - \breve{\mathcal{T}}^{\mathrm{j}}_1 \breve{\mathcal{T}}^{\mathrm{j}}_2, \breve{\mathcal{D}}^{\mathrm{j}}_1 + \breve{\mathcal{D}}^{\mathrm{j}}_2 - \breve{\mathcal{D}}^{\mathrm{j}}_1 \breve{\mathcal{D}}^{\mathrm{j}}_2, \breve{\mathcal{Y}}^{\mathrm{j}}_1 \breve{\mathcal{Y}}^{\mathrm{j}}_2, \breve{\mathcal{F}}^{\mathrm{j}}_1 \breve{\mathcal{F}}^{\mathrm{j}}_2)$

(ii) $\breve{\mathcal{K}}_1 \otimes \breve{\mathcal{K}}_2 = (\breve{\mathcal{T}}^{\mathrm{j}}_1 \breve{\mathcal{T}}^{\mathrm{j}}_2, \breve{\mathcal{D}}^{\mathrm{j}}_1 \breve{\mathcal{D}}^{\mathrm{j}}_2, \breve{\mathcal{Y}}^{\mathrm{j}}_1 + \breve{\mathcal{Y}}^{\mathrm{j}}_2 - \breve{\mathcal{Y}}^{\mathrm{j}}_1 \breve{y}^{\mathrm{j}}_2, \breve{\mathcal{F}}^{\mathrm{j}}_1 + \breve{\mathcal{F}}^{\mathrm{j}}_2 - \breve{\mathcal{F}}^{\mathrm{j}}_1 \breve{\mathcal{F}}^{\mathrm{j}}_2)$

(iii) $\lambda \breve{\mathcal{K}}_1 = ((1 - (1 - \breve{\mathcal{T}}^{\mathrm{j}}_1))^{\lambda}, (1 - (1 - \breve{\mathcal{D}}^{\mathrm{j}}_1))^{\lambda}, (\breve{\mathcal{Y}}^{\mathrm{j}}_1)^{\lambda}, (\breve{\mathcal{F}}^{\mathrm{j}}_1)^{\lambda})$

(iv) $(\breve{\mathcal{K}}_1)^{\lambda} = (\breve{\mathcal{T}}^{\mathrm{j}}_1)^{\lambda}, (\breve{\mathcal{D}}^{\mathrm{j}}_1)^{\lambda}, (1 - (1 - \breve{\mathcal{Y}}^{\mathrm{j}}_1))^{\lambda}, (1 - (1 - \breve{\mathcal{F}}^{\mathrm{j}}_1))^{\lambda})$

## 16.3 Shortest Path Determination in Networks Using $\mathcal{QSVNR}$ Algorithm

This section presents an algorithm for finding the shortest path and the minimum distance of each node from a specified source node. The method is based on a labeling framework, making it straightforward to understand and practical for addressing shortest path problems in practical applications.

**Step 1:** Let $\breve{\mathfrak{d}}_1 = \langle (0, 0, 1, 1), (0, 0, 1, 1), (0, 0, 1, 1) \rangle$ and assign a label to the source node as $[\langle (0, 0, 1, 1), (0, 0, 1, 1), (0, 0, 1, 1) \rangle, -]$.

**TABLE 16.1**: Edge weighting of the graph.

| Edges | $\mathcal{QSVNR}$ distance |
|---|---|
| 1-2 | $\langle(0.4, 0.8, 0.6, 0.4)\ (0.3, 0.5, 0.4, 0.2)\ (0.5, 0.3, 0.4, 0.2)\rangle$ |
| 1-3 | $\langle(0.5, 0.5, 0.6, 0.3)\ (0.4, 0.6, 0.3, 0.2)\ (0.5, 0.3, 0.5, 0.3)\rangle$ |
| 2-3 | $\langle(0.6, 0.4, 0.7, 0.4)\ (0.3, 0.5, 0.7, 0.2)\ (0.5, 0.3, 0.4, 0.2)\rangle$ |
| 2-5 | $\langle(0.5, 0.8, 0.2, 0.3)(0.4, 0.5, 0.5, 0.2)(0.5, 0.4, 0.6, 0.3)\rangle$ |
| 3-4 | $\langle(0.4, 0.7, 0.3, 0.2)(0.2, 0.4, 0.5, 0.4)(0.6, 0.40.6, 0.4)\rangle$ |
| 3-5 | $\langle(0.5, 0.6, 0.4, 0.3)(0.4, 0.6, 0.3, 0.4)(0.5, 0.5, 0.2, 0.4)\rangle$ |
| 4-6 | $\langle(0.5, 0.6, 0.2, 0.3)(0.5, 0.5, 0.2, 0.4)(0.3, 0.6, 0.3, 0.5)\rangle$ |
| 5-6 | $\langle(0.4, 0.4, 0.3, 0.3)(0.7, 0.2, 0.3, 0.2)(0.5, 0.4, 0.6, 0.3)\rangle$ |

**Step 2:** Compute $\breve{\mathfrak{d}}_j$ = minimum $(\breve{\mathfrak{d}}_j \oplus \breve{\mathfrak{d}}_{ij})$ ; j = 2,3,. . . ,n.

**Step 3:** If the minimum value corresponds to $i$ (i.e., $i = r$), then assign the label of node $j$ as $[\breve{\mathfrak{d}}_j, r]$. In cases where the minimum occurs for more than one value of $i$, the position with the smallest $i$ is selected.

**Step 4:** If the destination node $n$ is denoted by label $[\breve{\mathfrak{d}}_n, l]$, then the $\mathcal{QSVNR}$ shortest distance from the source node equals $\breve{\mathfrak{d}}_n$.

**Step 5:** When the destination node is assigned the label $[\breve{\mathfrak{d}}_n, l]$, the QSVNR shortest path from the source to the destination is derived by performing a backtracking procedure through the sequence of node labels. The process begins by inspecting the label of node $l$. If this label is $[\breve{\mathfrak{d}}_n, q]$, the label of node $q$ is then examined. This iterative tracing continues until the source node (node 1) is reached, thereby reconstructing the complete shortest path.

**Step 6:** The sequence of nodes obtained in Step 5, when combined, gives the $\mathcal{QSVNR}$ shortest path.

## 16.4 Illustrative Example

To illustrate and validate the proposed method, let us consider the network depicted in Figure 16.1, where the distance between each pair of vertices is expressed as a $\mathcal{QSVNR}$ number. The objective is to compute the shortest path and the corresponding shortest distance from the source to the destination node in this network.

Table 16.1 shows weights of the graphs.

Applying the algorithm described in Section 16.3, the computational results are derived as follows. As node 6 is the destination, we take $n = 6$.

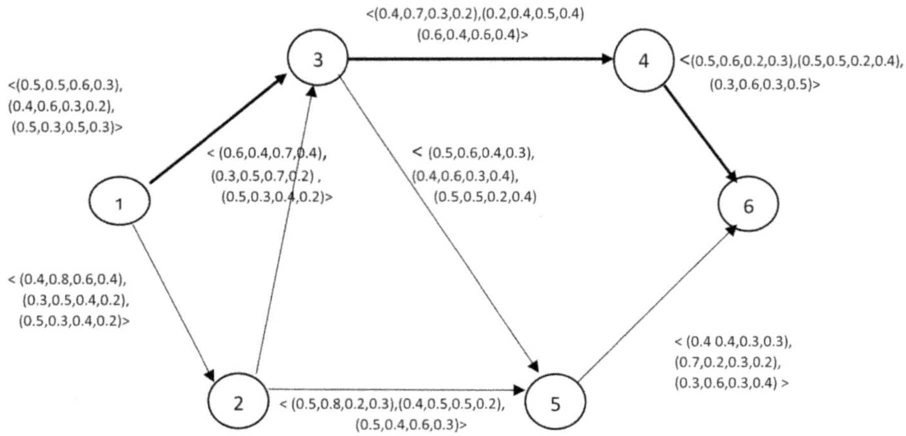


**FIGURE 16.1**: $\mathcal{QSVNR}$-based shortest distance network.

Let $\breve{\mathfrak{d}}_j = \langle(0{,}0{,}1{,}1),(0{,}0{,}1{,}1),(0{,}0{,}1{,}1)\rangle$. The source node (node 1) is labeled as $[\langle(0{,}0{,}1{,}1),(0{,}0{,}1{,}1),(0{,}0{,}1{,}1)\rangle, -]$. Then the values of $\breve{\mathfrak{d}}_n$ for $j$= 2, 3, 4, 5, 6 are determined as follows:

**Iteration 1:** As node 1 is the only predecessor of node 2, we substitute $i$= 1 and $j$= 2 into Step 2 of the proposed algorithm. Thus,

$$\begin{aligned}\breve{\mathfrak{d}}_j &= minimum\{\breve{\mathfrak{d}}_1 \oplus \breve{\mathfrak{d}}_{12}\} = minimum\{\langle(0,0,1,1),(0,0,1,1),(0,0,1,1)\rangle\oplus\\ &\qquad \langle(0.4,\ 0.8,\ 0.6,\ 0.4),(0.3,\ 0.5,\ 0.4,\ 0.2),(0.5,\ 0.3,\ 0.4,\ 0.2)\rangle\}\\ &= \langle(0.4,\ 0.8,\ 0.6,\ 0.4),(0.3,\ 0.5,\ 0.4,\ 0.2),(0.5,\ 0.3,\ 0.4,\ 0.2)\rangle.\end{aligned}$$

Since the minimum value corresponds to $i = 1$, node 2 is assigned the label as $[\langle(0.4{,}0.8{,}0.6{,}0.4),(0.3{,}0.5{,}0.4{,}0.2),0.5{,}0.3{,}0.4{,}0.2)\rangle{,}1]$.

**Iteration 2:** Node 3 has nodes 1 and 2 as its predecessors. Hence, by setting ($i$ = 1, 2) and ($j$ = 3) in Step 2 of the suggested procedure, the value of $\breve{\mathfrak{d}}_3$ is determined as

$$\begin{aligned}&\breve{\mathfrak{d}}_3 = minimum\{\breve{\mathfrak{d}}_1 \oplus \breve{\mathfrak{d}}_{13}, \breve{\mathfrak{d}}_2 \oplus \breve{\mathfrak{d}}_{23}\}\\ &= minimum\{\{\langle(0,0,1,1),(0,0,1,1),(0,0,1,1)\oplus\\ &\quad\langle(0.5,\ 0.5,\ 0.6,\ 0.3),(0.4,\ 0.6,\ \ 0.3,\ 0.2),(0.5,\ 0.3,\ 0.5,\ 0.3)\rangle\},\\ &\quad\{\langle(0.4,\ 0.8,\ 0.6,\ 0.4),(0.3,0.5,0.4,0.2),(0.5,0.3,0.1,0.4)\rangle\oplus\\ &\quad\langle(0.6,0.4,0.7,0.4),(0.3,0.5,0.7,0.2),(0.5,0.3,0.1,0.4)\rangle\}\}\\ &\quad= \ minimum\{\langle(0.5,0.5,0.6,0.3),(0.4,0.6,0.3,0.2),(0.5,0.3,0.5,0.3)\rangle,\\ &\quad\langle(0.76,0.88,0.42,0.16),(0.51,0.74,0.28,0.04),(0.75,0.51,0.0.4,0.08)\rangle\}\\ &\quad\mathcal{S}_{\breve{\mathcal{K}}_{\mathcal{QNR}}}\{\langle(0.5,0.5,0.6,0.3),(0.4,0.6,0.3,0.2),(0.5,0.3,0.5,0.3)\rangle\} = 0.583\\ &\quad\mathcal{S}_{\breve{\mathcal{K}}_{\mathcal{QNR}}}\{\langle(0.76,0.88,0.42,0.16),(0.51,0.74,0.28,0.04),(0.75,0.51,0.0.4,0.08)\rangle\}\\ &\quad= 0.655\end{aligned}$$

So

$$minimum\{\langle(0.5,0.5,0.6,0.3),(0.4,0.6,0.3,0.2),(0.5,0.3,0.5,0.3)\rangle,$$
$$\langle(0.76,0.88,0.42,0.16),(0.51,0.74,0.28,0.04)\rangle,$$
$$(0.75,0.51,0.0.4,0.08)\} = \langle(0.5,0.5,0.6,0.3),(0.4,0.6,0.3,0.2),$$
$$(0.5,0.3,0.5,0.3)\rangle$$

Since the minimum value corresponds to i = 1, node 3 is labeled as

$$[\langle(0.5,0.5,0.6,0.3),(0.4,0.6,0.3,0.2),(0.5,0.3,0.5,0.3)\rangle,1].$$

**Iteration 3:** Node 3 is the predecessor of node 4. Therefore, by substituting $i = 3$ and $j = 4$ into Step 2 of the suggested procedure, the value of $\breve{\mathfrak{d}}_4$ is

$$\breve{\mathfrak{d}}_4 = minimum\{\breve{\mathfrak{d}}_3 \oplus \breve{\mathfrak{d}}_{34}\} = minimum\{\langle 0.5,\ 0.5,\ 0.6,\ 0.3),(0.4,\ 0.6,\ 0.3,\ 0.2),$$
$$(0.5,\ 0.3,\ 0.5,\ 0.3)\}\rangle\oplus$$
$$\langle(0.4,\ 0.7,\ 0.3,\ 0.2),0.2,\ 0.4,\ 0.5,\ 0.4),(0.6,\ 0.4,\ 0.6,\ 0.4)\rangle\}$$
$$=\langle(0.7,\ 0.85,\ 0.18,\ 0.06),(0.52,0.76,0.15,0.08),(0.8,0.58,0.3,0.12)\rangle.$$

Since the minimum value corresponds to $i = 3$, node 4 is labeled as $[\langle(0.7,\ 0.85,\ 0.18,\ 0.06),(0.52,\ 0.76,\ 0.15,\ 0.08),(0.8,\ 0.58,\ 0.3,\ 0.12)\rangle,\ 3]$.

**Iteration 4:** Node 5 has nodes 2 and 3 as its predecessors. Thus, by setting ($i = 2$, 3) and ($j = 5$) in Step 2 of the suggested procedure, the value of

$$\breve{\mathfrak{d}}_5 = minimum\{\breve{\mathfrak{d}}_2 \oplus \breve{\mathfrak{d}}_{25}, \breve{\mathfrak{d}}_3 \oplus \breve{\mathfrak{d}}_{35}\}$$
$$= minimum\{\{\langle(0.4,\ 0.8,\ 0.6,\ 0.4),(0.3,\ 0.5,\ 0.4,\ 0.2),$$
$$(0.5,\ 0.3,\ 0.4,\ 0.2)\rangle\oplus$$
$$\langle(0.5,0.8,0.2,0.3),(0.4,0.5,0.5,0.2),(0.5,0.4,0.6,0.3)\rangle\},$$
$$\{\langle(0.5,0.5,0.6,0.3),(0.4,0.6,0.3,0.2),(0.5,0.3,0.5,0.3)\rangle\oplus$$
$$\langle(0.5,0.6,0.4,0.3),(0.4,0.6,0.3,0.4),(0.5,0.5,0.2,0.4)\rangle\}\}$$
$$= minimum\{\langle(0.7,0.96,0.12,0.12),(0.58,0.75,0.20,0.04),$$
$$(0.75,0.58,0.24,0.06)\rangle,$$
$$\langle(0.75,0.8,0.24,0.09),(0.64,0.84,0.09,0.08),0.75,0.65,0.10,0.12)\rangle$$
$$\mathcal{S}_{\breve{\mathcal{K}}_{\mathcal{QNR}}}\{\langle(0.7,\ 0.96,\ 0.12,\ 0.12),(0.58,\ 0.75,\ 0.20,\ 0.04),$$
$$(0.75,\ 0.58,\ 0.24,\ 0.06)\rangle\} = 0.664$$
$$\mathcal{S}_{\breve{\mathcal{K}}_{\mathcal{QNR}}}\{\langle(0.75,\ 0.8,\ 0.24,\ 0.09),(0.64,\ 0.84,\ 0.09,\ 0.08),$$
$$0.75,\ 0.65,\ 0.10,\ 0.12)\rangle\} = 0.678$$

So

$$minimum\{\langle(0.7,\ 0.96,\ 0.12,\ 0.12),(0.58,\ 0.75,\ 0.20,\ 0.04),\\ (0.75,\ 0.58,\ 0.24,\ 0.06)\rangle,\\ \langle(0.75,\ 0.8,\ 0.24,\ 0.09),(0.64,\ 0.84,\ 0.09,\ 0.08),\\ 0.75,\ 0.65,\ 0.10,\ 0.12)\rangle =\\ \langle(0.7,\ 0.96,\ 0.12,\ 0.12),(0.58,\ 0.75,\ 0.20,\ 0.04),\\ (0.75,\ 0.58,\ 0.24,\ 0.06)\rangle$$

Since the minimum value corresponds to $i = 2$, node 5 is labeled as

$$[\langle(0.7,\ 0.96,\ 0.12,\ 0.12),(0.58,\ 0.75,\ 0.20,\ 0.04),\\ (0.75,\ 0.58,\ 0.24,\ 0.06)\rangle,2].$$

**Iteration 5:** Node 6 has nodes 4 and 5 as its predecessors. Therefore, by setting ($i = 4,\ 5$) and ($j = 6$) in Step 2 of the suggested procedure, the value of

$$\breve{\mathfrak{d}}_6 = minimum\{\breve{\mathfrak{d}}_4 \oplus \breve{\mathfrak{d}}_{46}, \breve{\mathfrak{d}}_5 \oplus \breve{\mathfrak{d}}_{56}\}\\ = minimum\{\{\langle(0.7,0.85,0.18,0.06),(0.52,0.76,0.15,0.08),\\ (0.8,0.58,0.3,0.12)\rangle\oplus\\ \langle(0.5,0.6,0.2,0.3),(0.5,0.5,0.2,0.4),\\ (0.3,0.6,0.3,0.5)\rangle\},\\ \langle(0.7,0.96,0.12,0.12),(0.58,0.75,0.20,0.04),(0.75,0.58,0.24,0.06)\rangle\\ \oplus \langle(0.4,0.4,0.3,0.3),(0.7,0.2,0.3,0.2),(0.3,0.6,0.3,0.4)\rangle\}\\ = minimum\{\langle(0.85,0.94,0.036,0.018),(0.76,0.88,0.03,0.032),\\ (0.86,0.832,0.09,0.06)\rangle,\langle(0.82,0.984,0.036,0.036),\\ (0.874,0.8,0.06,0.008),(0.825,0.832,0.072,0.024)\rangle\\ \mathcal{S}_{\breve{\mathcal{K}}_{\mathcal{QNR}}}\{\langle(0.85,0.94,0.036,0.018),(0.76,0.88,0.03,0.032),\\ (0.86,0.832,0.09,0.06)\rangle\} = 0.712\\ \mathcal{S}_{\breve{\mathcal{K}}_{\mathcal{QNR}}}\{\langle(0.82,0.984,0.036,0.036),(0.874,0.8,0.06,0.008),\\ (0.825,0.832,0.072,0.024)\rangle\\ = 0.722$$

So

$$minimum\{\langle(0.85,0.94,0.036,0.018),(0.76,0.88,0.03,0.032),\\ (0.86,0.832,0.09,0.06),\\ \langle(0.82,0.984,0.036,0.036),(0.874,0.8,0.06,0.008),\\ (0.825,0.832,0.072,0.024)\rangle =\\ \langle(0.85,0.94,0.036,0.018),(0.76,0.88,0.03,0.032),\\ (0.86,0.832,0.09,0.06)\rangle$$

**TABLE 16.2**: Tabulated results of different shortest paths determined by $\mathcal{QSVNR}$.

| **Node No. ($j$)** | $\widetilde{d_j}$ | $\mathcal{QSVNR}$ **shortest path between $j$th and first node** |
|---|---|---|
| 2 | ⟨(0.4, 0.8, 0.6, 0.4) (0.3, 0.5, 0.4, 0.2) (0.5, 0.3, 0.4, 0.2)⟩ | 1-2 |
| 3 | ⟨(0.5, 0.5, 0.6, 0.3) (0.4, 0.6, 0.3, 0.2) (0.5, 0.3, 0.5, 0.3)⟩ | 1-3 |
| 4 | ⟨(0.7, 0.85, 0.18, 0.06), (0.52, 0.76, 0.15, 0.08), (0.8, 0.58, 0.3, 0.12)⟩ | 1-3-4 |
| 5 | ⟨(0.7, 0.96, 0.12, 0.12), (0.58, 0.75, 0.20, 0.04), (0.75, 0.58, 0.24, 0.06)⟩ | 1-2-5 |
| 6 | ⟨(0.85, 0.94, 0.036, 0.018), (0.76, 0.88, 0.03, 0.032), (0.86, 0.832, 0.09, 0.06)⟩ | 1-3-4-6 |

Since the minimum value corresponds to $i = 4$, node 6 is labeled as

$$[\langle(0.85, 0.94, 0.036, 0.018), (0.76, 0.88, 0.03, 0.032), (0.86, 0.832, 0.09, 0.06)\rangle, 4]$$

Given that node 6 is the destination in the network, the shortest path from node 1 to node 6 is obtained using the $\mathcal{QSVNR}$ method.

$$\langle(0.85, 0.94, 0.036, 0.018), (0.76, 0.88, 0.03, 0.032), (0.86, 0.832, 0.09, 0.06)\rangle.$$

The $\mathcal{QSVNR}$ shortest path from node 1 to node 6 is identified as follows:

Node 6 is assigned the label $[\langle(0.85, 0.94, 0.036, 0.018), (0.76, 0.88, 0.03, 0.032), (0.86, 0.832, 0.09, 0.06)\rangle, 4]$, indicating that it is reached from node 4. Node 4 is labeled as [⟨(0.7, 0.85, 0.18, 0.06), (0.52, 0.76, 0.15, 0.08), (0.8, 0.58, 0.3, 0.12)⟩, 3], showing that it is reached from node 3. Similarly, node 3 is labeled as $[\langle(0.5, 0.5, 0.6, 0.3), (0.4, 0.6, 0.3, 0.2), (0.5, 0.3, 0.5, 0.3)\rangle, 1]$, which indicates that it is reached from node 1. Thus, the $\mathcal{QSVNR}$ shortest path from node 1 to node 6 is obtained by tracing these connections, yielding the path 1–3–4–6 with the $\mathcal{QSVNR}$ value $\langle(0.85, 0.94, 0.036, 0.018), (0.76, 0.88, 0.03, 0.032), (0.86, 0.832, 0.09, 0.06)\rangle$. In Figure 16.2, the darkened lines represent the identified optimal route between the source and destination nodes. The QSVNR shortest distance and path for all nodes from node 1 are summarized in Table 16.2, and the corresponding node labels are depicted in Figure 16.2.

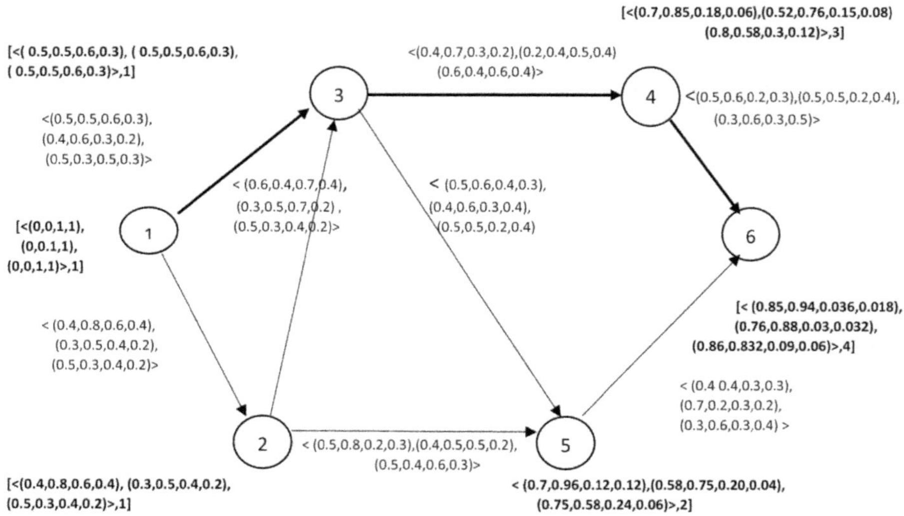


**FIGURE 16.2**: $\mathcal{QSVNR}$ network depicting the shortest distances from node 1 to all other nodes.

## 16.5 Conclusion

In this chapter, we have formulated a $\mathcal{QSVNR}$ shortest path technique for determining the shortest path and its length between a source and a destination node in a network, where the edge weights are expressed as $\mathcal{QSVNR}$ numbers. The methodology has been clearly described and illustrated with an example, emphasizing both its applicability and ease of implementation. Unlike intuitionistic approaches, which are limited to truth and falsity, the proposed method incorporates truth, contradiction, unknown, and falsity, thus providing a more thorough representation of uncertainty. Moreover, the algorithm can be generalized to address a broad spectrum of shortest path problems involving soft, rough, or interval quadripartitioned neutrosophic refined numbers. Overall, the proposed approach provides an efficient and effective tool for solving shortest path problems under uncertainty, contributing significantly to the broader field of $\mathcal{QSVNR}$-based decision-making and optimization.

## Bibliography


[1] Atanasov K.T., Intuitionistic Fuzzy Sets, Fuzzy Sets Syst., 1986, 20, 87–96.

[2] Arokia Pratheesha S.V. and Mohana, K., Quadripartitioned single valued neutrosophic refined sets and its topological spaces, Int. J. Creat. Res. Thoughts, March 2022, 10(3), d937–d948.

[3] Arokia Pratheesha, S.V., Mohana, K., and Radha, R., Softmax Function-based on Quadripartitioned Single Valued Neutrosophic Refined and Application in Multi-Attribute Decision Making Problem, Revolutionary Topologies and Their Applications, IGI Global, 2026.

[4] Blizard, W.D., Multiset theory, Notre Dame J. Form. Log., 1989, 30(1), 36–66.

[5] Bellman, R.E. and Zadeh, L.A., Decision-making in a fuzzy environment. Manag. Sci. 1970, 17, B-141.

[6] Belnap Jr., A useful four valued logic, Modern Uses of Multiple Valued Logic Springer, Dordrecht, 1977, pp. 5–37.

[7] Broumi, S., Talea, M., Bakali, A., and Smarandache, F., Single valued neutrosophic graphs, J. New Theory, 2016, N10, 86–101.

[8] Broumi, S., Talea, M., Bakali, A., and Smarandache, F., On bipolar single valued neutrosophic graphs, J. New Theory, 2016, N11, 84–102.

[9] Broumi, S., Talea, M., Bakali, A., and Smarandache, F., Interval Valued Neutrosophic Graphs, Critical Review, A Publication of Society for Mathematics of Uncertainty, 2016, 10, 5–33.

[10] Broumi, S., Bakali, A., Talea, M., and Smarandache, F., Isolated single valued neutrosophic graphs. Neutrosophic Sets Syst., 2016, 11, 74–78.

[11] Broumi, S., Bakali, A., Talea, M., and Smarandache, F., Computation of Shortest Path Problem in a Network with SV-Trapezoidal Neutrosophic Numbers, IEEE International Conference on Advanced Mechatronic Systems (ICAMechS) 2016, pp. 417–422.

[12] Broumi, S., Bakali, A., Talea, M., Smarandache, F., and, Ali, M., Shortest Path Problem under Bipolar Neutrosophic Setting, Appl. Mech. Mater., 2016, 859, 59–66.

[13] Broumi, S., Bakali, A., Talea, M., and Smarandache, F., Shortest Path Problem under Interval Valued Neutrosophic Setting, Int. J. Adv. Trends Comput. Sci. Eng., 2019, 8, 216–222.

[14] Broumi, S., Smarandache, F., Talea, M., and Bakali, A., An introduction to bipolar single valued neutrosophic graph theory. Appl. Mech. Mater., 2016, 841, 184–191.

[15] Broumi, S. Talea, M., Smarandache, F., and Bakali, A., Single valued neutrosophic graphs: Degree, order and size, 2016 IEEE International Conference on Fuzzy Systems (FUZZ), 2016, pp. 2445–2451.

[16] Calude, C.S., Paun, G., Rozenberg, G., and Salomaa, A., Lecture notes in computer science: Multiset processing mathematical, computer science, and molecular computing points of view, Springer-New York, 2001, 2235.

[17] Deli, I. and Broumi, S., Neutrosophic refined sets and their applications in medical diagnosis, J. New Theory, 2015, 6, 88–98.

[18] Dijkstra, E.W., A note on two problems in connexion with graphs, Numer. Math., 1959, 1, 269–271.

[19] Jayagowri, P. and Geetha Ramani, G., Using trapezoidal intuitionistic fuzzy number to find optimized path in a network, Adv. Fuzzy Syst., 2014, 183607, 1–6.

[20] Kumar, A. and Kaur, M., A new algorithm for solving shortest path problem on a network with imprecise edge weight, Appl. Appl. Math., 2011, 6(2), 602 – 619.

[21] Kumar, A. and Kaur, M., Solution of fuzzy maximal flow problems using fuzzy linear programming, Int. J. Math. Comput. Methods, 2011, 5(6), 811–815.

[22] Majumder, S. and Pal, A., Shortest path problem on intuitionistic fuzzy network, Ann. Pure Appl. Math., November 2013, 5(1), 26–36.

[23] Okada, S. and Gen, M., Fuzzy shortest path problems, Comput. Ind. Eng., 1994, 27(1-4), 465–468.

[24] Chatterjee, R., Majumdar, P., and Samanta, S.K., Single valued neutrosophic multisets, Ann. Fuzzy Math. Inform., 2015, 10(3), 499–514.

[25] Chatterjee, R., Majumdar, P., and Samanta, S.K., On some similarity measures and entropy on quadripartitioned single valued neutrosophic sets, J. Intell. Fuzzy Syst., 2016, 30, 2475–2485.

[26] Smarandache, F., A Unifying Field in Logics: Neutrosophic Logic. Neutrosophy, Neutrosophic Set, Neutrosophic Probability, American Research Press, Rehoboth, NM, 1999.

[27] Smarandache, F., Symbolic Neutrosophic Theory, EuropaNova Asbl, Brussels, 2015, 195.

[28] Smarandache, F., Neutrosophy and neutrosophic logic, First International Conference on Neutrosophy, Neutrosophic Logic, Set, Probability, and Statistics University of New Mexico, Gallup, NM 87301, USA, 2002.

[29] Shinoj, T.K and John, S.J., Intuitionistic fuzzy multisets and its application in medical diagnosis, World Acad. Sci. Eng. Technol., 2012, 6, 1–28.

[30] Smarandache, F., n-valued refined neutrosophic logic and its applications in physics, Prog. Phys., 2013, 4, 143–146.

[31] Wang, H., Smarandache, F., Zhang, Y.Q., and Sunderraman, R., Single valued neutrosophic sets, Multispace and Multistruct, 2010, 4, 410–413.

[32] Yager, R.R., On the theory of bags (multi sets), Int. J. Gen. Syst., 1986, 13, 23–37.

[33] Yoon, K. A reconciliation among discrete compromise solutions. J. Oper. Res. Soc. 1987, 38, 277–286.

[34] Ye, S., Fu, J., Ye, J., Medical diagnosis using distance-based similarity measures of single valued neutrosophic multisets, Neutrosophic Sets and Syst., 2015, 7, 47–52.

[35] Zadeh, L.A., Fuzzy sets, Inform. and Control, 1965, 8(3), 338–353.

# *Index*

**Note:** *Italic* and **Bold** page numbers refer to figures and tables, respectively.

www.ingramcontent.com/pod-product-compliance
Ingram Content Group UK Ltd.
Pitfield, Milton Keynes, MK11 3LW, UK
UKHW022215190926
475702UK00003B/369
* 9 7 8 1 0 4 1 1 6 2 0 2 5 *